ROTOR DYNAMICS

(SECOND EDITION)

J. S. RAO
Professor & Head
Department of Mechanical Engineering
Indian Institute of Technology, Delhi
India

JOHN WILEY & SONS
New York   Brisbane   Chichester   Toronto   Singapore

Second published in 1991 by
WILEY EASTERN LIMITED
4835/24 Ansari Road, Daryaganj
New Delhi 110 002, India

Distributors:

Australia and New Zealand:
JACARANDA WILEY LTD.,
PO Box 1226, Milton Old 4064, Australia

Canada:
JOHN WILEY & SONS CANADA LIMITED
22 Worcester Road, Rexdale, Ontario, Canada

Europe and Africa:
JOHN WILEY & SONS LIMITED
Baffins Lane, Chichester, West Sussex, England

South East Asia:
JOHN WILEY & SONS, (PTE) LIMITED
05-04, Block B, Union Industrial Building
37 Jalan Pemimpin, Singapore 2057

Africa and South Asia:
WILEY EASTERN LIMITED
4835/24 Ansari Road, Daryaganj
New Delhi 110 002, India

North and South America and rest of the World:
JOHN WILEY & SONS, INC.
605 Third Avenue, New York, N.Y. 10158, USA

Copyright © 1991, WILEY EASTERN LIMITED
New Delhi, India

Library of Congress Cataloging-in-Publication Data

Rao, J.S.
   Rotor dynamics/J.S. Rao—2nd ed.
     p. cm.
   Includes bibliographical references and index.
    1. Rotor—Dynamics. I. Title.   91-3169

TJ105.R36   1991   CIP
621.8'2—dc20

ISBN 0-470-21787-1 John Wiley & Sons, Inc.
ISBN 81-224-0314-X Wiley Eastern Limited

Printed in India at Urvashi Press, Meerut.

To
My Wife Indira
and
To the memory of my parents Jammi Chikka Rao and Ramanamma

Preface to the Second Edition

The opportunity of preparing a second edition is taken to include certain useful additions to the text. Firstly, a transfer matrix analysis using distributed elements is introduced in Chapter 8. This method is useful in modelling large rotors with several stations to determine the unbalance response considering the fluid film bearing supports. In view of developments in aircraft propulsion systems, wherein a two spool rotor system with intershaft bearings is becoming a standard layout, a transfer matrix procedure to determine the unbalance response of dual rotor systems is also included in Chapter 8. The concept of an optimum design of bearings for minimum unbalance response of a rotor is also introduced.

In recent years, transient response studies have taken an important place in the subject of Rotor Dynamics. This has become necessary in view of the operating speeds approaching or crossing the stability threshold speeds predicted by linear theories, besides the need for analysis of rotors of machinery subjected to earthquake excitations. Another important development is the study of cracked rotors, which is assuming significance in the understanding of dynamic behaviour of rotors with a transverse crack developed during its operation. Analysis based on Jeffcott models using time marching techniques, is introduced in Chapter 9. The instability due to a transverse crack is introduced in Chapter 9 and the instability due to the negative cross-coupled stiffness is introduced in Chapter 10. The transient orbital analysis by transfer matrix method is given in Chapter 10, with several examples.

This material is prepared, while giving a course on Rotor Dynamics to the post graduate students of the Institute of Mechanics at the Gesamthochschule of Kassel University. The author wishes to thank Professor H. Irretier and his Institute who organized these presentations and helped in preparing the second edition of this text.

New Delhi J.S. RAO
August 20, 1991

Preface to the First Edition

With ever increase in demand for larger size and velocity in modern machines, Rotor Dynamics became more and more an important subject in the mechanical engineering design. It is well known that torsional vibrations in rotating machines, reciprocating machine installations and geared systems, whirling of rotating shafts, the effect of flexible bearings, instabilities due to asymmetric cross-section shafts, hydrodynamic bearings, hysteresis, balancing of rigid and flexible rotors can be thoroughly understood only on the basis of Rotor Dynamic studies. In the recent years, increasing dependence is observed in the maintenance of heavy duty and high speed machines, based on the vibration signature analysis. Thus in the recent years some technical institutions have introduced a course, either at final year undergraduate level or at postgraduate level, on Rotor Dynamics. This book grew out of such lectures given to the students at the Indian Institute of Technology, Delhi. The formal script was written during the courses delivered in National Institute of Applied Sciences, Lyon, France and Concordia University, Montreal, Canada.

The present book assumes a background of at least one course on theory of vibration. It aims to be simple with a reasonable complete treatment of the subject. The mathematical approach has been quite simple in its treatment, leading to modern computer application using matrix algebra. The solution of technical problems is illustrated by various examples, taken in many cases from actual experience with vibration of machines in service.

The author owes a very large debt of gratitude to the management of Indian Institute of Technology, New Delhi, which made it possible for him to develop this subject in the last 8 years. He also takes this opportunity to thank his friends, Professor M. Lalanne, Professor T. S. Sankar and Professor N. F. Rieger for their criticism and valuable suggestions. He is indebted, also, to M/s Wiley Eastern for their care in publication of this book.

New Delhi   J. S. Rao
July 20, 1983

Contents

Preface to the Second Edition   v

Preface to the First Edition   vi

Chapter 1. Introduction   3

Chapter 2. Torsional Vibrations in Rotating Machinery   9
   2.1 Modelling of rotating machinery shafting   *9*
   2.2 Transfer matrix analysis for free vibrations   *10*
   2.3 Equivalent discrete system   *15*
   2.4 Excitation torque   *17*
   2.5 Transient response in torsional vibration   *18*
   2.6 Branched systems   *20*

Chapter 3. Torsional Vibrations in Reciprocating Machinery   25
   3.1 Modelling to the reciprocating machine systems   *25*
   3.2 Free vibration calculations   *29*
   3.3 Excitation torque   *32*
   3.4 Forced vibration and resonant stresses   *37*
   3.5 Cyclic irregularity   *40*

Chapter 4. Bending Critical Speeds of Simple Shafts   47
   4.1 Whirling of an unbalanced simple elastic rotor   *47*
   4.2 Simple shafts with several disks   *51*
   4.3 Transfer matrix analysis for bending critical speeds   *52*
   4.4 Shafts with overhangs   *56*

Chapter 5. Out-of-Balance Response of Rotors in Rigid Supports   62
   5.1 Field matrix   *62*
   5.2 Point matrix for unbalance mass   *66*
   5.3 Out-of-balance response of rotors with rigid end supports   *68*
   5.4 Simply supported rotor with overhangs   *71*

Chapter 6. Gyroscopic Effects   79
   6.1 Gyroscopics of a spinning disk   *79*
   6.2 Synchronous whirl of an overhung rotor   *82*
   6.3 Nonsynchronous whirl   *84*
   6.4 Rotor system with a coupling   *86*

Chapter 7. Fluid Film Bearings   91
   7.1 Steady state characteristics of plain cylindrical hydrodynamic bearings   *91*
   7.2 Stiffness and damping coefficients of a journal bearing   *96*

Chapter 8. Rotors Mounted on Fluid Film Bearings 107
- 8.1 A simple rotor in fluid film bearings *107*
- 8.2 Transfer matrix analysis of rotors in fluid film bearings *125*
- 8.3 Transfer matrix analysis of turbine rotors by distributed elements *131*
- 8.4 Dual rotor system analysis *142*
- 8.5 Optimum design of bearings for minimum unbalance response *154*

Chapter 9. Shafts with Dissimilar Moments of Area 163
- 9.1 Stability of a shaft with dissimilar stiffness *163*
- 9.2 Whirling of a shaft with dissimilar stiffnesses *171*
- 9.3 Effect of disk unbalance *174*
- 9.4 Effect of gravity on a balanced disk *175*
- 9.5 Transient response by time marching technique *177*
- 9.6 Instability due to a transverse crack *187*

Chapter 10. Instability Due to Fluid Film Forces and Hysteresis 199
- 10.1 Instability of rotors mounted on fluid film bearings *199*
- 10.2 Rigid rotor instability *202*
- 10.3 Instability of a flexible rotor *205*
- 10.4 Instability threshold by transfer matrix method *209*
- 10.5 Internal hysteresis of shafts *210*
- 10.6 Instability due to negative cross-coupled stiffness *214*
- 10.7 Orbital analysis by transfer matrix method *219*

Chapter 11. Instability in Torsional Vibration 236
- 11.1 Variable stiffness in torsional vibrations *236*
- 11.2 System with variable inertia *239*

Chapter 12. Balancing of Rotors 247
- 12.1 Classification of rotors *248*
- 12.2 Rigid rotor classification and balancing criteria *248*
- 12.3 Balancing of rigid rotors *253*
- 12.4 Balancing of flexible rotors *259*
- 12.5 Balance criteria for flexible rotors *269*

Chapter 13. Condition Monitoring Using Vibration Measurements 275
- 13.1 Vibration generating mechanisms *276*
- 13.2 Condition monitoring *281*
- 13.3 Noise spectrum *287*
- 13.4 Real time analysis *289*

Index 299

120 MW LP Rotor (Courtesy BHEL, Hyderabad)

Chapter 1

Introduction

Rotating shafts are employed in industrial machines such as steam and gas turbines, turbogenerators, internal combustion engines, reciprocating and centrifugal compressors for power transmission. On account of the ever increasing demand for power and high speed transportation, the rotors of these machines are made extremely flexible, which makes the study of vibratory motion an essential part of design. The shafting of these machine installations is subjected to torsional and bending vibration and in some cases unstable conditions of operation. This book is concerned with the dynamic analysis of rotors in rotating and reciprocating machinery.

Rotating machines such as turbo-alternator sets, propulsion systems of ships, drive train of a hot strip steel mill, can develop excessive stresses in torsion, because of low torsional natural frequencies of the system involving flexible couplings. Determination of natural frequencies and mode shapes is thus important from a design point of view and a transfer matrix method is given in Chapter 2, taking into account the distributed inertia of the shaft. Dynamic stresses induced in such installations due to transient loads, such as electrical unbalance in generators, input loading in a steel mill etc., are important design considerations, as these stresses can lead to low cycle fatigue. Modal analysis is presented in this chapter to determine the transient response of rotating machine installations. An example of a 6 MW turbo-alternator set is considered to determine the natural frequencies and mode shapes and transient response due to a typical short circuiting torque acting on the generator. In several units such as ship propulsion drives, steel mill stands, there are one or two branch points and the transfer matrix method for the branched systems is also outlined in Chapter 2, with an illustrative example.

In reciprocating machine installations, internal combustion engines are used to drive units such as generators, pumps etc., or conversely, reciprocating pumps and compressors are driven by prime movers such as turbines, electrical motors. The inherent disadvantage of these installations is that the drive torque is highly pulsating, with several harmonics of the engine speed inducing large amplitude vibrations at or near the critical speeds of operation. The analysis for free vibrations of such installations using matrix approach for a discrete system is given in Chapter 3. The excitation torque for an in-line engine is described and the procedure to determine resonant stresses is given. It may be noted that the matrix approach has been adopted here, which is more suitable for computer applications than the conventional Holzer procedure. The fluctuations in speed that occur in the system due to torsional vibration are also discussed in this chapter. Several case studies are presented.

Lateral bending and consequently whirling of a rotor may arise out of a number of possible sources, the most important and of high practical significance being the residual unbalance. The simple circular shaft on rigid supports carrying concentrated masses, experiences synchronous whirl due to the residual unbalance and when the shaft runs at a speed equal to its natural frequency in lateral bending, the whirling becomes predominant. These speeds are called critical speeds in bending and the phenomenon is different from conventional resonance, since large amplitude vibrations cannot be controlled by additional damping. The whirling of a simple Jeffcott type rotor is first described in Chapter 4, along with Dunkerley and Rayleigh methods for a quick estimation of the critical speed of rotors carrying several concentrated masses. Transfer matrix method is then discussed to determine the critical speeds and mode shapes of simple rotors supported on rigid end supports or overhung on both ends. Some examples are presented. The procedure to determine the out-of-balance response of rotors on rigid supports, is outlined in Chapter 5. General theory for the elliptical orbit whirl of an unbalanced rotor is given, with an example of an overhung rotor.

Rotors of turbines, compressors and pumps carry one or more disks, and these disks besides contributing to a lumped mass at stations where they are located, also give rise to a gyroscopic couple. The effect of gyroscopic couple is predominant, if the disk is located at a nodal point or at a free end of the rotor. The general transfer matrix of a disk for synchronous whirl is derived in Chapter 6, and an example of overhung shaft is considered. The case of nonsynchronous whirl is also considered for an overhung shaft in this chapter. A typical case study of a system of rotors connected by a coupling is presented to determine the effect of gyroscopic couple on the natural frequencies and mode shapes.

Fluid film bearings play a very significant role in the dynamic behaviour of rotors. The subject of fluid film lubrication is very wide and a full study of this topic is beyond the scope of this book. The essential characteristics of steady state behaviour and linear stiffness and damping coefficients of such bearings are presented in Chapter 7, without any formal solution of the governing differential equations. This gives the basic understanding necessary for a study of the effect of journal bearing properties on the behaviour of rigid and flexible rotors.

The effect of asymmetric and cross-coupled spring coefficients of a hydrodynamic bearing on the out-of-balance response of a Jeffcott rotor is given in Chapter 8. It has been demonstrated that the rotor can exhibit two distinct critical speeds or does not exhibit any critical speed corresponding to the rigid bearing critical speed. For a rotor which exhibits two critical speeds; the backward whirl between these two critical speeds is demonstrated by plotting the whirl orbits. The effect of bearing damping is next considered and its influence on the dynamic behaviour of a Jeffcott rotor is discussed in detail. The general transfer matrix for the bearing station is then derived and an illustrative example presented to determine the out-of-balance response and the whirl mode shape at a speed where the response is high.

A transfer matrix analysis, using distributed elements is next derived in Chapter 8. Some rules in modelling the rotor by this method are presented. An example of a 6 MW rotor is illustrated. The transfer matrix method is then extended to the case of a two spool rotor system with an intershaft bearing, the layout of which is becoming a standard for aircraft propulsion systems. The steady state unbalance response of such rotors is shown to exhibit periodicity in the whirl orbits. The concept of an optimum design procedure of the bearings of a rotor for minimum unbalance response is then introduced with some examples of Jeffcott rotors.

Chapters 9, 10 and 11 deal with instability phenomenon that occurs in rotor bending and torsional vibrations. Double pole electric generator rotors have two different principal moments of area. The phenomenon of secondary resonance and the instability of such a horizontal rotor at half critical speed is discussed in Chapter 9. Using rotating coordinates, the behaviour of a rotor mounted on a rectangular cross-section shaft was next considered taking into account the disk unbalance and gravity effects.

The influence of hydrodynamic bearings on the dynamic behaviour of asymmetric shafts is next considered by a transient analysis using time marching techniques. Transient whirl orbits of Jeffcott rotors, including the effects of gravity, unbalance, asymmetry and the fluid film bearings are presented. The instability that arises from the presence of a transverse crack is then illustrated in Chapter 9.

The instability phenomenon of rigid and flexible rotors mounted on fluid film bearings, which is essentially a self-excited vibration is discussed in Chapter 10. The conditions for the threshold instability of rigid and flexible rotors are derived in a unified manner and simple design charts presented for Jeffcott models. Root searching techniques generally practiced have been eliminated. The instability phenomenon due to internal hysteresis is also discussed in this chapter, and the threshold instability condition for a Jeffcott rotor is derived. The procedure for determining the instability threshold speed of a general rotor using the transfer matrix theory is next discussed, with a case study.

In Chapter 8, rotors mounted on bearings with a negative cross-coupled stiffness were considered to determine the steady state unbalance response. When two distinct critical speeds do not occur, the rotor is actually unstable. In Chapter 10, the instability that arises under such conditions is studied. A general time dependent transfer matrix method is then derived in Chapter 10, to study the transient whirl orbit analysis of a multistation rotor. The example of 6 MW rotor considered in Chapter 8 was then studied to observe the transient whirl orbits before attaining the steady response.

In Chapter 11, the instabilities that occur due to variable stiffness and inertia in torsional systems are presented. The variable stiffness problem in torsional vibration is similar to the variable stiffness problem of a shaft with dissimilar moments of area, but the major instability region is around half the critical speed. This is illustrated with an example of side rod locomotive drive unit. The variable inertia problem in torsional vibration that occurs in reciprocating engines with heavy pistons, is next considered in Chapter 11. As the governing equations of motion are not of Mathieu or Hill type, modified Euler's method is used to obtain a numerical solution, to determine the instability boundaries of a reciprocating single cylinder engine. An example of a marine engine is considered to illustrate the instability regions of operation.

As the major cause of vibrations in a rotor is the residual unbalance, removal of the same is an important step to achieve smooth operation. Balancing of rigid rotors (rotors operating at speeds well below the first critical speed), is discussed in Chapter 12. Thearle's balancing procedure in the form of influence coefficients is illustrated with a case study. The balance quality levels for rigid rotors, as per the international standards are presented in tabular and graphical form. The balancing procedure for flexible rotors is complicated. The techniques of modal balancing and influence coefficients are outlined in this Chapter 12, and a case study for influence coefficient method is illustrated. The balance qualities for flexible rotors are also presented.

The rotating and reciprocating parts of a machine cause vibrations and noise and therefore the analysis of these records gives valuable information to assess the performance of a machine and continuously monitor its health. The use of vibration and noise measurements and their analysis is rapidly becoming a standard form of maintenance of rotating and reciprocating machinery. Such records also help in the diagnosis of an existing fault in a machine installation. In Chapter 13, different vibration generating mechanisms are discussed, which help in identifying the source of trouble. Different methods of condition monitoring are then described. International standards for quality judgement of vibration severity of different types of machines are also presented. Typical case studies involving failure of machinery are discussed.

This book assumes a basic knowledge of theory of vibrations and mechanics of machinery. The knowledge of fluid film lubrication is helpful but not essential. The material in this book can form an advanced undergraduate or a graduate course for students in mechanical and aerospace engineering.

A 3.5 MW Turbo Generator plant (Courtsey BHEL, Hyderabad)

Chapter 2

Torsional Vibrations in Rotating Machinery

Rotating machines such as steam turbines, compressors, generators can develop excessive dynamic stresses, if they are run at speeds near their natural frequencies in torsional vibration. When the driving and driven units of such large rotating machines are coupled through flexible couplings, their natural frequencies in torsional vibration are low enough to be in the operational speed ranges of their installations. In rotating machines the driving torque is practically uniform unlike in reciprocating machines, and hence the harmful effects harmonics of the engine speed occurring in the operational speed is remote. However, a system can develop excessive stresses at the critical speeds due to misalignment, gear meshing frequencies, and the design should avoid all such possibilities.

An important factor in the design of rotating machines, in the possibility of transient loads acting on a rotor, for example, the electrical unbalance in a generator. These transient loads can induce excessive stresses, though for a short period, which can lead to low cycle fatigue and premature failure of some parts such as couplings. In such cases the determination of transient stresses is an important design consideration of rotating machine shafts and couplings.

2.1 Modelling of Rotating Machinery Shafting

All identifiable rigid bodies such as flywheels, inner and outer parts of a flexible coupling, a turbine disk, can be considered as rigid disks, whose mass moments of inertia can be determined easily. Flexible couplings and thin shafts whose polar mass moments of inertia are small, can be considered as massless shafts, whose torsional stiffness can be determined in the normal way. Where the shaft diameter is large and its polar mass moment of inertia cannot be ignored as in the case of steam turbine of generator shaft, we can either consider a large number of stations and lump the inertia of each section as a rigid disk while retaining the elasticity of the shaft as in massless torsional shafts, or alternatively we can consider a distributed inertia and stiffness of the shaft between sections, where the shaft diameter is large and that its inertia cannot be ignored. Hence the system reduces to either several rigid disks connected by massless elastic shifts or distributed mass and elastic shafts. If one part of the system is coupled to another part through gears, belts or chains, the system inertias and stiffnesses should be reduced to one reference speed.

The number of stations to be considered in a large machine can be quite large, as for example 79 in a turbo-alternator set shown in Fig. 2.1, and a transfer matrix procedure will be preferred to determine the free vibration characteristics, see Pestel and Leckie [1]. The procedure for a continuous system model is first given by Eshelman [2], to determine the torsional response of an internal combustion engine. Rao et al. [3] adopted such a procedure for the analysis of torsional response in rotating machines, as described below.

2.2 Transfer Matrix Analysis for Free Vibrations

Figure 2.2a gives a continuous system model representation of the turbo-alternator set given in Fig. 2.1, with n number of elements. The ith element

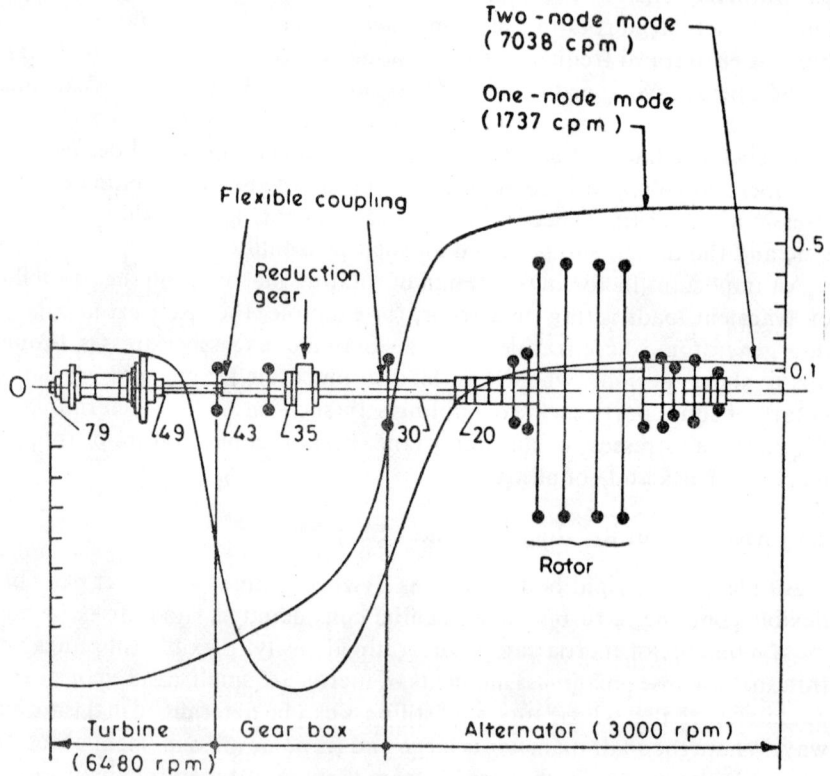

Fig. 2.1 6-MW turbo-alternator, torsional frequencies and mode shapes

is shown in Fig. 2.2b, with the notations adopted for the analysis given below. The differential equations of motion of the shaft elements are [4]

$$\rho J_i \ddot{\phi}_i - G J_i \phi_i'' = 0; \quad i = 1, 2, \ldots, n \qquad (2.2.1)$$

where ρ is mass density, J is polar second moment of area of ith shaft, G

is modulus of rigidity, ϕ is the angle of twist, ′ and · represent partial

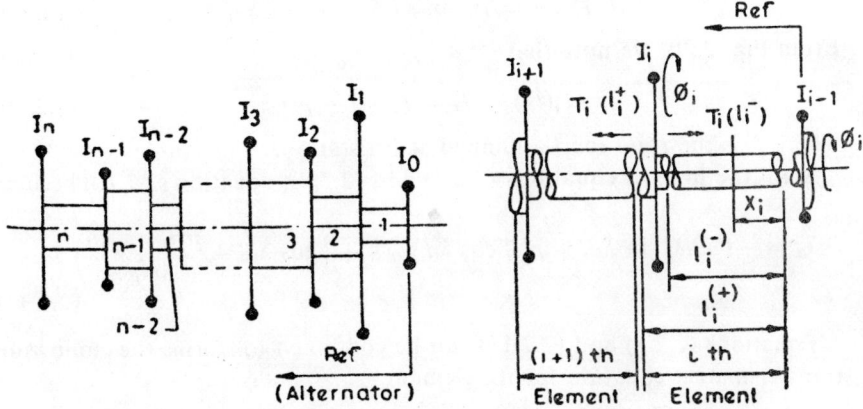

Fig. 2.2 (a) Continuous model, (b) notations used in ith element

derivatives with respect to the axial coordinate x and time t respectively.
The free vibration solution of equation (2.2.1) is

$$\phi_i(x_i, t) = \phi_i(x_i) \sin pt$$

where

$$\phi_i(x_i) = A_i \sin \lambda_i x_i + B_i \cos \lambda_i x_i \qquad (2.2.2)$$

and λ is non-dimensional value of natural frequency, p, given by

$$\lambda = p \sqrt{\frac{\rho}{G}} \qquad (2.2.3)$$

At $x_i = 0$, let

$$\phi_i(x_i = 0) = \phi_i(0)$$
$$T_i(x_i = 0) = GJ_i \phi'_i(0) = T_i(0) \qquad (2.2.4)$$

where T is the torque in shaft.

Using equations (2.2.4) in (2.2.2), we obtain

$$B_i = \phi_i(0)$$
$$A_i = \frac{T_i(0)}{GJ_i \lambda_i} \qquad (2.2.5)$$

Hence

$$\phi_i(x_i) = \phi_i(0) \cos \lambda_i x_i + \frac{T_i(0)}{GJ_i \lambda_i} \sin \lambda_i x_i \qquad (2.2.6)$$

$$T_i(x_i) = -GJ_i \lambda_i \phi_i(0) \sin \lambda_i x_i + T_i(0) \cos \lambda_i x_i \qquad (2.2.7)$$

Substituting $x_i = l_i$, equation (2.2.6) becomes

$$\phi_i = \cos \lambda_i l_i \phi_{i-1} + \frac{\sin \lambda_i l_i}{GJ_i \lambda_i} T_{i-1} \qquad (2.2.8)$$

For notation of ϕ and T, see Fig. 2.2b.

Similarly equation (2.2.7) becomes

$$T_i(l_i^-) = -GJ_i\lambda_i \sin \lambda_i l_i \phi_{i-1} + \cos \lambda_i l_i T_{i-1} \qquad (2.2.9)$$

From Fig. 2.2b, we note that

$$T_i(l_i^+) = T_i = T_i(l_i^-) - p^2 I_i \phi_i \qquad (2.2.10)$$

where I_i is the rigid inertia lumped at ith station.

With the help of equations (2.2.8) and (2.2.9), equation (2.2.10) become

$$T_i = -(GJ_i\lambda_i \sin \lambda_i l_i + p^2 I_i \cos \lambda_i l_i)\phi_{i-1} + \left(\cos \lambda_i l_i + \frac{p^2 I_i \sin \lambda_i l_i}{GJ_i\lambda_i}\right) T_{i-1} \qquad (2.2.11)$$

Equations (2.2.8) and (2.2.11) can be combined to form the following transfer matrix equation for ith element.

$$\{S\}_i = [T]_i \{S\}_{i-1} \qquad (2.2.12)$$

where $\{S\}$ is the state vector defined by

$$\{S\} = \begin{Bmatrix} \phi \\ T \end{Bmatrix} \qquad (2.2.13)$$

and $[T]$ is the ith element transfer matrix given by

$$[T]_i = \begin{bmatrix} \cos \lambda l & \dfrac{\sin \lambda l}{GJ\lambda} \\ -(p^2 I \cos \lambda l + GJ\lambda \sin \lambda l) & \left(\cos \lambda l - \dfrac{p^2 I \sin \lambda l}{GJ\lambda}\right) \end{bmatrix}_i \qquad (2.2.14)$$

For a discrete system, with no distributed inertia, i.e., the shaft is massless and of stiffness K, the above transfer matrix becomes

$$[T]_i = \begin{bmatrix} 1 & \dfrac{1}{K} \\ -p^2 I & 1 - \dfrac{p^2 I}{K} \end{bmatrix}_i \qquad (2.2.15)$$

The procedure to determine the natural frequencies and modes is as follows. Starting from element 1, we can apply equation (2.2.12) repeatedly to obtain the following:

$$\{S\}_1 = [T]_1 \{S\}_0$$
$$\{S\}_2 = [T]_2 \{S\}_1 = [T]_2 [T]_1 \{S\}_0$$
$$\vdots$$
$$\{S\}_n = [T]_n [T]_{n-1} \ldots [T]_2 [T]_1 \{S\}_0$$

i.e.,

$$\{S\}_n = [U]\{S\}_0 \qquad (2.2.16)$$

In the above $[U]$ is the overall transfer matrix of the system. Equation (2.2.16) in expanded form is

$$\begin{Bmatrix} \phi \\ T \end{Bmatrix}_n = \begin{bmatrix} u_{11} & u_{12} \\ u_{21} & u_{22} \end{bmatrix} \begin{Bmatrix} \phi \\ T \end{Bmatrix}_0 \tag{2.2.17}$$

Since T_n and T_0 at both the free ends are zero, we get from the above equation

$$u_{21} = 0 \tag{2.2.18}$$

The above condition will be satisfied, if p in equations (2.2.14) and (2.2.15) is a natural frequency. To determine this natural frequency, we can adopt the following procedure: Starting from an initial guess of p for the natural frequency, the transfer matrix for each element is first set up and the product indicated in equation (2.2.16) is taken to obtain the overall transfer matrix in equation (2.2.17). If the element u_{21} is zero according to equation (2.2.18), then the assumed p is a natural frequency. If u_{21} is not zero, then p is incremented and the procedure repeated until either u_{21} is zero or it changes sign. When the sign of u_{21} changes, a linear interpolation technique can be used to determine the natural frequency. This root searching technique is shown diagrammatically in Fig. 2.3. The interpolations indicated in this figure are carried until the required accuracy for the natural frequency is obtained. This value of the natural frequency is substituted in equation (2.2.16) with $\phi_0 = 1$, and $T_0 = 0$ in $\{S\}_0$ and the mode shape obtained.

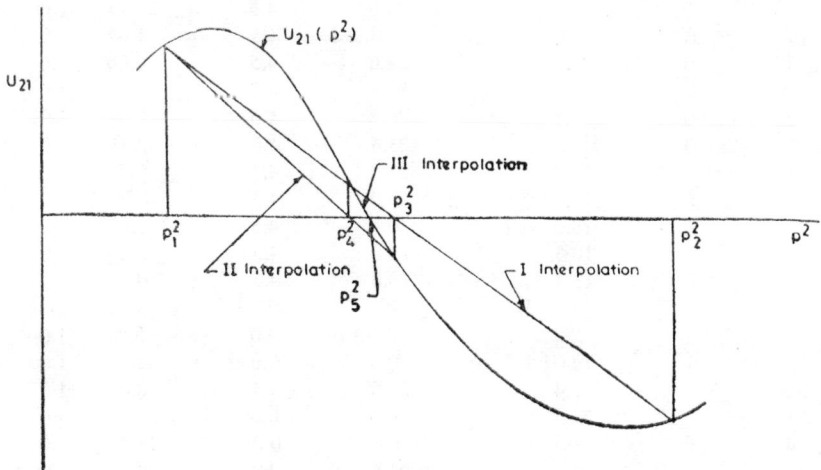

Fig. 2.3 Linear interpolation root searching technique

Example 2.1

The data for a 6 MW turbo-alternator set is given in Table 2.1. The natural frequencies and corresponding mode shapes can be obtained by the computer program [3] developed for this purpose using the above analysis. They are given in Fig. 2.1, for the first two modes.

Table 2.1 Input data for a 6 MW turbo-alternator

L—length of the shaft; D_0—outer diameter of the shaft; D_i—inner diameter of the shaft; I—lumped mass moment of inertia; TR—transmission ratio; LC—location where the masses have to be lumped; CN—coupling stiffness; $I1$—Component code; 0 for turbine or alternator 1 for coupling

Element No.	$I1$	L (cm) CN(Nm/rad)	D_0 (cm)	D_i (cm)	I (kg m²)	TR LC
1	2	3	4	5	6	7
1	0	14.0	22.0	4.5	0.00	1.0
2	0	16.5	20.0	4.5	0.00	1.0
3	0	16.5	20.0	4.5	0.00	1.0
4	0	11.5	22.0	4.5	0.00	1.0
5	0	3.0	23.5	4.5	0.00	1.0
6	0	14.1	23.6	4.5	2.11	1.0
7	0	22.5	30.8	4.5	0.01	1.0
8	0	11.5	32.0	4.5	7.26	1.0
9	0	19.9	34.0	4.5	0.00	1.0
10	0	7.0	35.0	4.5	5.00	1.0
11	0	36.5	35.0	4.5	27.80	1.0
12	0	3.1	35.0	4.5	6.81	1.0
13	0	37.7	55.3	4.5	48.01	1.0
14	0	37.7	55.3	4.5	48.01	1.0(1)
15	0	37.7	55.3	4.5	48.01	1.0
16	0	37.7	55.3	4.5	48.01	1.0
17	0	3.1	35.0	4.5	6.81	1.0
18	0	36.5	35.0	4.5	27.80	1.0
19	0	7.0	35.0	4.5	5.00	1.0
20	0	19.9	34.0	4.5	0.00	1.0
21	0	11.5	32.0	4.5	7.26	1.0
22	0	22.5	30.8	4.5	0.01	1.0
23	0	14.1	23.6	4.5	2.11	1.0
24	0	3.0	23.5	4.5	0.00	1.0
25	0	10.0	22.0	4.5	0.00	1.0
26	0	18.0	20.0	4.5	0.00	1.0
27	0	18.5	20.0	4.5	0.00	1.0
28	0	13.4	20.0	4.5	0.00	1.0
29	1	15.4×10^7	0.0	0.0	0.00	1.0
30	1	39.8×10^7	0.0	0.0	3.82	1.0
31	1	13.0×10^7	0.0	0.0	3.83	1.0
32	0	13.4	20.0	0.0	0.00	1.0
33	0	39.8	20.0	0.0	0.00	1.0
34	0	9.5	26.0	0.0	11.20	1.0
35	0	9.7	61.6	0.0	0.00	1.0(2)
36	0	9.7	61.6	0.0	0.00	1.0
37	0	28.5	28.5	0.0	0.00	2.16
38	0	1.5	16.0	0.0	0.02	2.16
39	0	33.0	16.0	0.0	0.02	2.16
40	0	2.0	17.0	0.0	0.00	2.16
41	0	5.0	15.0	0.0	0.00	2.16
42	0	10.5	14.5	0.0	0.00	2.16

1	2	3	4	5	6	7
43	1	5.0×10^7	0.0	0.0	0.00	2.16
44	1	14.5×10^7	0.0	0.0	1.36	2.16
45	1	5.0×10^7	0.0	0.0	1.36	2.16
46	0	10.5	14.5	0.0	0.00	2.16
47	0	17.8	16.0	0.0	0.00	2.16
48	0	1.2	13.0	0.0	0.00	2.16
49	0	11.2	13.0	0.0	0.00	2.16
50	0	1.5	13.0	0.0	0.01	2.16
51	0	3.8	18.0	0.0	0.00	2.16
52	0	1.4	18.0	0.0	0.03	2.16
53	0	14.8	24.6	0.0	0.00	2.16
54	0	2.5	24.6	0.0	0.48	2.16
55	0	2.9	39.4	0.0	0.46	2.16
56	0	5.3	34.5	0.0	4.37	2.16
57	0	3.0	38.5	0.0	0.53	2.16
58	0	3.8	37.7	0.0	2.50	2.16
59	0	4.5	42.0	0.0	0.44	2.16
60	0	11.7	39.0	0.0	3.97	2.16
61	0	10.2	43.0	0.0	0.00	2.16
62	0	10.2	38.5	0.0	2.83	2.16
63	0	13.7	42.5	0.0	0.00	2.16(3)
64	0	6.1	38.9	0.0	0.00	2.16
65	0	1.4	34.4	0.0	0.05	2.16
66	0	11.2	33.2	0.0	1.05	2.16
67	0	12.7	33.2	0.0	1.17	2.16
68	0	12.7	33.2	0.0	1.12	2.16
69	0	8.6	37.0	0.0	0.00	2.16
70	0	4.5	61.4	0.0	0.00	2.16
71	0	19.7	44.0	0.0	0.00	2.16
72	0	5.3	25.0	0.0	0.00	2.16
73	0	5.8	18.0	0.0	0.00	2.16
74	0	9.6	11.0	0.0	0.00	2.16
75	0	6.3	9.0	0.0	0.00	2.16
76	0	2.3	18.0	0.0	0.00	2.16
77	0	9.0	8.0	0.0	0.00	2.16
78	0	2.3	18.0	0.0	0.00	2.16
79	0	4.2	3.0	0.0	0.00	2.16

The free vibrational characteristics thus obtained will be used in forced vibration analysis, given in next sections.

2.3 Equivalent Discrete System

To determine the transient response due to excitation torque, which acts on one rotor in the system, like the generator rotor in a turbo-alternator system, it is convenient to reduce the continuous system model considered earlier to an equivalent discrete system with a fewer number of rotors. To obtain such a model, the mass moments of inertia of both the attached masses and shaft elements are summed up between two nodal points and lumped at chosen locations corresponding to the approximate centres of gravity [5]. The approximate stiffnesses, k_j, of the shafting between these locations of masses are then evaluated, which are refined to obtain the dynamically equivalent discrete model of the continuous system as explained below. For the discrete system, the equations of motion in matrix form are (see section 3.2):

$$[M]\{\ddot{\phi}\} + [K]\{\phi\} = 0 \qquad (2.3.1)$$

Equation (2.3.1) reduces to standard eigen value problem,

$$[K] - p^2[M] = 0 \qquad (2.3.2)$$

from which we can determine the natural frequencies and mode shapes of the discrete system with given mass moments of inertia, m_j, and approximate stiffnesses. We shall denote them as p_i^* with the corresponding mode shapes $\{\phi^*\}_i$. In ith mode vibration, the potential energy of the discrete system is (N — number of rotors in the discrete system)

$$V_i = \sum_{j=1}^{N-1} \tfrac{1}{2} k_j (\phi_j - \phi_{j+1})_i^2 \qquad (2.3.3)$$

The expression for the kinetic energy is given by

$$T_i = \sum_{j=1}^{N} \tfrac{1}{2} m_j (\phi_j^2)_i p_i^2 \qquad (2.3.4)$$

Since in any normal mode, the kinetic and potential energies are equal, we have

$$\sum_{j=1}^{N-1} k_j (\phi_j - \phi_{j+1})_i^2 = p_i^2 \sum_{j=1}^{N} m_j (\phi_j^2)_i; \quad i = 2, 3, \ldots, N \qquad (2.3.5)$$

In the above system of equations, we substitute exact p_i and m_j and approximate mode shape vectors $\{\phi^*\}_i$ to evaluate modified stiffnesses. With these modified stiffnesses, we return to equation (2.3.2) and repeat the process until we get the desired accuracy for the stiffnesses.

Example 2.2

Consider the turbo-alternator set given in Fig. 2.1. The locations where the masses are to be lumped are indicated in Table 2.1. These masses, are

Generator, $m_1 = 380$ kg m²

Gear box, $m_2 = 71.5$ kg m²

Turbine, $m_3 = 230$ kg m²

The values of stiffness determined from a computer program for an equivalent discrete system of this set are

Stiffness between generator and gear box, $k_1 = 0.644 \times 10^7$ Nm/rad,

Stiffness between gear box and turbine, $k_2 = 2.563 \times 10^7$ Nm/rad.

This discrete system can be used to determine the transient response of the turbo-alternator set in Fig. 2.1, as outlined later in this chapter.

2.4 Excitation Torque

In rotating machine installations, having driver and driven units, the two units can be connected by suitable couplings to avoid the possibility of resonance in the operating range of the system. This avoids the necessity of forced vibration analysis under steady state conditions, as in the case of reciprocating machine installations, where many critical speeds can exist in the operating zone. However, transient response analysis is often necessary to avoid possible low cycle fatigue caused by suddenly applied loads acting for a short duration. Two typical cases of excitation forces which are transient in nature are discussed below.

2.4.1 Electrical unbalance load in a generator

The most important disturbances creating abnormal pulsating torques in a turbo-alternator shafting system are: short circuit at generator terminals, faulty synchronising, short circuit clearing, and line switching. Sudden short circuit at generator terminals considered as the most unfavourable condition, sets up large torques in the rotor and induces severe stresses of the order of four to five times the normal value [6]. Short circuiting in a generator terminal occurs in a small interval of time, about 0.35 seconds, and is removed by automatic controls. This short circuiting period can be divided into two parts, the first of which lasts about 0.05 seconds known as subtransient period. In this period, the current across the terminals rises up to 10-15 times the rated value. In the second part known as transient period, the current fluctuations are 3-5 times the rated current.

The short circuiting torque consists of two components:
1. Air gap torque T_a.
2. Armature torque T_s.

The air gap torque decays depending on the time constant of the generator and is more predominant in the subtransient period. The equation for this torque T_a, in terms of the normal torque T_n is given by [7]

$$T_a = T_n m_a e^{-\alpha t} \sin \omega t \qquad (2.4.1)$$

where α = subtransient saliency constant; m_a = oscillating moment; and ω = rotational frequency.

The second component comes into action in the transient period and the equation for this torque is

$$T_s = T_n m_s e^{-\beta t} \sin 2\omega t \qquad (2.4.2)$$

where m_s and β are transient saliency constants. It may be noted that the armature torque occurs at twice the rotational frequency of the shaft. For a more detailed description on the short circuit torque see [6, 7 and 8].

2.4.2 Transient electric motor torque of a steel mill stand

Consider the hot strip steel mill stand as given by Smalley [9] shown in Fig. 2.4 schematically. There are two rolls 8 and 9, one vertically above the other which are driven through a series of couplings and a pinion stand by a pair of D.C. motors 1 and 2. Initially the rolls run at constant speed, as the mill is essentially unloaded and the only torque output from the motors is that required to overcome friction at the bearings in the system. When a hot strip of steel enters the rolls, the rolls slow down and transmit a torque back to the motors. Thus the motors themselves slow down, though to a limited extent, because of their very large inertia. Responding to this deceleration, the motor windings impose an accelerating torque on the armatures, part of which is transmitted back to the rolls. This results in an electrical torque characteristic of the motors with a response time of about 0.3 seconds.

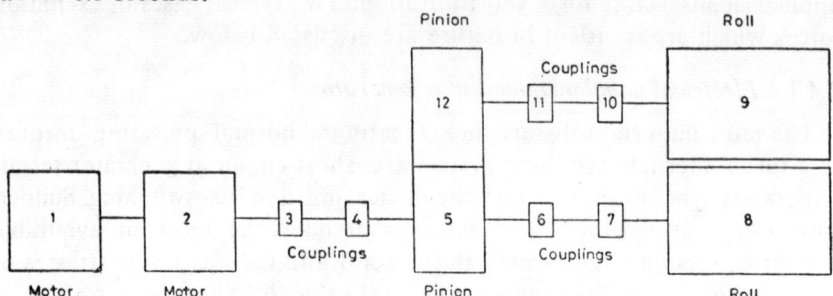

Fig. 2.4 Schematic of mill stand drive train [9]

To determine the angles of twist to each rotor due to the transient excitation torques, the well-known method of modal expansion can be used, as outlined in the next section.

2.5 Transient Response in Torsional Vibration

The equations of motion for the discrete system in forced vibration are

$$[M]\{\ddot{\phi}\} + [K]\{\phi\} = \{f\} \tag{2.5.1}$$

where $\{f\}$ is the excitation vector, which is a time dependent function. Using normal mode expansion for the response, we write

$$\{\phi\} = [X]\{q\} \tag{2.5.2}$$

where $[X]$ is twist modal matrix and $\{q\}$ is generalised coordinate vector. Substituting equation (2.5.2) in equation (2.5.1), premultiplying this equation with the transpose of modal matrix $[X]$, and using the orthogonal properties of normal mode shapes, we obtain the modal equations as follows [4]:

$$\overline{m}_k \ddot{q}_k + \overline{k}_k q_k = \bar{f}_k; \quad k = 1, 2, \ldots, N \tag{2.5.3}$$

where $\overline{m}_k$ = generalised mass in kth mode of vibration

$$= \sum_{i=1}^{N} m_i x_{ik}^2 \tag{2.5.4}$$

$\overline{k}_k$ = generalised stiffness in kth mode of vibration

$$= p_k^2 \overline{m}_k \tag{2.5.5}$$

and $\bar{f}_k$ = generalised force in kth mode of vibration

$$= \sum_{i=1}^{N} f_i x_{lx} \tag{2.5.6}$$

The solution of equation (2.5.3) is [4],

$$q_k(t) = q_k(0) \cos p_k t + \dot{q}_k(0) \frac{\sin p_k t}{p_k} + \int_0^t \frac{\bar{f}_k(\tau) \sin p_k(t-\tau)}{p_k m_k} d\tau \quad (2.5.7)$$

For the rigid body mode ($k=1$, $p_1=0$), in the limit, the above becomes

$$q_1(t) = q_1(0) + \dot{q}_1(0)t + \frac{1}{m_k} \int_0^t \bar{f}_k(\tau)(t-\tau) d\tau \quad (2.5.8)$$

To evaluate $q_k(0)$, consider the state of the turbo-alternator just before short circuiting. The angles of twist to all the rotors will not depend on time, hence

$$[K]\{\phi(0)\} = \{f(0)\} \quad (2.5.9)$$

Knowing $f_i(0)$, $\phi_i(0)$ can be determined by a solution of the above equations. However, the stiffness matrix $[K]$ is singular and it cannot be inverted. We replace the first equation in the N equations of (2.5.9) by

$$\sum_{i=1}^N m_i \phi_i(0) = 0 \quad (2.5.10)$$

which is merely the sum of all equations (2.5.1), for free vibration.

Equation (2.5.10) and the last $N-1$ equations of (2.5.9) can be solved to determine $\phi_i(0)$. Then $q_i(0)$ are determined from equation (2.5.2). A suitable numerical integration procedure can be used for evaluating $q_k(t)$ in equation (2.5.7), and $\phi_i(t)$ are determined from equation (2.5.2) again.

Example 2.3

A step input torque is given to the generator rotor of the equivalent discrete system obtained in example 2.2. For this input, the integration in

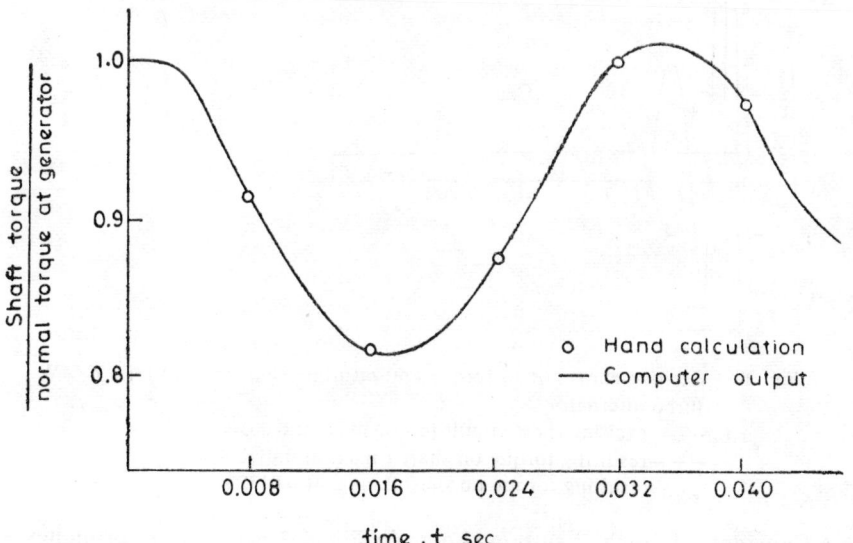

Fig. 2.5 Response to step input

equation (2.5.7) can be carried out by hand. The response obtained from a computer program [3] and the hand calculations are given in Fig. 2.5.

Example 2.4

A typical short circuit excitation torque for the generator is shown in Fig. 2.6. The resulting torques on shaft 1 between the generator and gear box and on shaft 2 between the gear box and turbine, which correspond to the flexible couplings, are determined and plotted in Fig. 2.6. The magnitude of these torques is as much as three to four times the rated value of the torque.

2.6 Branched Systems

In the analysis of previous sections, we considered straight torsional systems with no branches. In several systems like the strip steel mill stand of Fig. 2.4, ship propulsion systems, machine tool drives etc., there are one or two branch points, as they employ one or two drivers driving one or more driven members. In the example of the hot strip steel mill stand in Fig. 2.4, rotors 1 to 8 form a straight torsional system with a branch at station 5 consisting rotors 9 to 12. The procedure to determine natural frequencies and mode shapes is discussed in this section. Once the free vibrational characteristics are determined the transient analysis can be accomplished by following the modal expansion method described earlier in section 2.5.

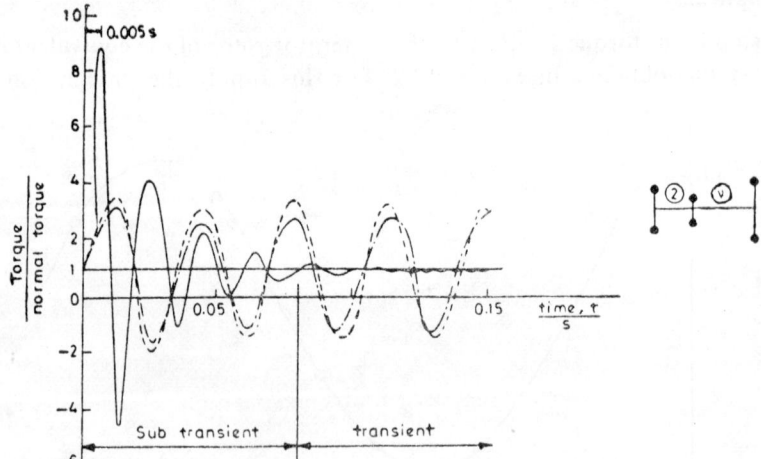

Fig. 2.6 Exciting short circuit torque and resulting shaft torques for a 6-MW turbo-alternator
——— exciting short circuit torque max. at 0.005 s
− − − resulting torque on shaft 1 max. at 0.0132 s
−·−·− resulting torque on shaft 2 max. at 0.0132 s

Consider a branched system shown in Fig. 2.7, with three branches A, B and C. The number of rotors in the branches A, B and C are denoted

Torsional Vibrations in Rotating Machinery

by n_A, n_B and n_C respectively. We first determine the overall transfer matrices for each individual branch as follows.

$$\{S\}_{nA}^R = [A]\{S\}_{A0} \qquad (2.6.1)$$

$$\{S\}_{nB}^R = [B]\{S\}_{B0} \qquad (2.6.2)$$

$$\{S\}_{nC}^R = [C]\{S\}_{C0} \qquad (2.6.3)$$

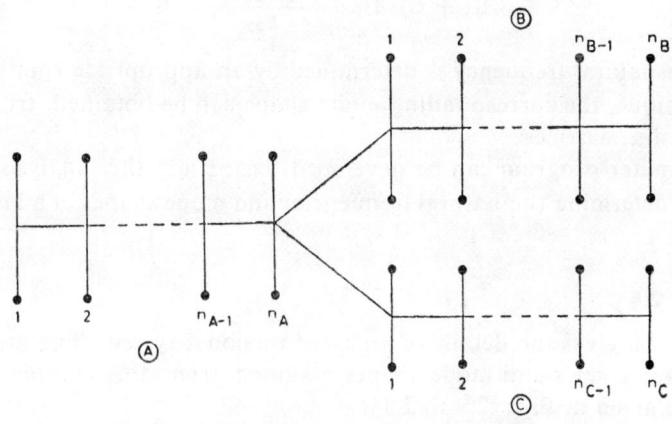

Fig. 2.7 Branched torsional system

We have the following conditions to be statisfied at the branch point:

$$\phi_{nA} = \phi_{B0} = \phi_{C0} \qquad (2.6.4)$$

$$T_{nA} = T_{B0} + T_{C0} \qquad (2.6.5)$$

Using the conditions at the free end of branch A, we have

$$\left\{\begin{matrix}\phi \\ T\end{matrix}\right\}_{nA}^R = \begin{bmatrix}A_{11} & A_{12} \\ A_{21} & A_{22}\end{bmatrix}\left\{\begin{matrix}1 \\ 0\end{matrix}\right\} \qquad (2.6.6)$$

i.e.,

$$T_{nA}^R = A_{21}$$

$$\phi_{nA}^R = A_{11} \qquad (2.6.7)$$

Equation (2.6.2) now becomes

$$\left\{\begin{matrix}\phi \\ 0\end{matrix}\right\}_{nB}^R = \begin{bmatrix}B_{11} & B_{12} \\ B_{21} & B_{22}\end{bmatrix}\left\{\begin{matrix}A_{11} \\ T\end{matrix}\right\}_{B0} \qquad (2.6.8)$$

Therefore

$$T_{B0} = -\frac{B_{21}A_{11}}{B_{22}} \qquad (2.6.9)$$

Using equations (2.6.7) and (2.6.9), in equation (2.6.5), we get

$$T_{C0} = A_{21} + \frac{B_{21}A_{11}}{B_{22}} \qquad (2.6.10)$$

22 *Rotor Dynamics*

Since $T_{nC}^R = 0$, equation (2.6.3) can now be written as

$$\begin{Bmatrix} \phi \\ 0 \end{Bmatrix}_{nC}^R = \begin{bmatrix} C_{11} & C_{12} \\ C_{21} & C_{22} \end{bmatrix} \begin{Bmatrix} A_{11} \\ A_{21} + \dfrac{B_{21} A_{11}}{B_{22}} \end{Bmatrix} \qquad (2.6.11)$$

The frequency equation is therefore

$$C_{21} A_{11} + C_{22} A_{21} + \dfrac{C_{22} B_{21} A_{11}}{B_{22}} = 0 \qquad (2.6.12)$$

Once a natural frequency is determined by an appropriate root searching technique, the corresponding mode shape can be obtained from station transfer matrices.

A computer program can be developed based on the analysis given above to determine the natural frequencies and mode shapes of a branched system.

Example 2.5

Figure 2.8 gives the details of a geared torsional system. The first three natural frequencies and mode shapes obtained from the computer program are given in Figs. 2.9 to 2.11.

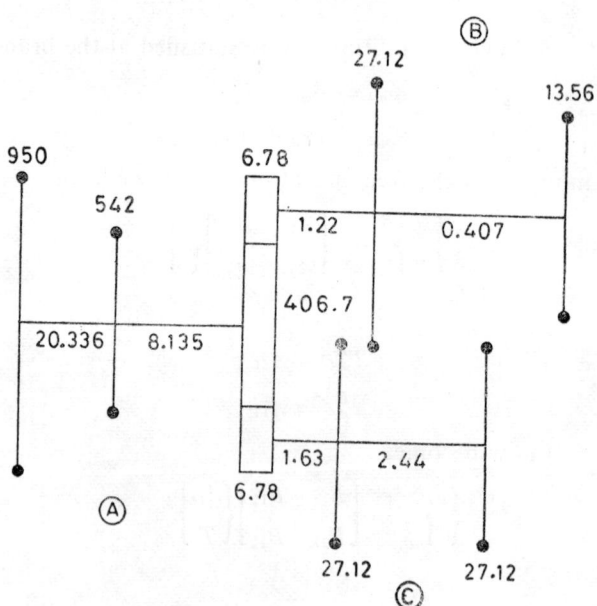

Fig. 2.8 Geared torsional system of Jacobsen and Ayre [10]
  Inertias—kg m²
  Stiffnesses—Nm/rad × 10⁶
  Gear ratio = $\dfrac{76}{24}$

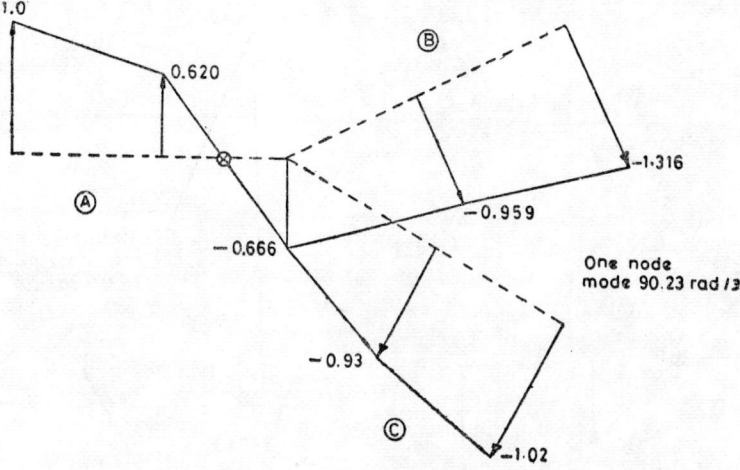

Fig. 2.9 One node mode of the geared torsional system

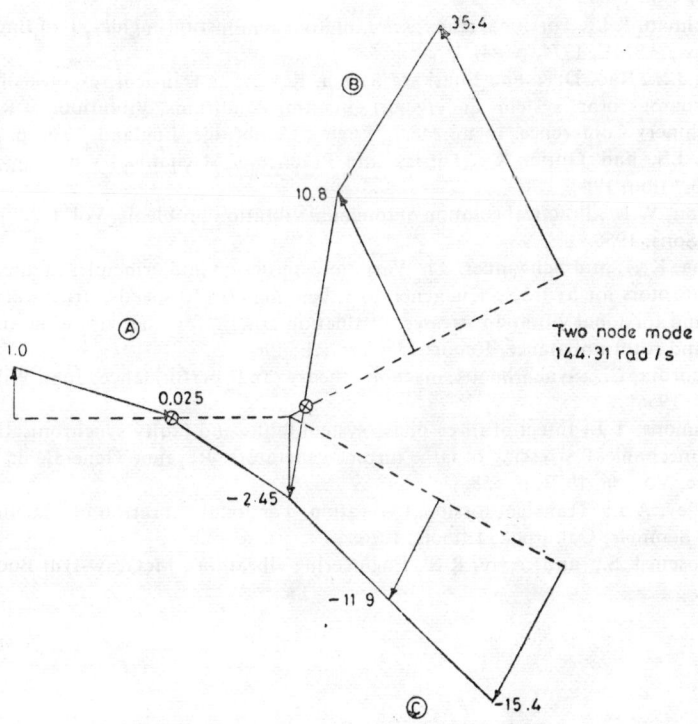

Fig. 2.10 Two node mode of geared torsional system

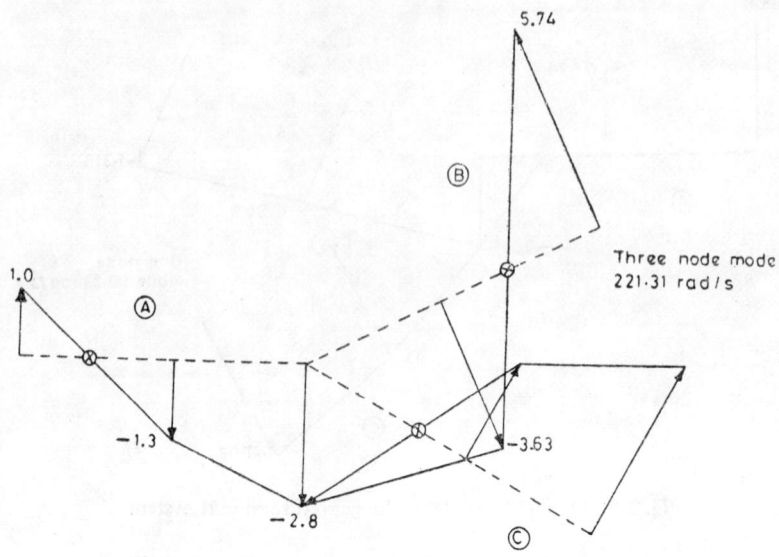

Fig. 2.11 Three node mode of geared torsional system

References

1. Pestel, E.C. and Leckie, F.A., Matrix methods in elastomechanics, McGraw-Hill Book Co., 1963.
2. Eshelman, R.L., Torsional response of internal combustion engines, J. of Engng, for Indus., ASME, 1974, p. 441.
3. Rao, J.S., Rao, D.K. and Bhaskara Sarma, K.V., The transient response of turbo-alrernator rotor systems under short circuiting conditions, Vibrations of Rotating Machinery Conference, Instn. Mech. Engrs., Cambridge, England, 1980, p. 271.
4. Rao, J.S. and Gupta, K., Theory and Practice of Mechanical Vibrations, Wiley Publication, 1984.
5. Wilson, W.K., Practical solution of torsional vibration problems, Vol. I, John Wiley and Sons, 1956.
6. Saling, K.H. and Schwinder, Th. Variations in design and principles of dimensioning of rotors for hydroelectric generators, bending critical speeds, structural resources and torsional vibration stresses; Evaluation criteria for the vibration of bearing and rotor unbalance, Report 138, CIGRE, 1966.
7. Concordia, C., Synchronous machine theory and performance, John Wiley and Sons, 1956.
8. Hammons, T.J., Effect of three-phase system faults and faulty synchronisation on the mechanical stressing of large turbine-generators, Revenue Generale de l'Electricite, Vol. 86, 1977, p. 558.
9. Smalley, A.J., Transient torsional vibration, Torsional Vibration of Machine Systems Seminar, Oakbrook, Illinois, 1976.
10. Jacobsen, L.S., and Ayre, R S., Engineering vibrations, McGraw-Hill Book Co., 1963.

Chapter 3

Torsional Vibrations in Reciprocating Machinery

Torsional oscillations in the crankshaft and in the shafting of driven machinery is a vibration phenomenon of practical importance in the design of reciprocating machines. The average torque delivered by a cylinder in a reciprocating machine, is a small fraction of the maximum torque which occurs during the firing period. Even though the torque is periodic the fact that it fluctuates so violently within the period, constitutes one of the inherent disadvantages of a reciprocating machine, from the dynamics point of view, as compared with a turbine where the torque is practically uniform. It is possible to express the torque delivered by a reciprocating engine into its harmonic components of several orders of the engine speed, and these harmonic components can excite the engine driven installations into forced torsional vibrations. The engine and the driven unit such as a generator or a pump are normally connected by a flexible coupling and thus the total installation has fairly low natural frequencies falling in the speed range of the engine and the harmonics of different orders.

It is a commonly known fact that failures can occur in reciprocating machine installations, when the running speed of the engine is at or near a dominant critical speed of the system. High dynamic stresses can occur in the main shafting of such engine installations and to avoid these conditions, it is essential that the torsional vibration characteristics of the entire installation be analysed before the unit is put into operation. Various classification societies, such as Lloyd's register of shipping, American bureau of shipping, Germanischer Lloyd etc., have framed recommendations, rules and guidance notes regarding the permissible torsional vibration stresses in any installation, see BICERA [1]. Any analysis of torsional vibration characteristics of reciprocating machinery should finally predict the maximum dynamic stresses or torque developed in the shafting and couplings of the system, as accurately as possible, so that they can be compared with the permissible values, to check the safety of the installation. The hand book of BICERA [1] and series of the volumes on torsional vibration problems by Wilson [2], contain very valuable practical information on this subject.

3.1 Modelling of the Reciprocating Machine Systems

A multicylinder reciprocating machine contains many reciprocating and rotating parts such as pistons, connecting rods, crankshaft, flywheel, damper and auxiliary drives. The system is so complicated that it is difficult, if not impossible, to undertake an exact analysis of its vibrational characteristics. The information that one can obtain from such a rigorous analysis, is also of doubtful value for the present-day speeds of the reciprocating machinery. The actual system is characterised by the presence of unpredictable effects like variable inertia, internal damping, misalignments in the transmission units, uneven firing intervals etc. The analysis can be

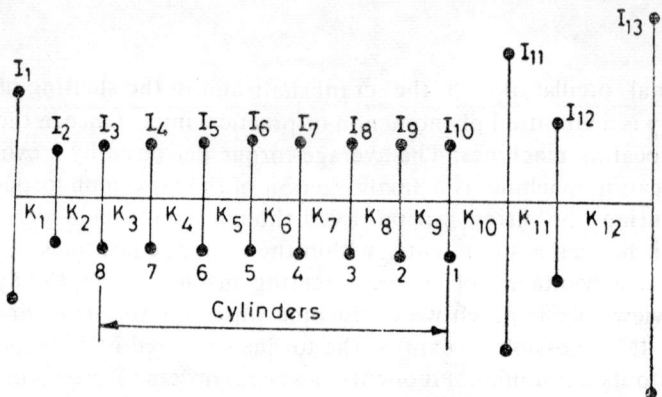

Fig. 3.1 N rotor model of a typical reciprocating engine installation

Table 3.1

Rotor No.	I		K	
	Description	kg m²	Description	N m/rad × 10⁶
1	Outer part of damper	3.2262		
2	Inner part of damper	0.3628	Sleeve spring stiffness	0.3334
3	Cyl No. 8	0.5482	Crankshaft up to Cyl 8	5.9817
4	Cyl No. 7	0.5482	Crankshaft Cyl 8 − 7	4.5794
5	Cyl No. 6	0.5482	Cyl 7 − 6	4.5794
6	Cyl No. 5	0.5482	Cyl 6 − 5	4.5794
7	Cyl No. 4	0.5482	Cyl 5 − 4	4.5794
8	Cyl No. 3	0.5482	Cyl 4 − 3	4.5794
9	Cyl No. 2	0.5482	Cyl 3 − 2	4.5794
10	Cyl No. 1	0.5482	Cyl 2 − 1	4.5794
11	Fly wheel + Primary of coupling	15.3563	Cyl 1 − fly wheel	3.6871
12	Secondary of coupling	0.5952	Vulcan coupling	0.08825
13	Generator	16.0525	Generator shaft	2.1966

best carried out, by lumping the inertias of rotating and reciprocating parts at discrete points on the main shaft. The problem then reduces to the forced vibration study of an N rotor system subjected to varying torques at different cylinder points.

The crankshaft and the other drive or driven shafts are generally flexible in torsion, but have low polar mass moments of inertia, unlike in the case of some large turbines or compressors. On the other hand, the parts mounted on the shafting, like the damper, flywheel, generator etc., are rigid and will have very high polar mass moments of inertia. The system containing the crankshaft, coupling, generator/auxiliary drive shaft/other driven shafts like pumps, and the mounted parts can then be reduced to a simple system with a series of rigid rotors (representing the inertias) connected by the massless flexible shafts as shown in Fig. 3.1, which shows a torsional vibration model of a typical diesel engine driven installation.

3.1.1 *Polar mass moments of inertia*

Determination of polar mass moments of inertia is a straightforward matter for rotating parts [1], however, it is not quite so simple in the case of reciprocating parts. Consider the piston shown in two different positions in Fig. 3.2 and imagine the crankshaft with a polar mass moment of inertia, I_{rot}, to be nonrotating but executing small torsional oscillations. In Fig. 3.2a, there is no motion for the piston, with small oscillations of the crank and hence the equivalent inertia of the piston is zero, whereas

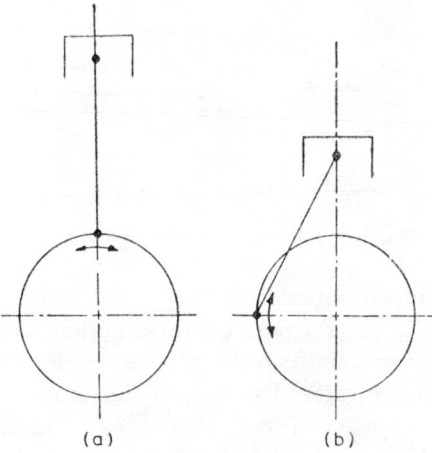

Fig. 3.2 Equivalent mass moment of inertia of a piston

in Fig. 3.2b, the piston has practically the same acceleration as that of the crank pin and the equivalent inertia is $m_{rec}r^2$, where m_{rec} is the mass of the reciprocating parts and r is the crank throw. Hence, the total polar mass moment of inertia varies from I_{rot} to $I_{rot} + m_{rec}r^2$, when the crankshaft is rotating. We consider as an approximation, the system to have an average inertia given by

$$I = I_{rot} + \tfrac{1}{2}m_{rec}r^2 \qquad (3.1.1)$$

The values of polar mass moments of inertia of the system in Fig. 3.1 are given in Table 3.1.

3.1.2 *Torsional stiffness of shafts connecting the rotors*

In determining the torsional stiffness of shafts connecting the rotors, the main difficulty arises from the crank webs. Consider the idealisation of a crank throw, into an ordinary shaft having the same flexibility as shown in Fig. 3.3. Though this is physically possible, the calculations involved are extremely difficult. This is because the crank webs are subjected to bending and the crank pin to twisting, if the main shaft is subjected to twisting. Moreover the beam formulae, if used will not be very accurate, because of short stubs involved rather than long beams usually considered. Further the torque applied at the free end also gives rise to sidewise displacement which is prevented in the machine. For high speed light weight engines, the crank webs are no more rectangular blocks and application of the theory becomes extremely difficult.

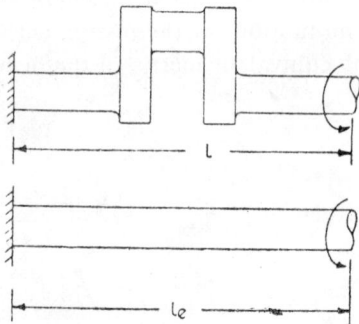

Fig. 3.3 Equivalent length of a crank

Because of these uncertainties in analytical calculation to estimate the torsional stiffness of crank throws, several experiments have been carried out on a number of crank shafts of large slow speed engines, which have shown [3] that the equivalent length l_e of Fig. 3.3 is nearly equal to the actual length, if the diameter of main shaft is equal to the crank pin diameter. In general the procedure that is adopted to reduce the reciprocating machine system to a mathematical model, is to use a basic diameter, which corresponds to the journal diameter of the crank shaft. The torsional stiffnesses are all calculated based on the basic diameter, irrespective of their actual diameter. For crank webs of different designs, there are several different formulae proposed, such as BICERA, Carter, Constant, Heldt, Jackson, Wilson, Timoshenko, Tuplin and Zimanenko, see [1, 2].

For the end rotors in Fig. 3.1, e.g., the generator rotor, the point of rigidity is to be calculated as given by BICERA [1], and compute the stiffness of the shaft from the coupling up to the point of rigidity. In case where one part of the system is connected to the other part through gears, or other transmission units, it is convenient to reduce all the inertias and stiffnesses to one reference speed. The numerical data for a typical reciprocating engine [4], given in Table 3.1, is used in illustrating the critical speed calculations, forced vibration response, and the coefficient of cyclic irregularity in the following sections.

3.2 Free Vibration Calculations

In the previous chapter, we adopted a transfer matrix procedure to determine the free vibration characteristics. In reciprocating machine installations, the number of stations are generally few in the mathematical model, e.g., 13 for the system in Fig. 3.1, and it is convenient to adopt the classical eigen value problem approach [5], rather than the transfer matrix procedure. This procedure is briefly described here. Let ϕ_i denote the angle of twist of ith rotor. The free torsional vibration equations of motion are

$$I_1\ddot{\phi}_1 + K_1\phi_1 - K_1\phi_2 = 0$$
$$I_2\ddot{\phi}_2 - K_1\phi_1 + (K_1 + K_2)\phi_2 - K_2\phi_3 = 0$$
$$\vdots$$
$$I_i\ddot{\phi}_i - K_{i-1}\phi_{i-1} + (K_{i-1} + K_i)\phi_i - K_i\phi_{i+1} = 0$$
$$\vdots$$
$$I_N\ddot{\phi}_N - K_{N-1}\phi_{N-1} + K_{N-1}\phi_N = 0 \tag{3.2.1}$$

In matrix notation the above can be written as

$$[M]\{\ddot{\phi}\} + [K]\{\phi\} = 0 \tag{3.2.2}$$

where

$$[M] = \begin{bmatrix} I_1 & & & & \\ & I_2 & & & \\ & & I_3 & & \\ & & & - & \\ & & & & - \\ & & & & & I_N \end{bmatrix} \tag{3.2.3}$$

$$[K] = \begin{bmatrix} K_1 & -K_1 & & & & \\ -K_1 & (K_1 + K_2) & -K_2 & & & \\ & -K_2 & (K_2 + K_3) & -K_3 & & \\ & & - & - & - & \\ & & & & & -K_{N-1} \\ & & & & -K_{N-1} & K_{N-1} \end{bmatrix}$$

$$\tag{3.2.4}$$

$$\{\phi\} = \begin{Bmatrix} \phi_1 \\ \phi_2 \\ - \\ - \\ \phi_N \end{Bmatrix} \qquad (3.2.5)$$

In the above $[M]$ and $[K]$ are the familiar mass and stiffness matrices and $\{\phi\}$ is the response vector.

The normal procedure adopted to find the natural frequencies and modes of vibration for the reciprocating machine systems is the Holzer method [1, 2, 3]. If the number of rotors is fairly small, as in the present example, matrix methods are most amenable for programming a digital computer.

For free harmonic vibrations,

$$\{\phi(t)\} = \{\phi\} \exp{(ipt)} \qquad (3.2.6)$$

Equation (3.2.2) now becomes

$$[[K] - p^2[M]]\{\phi\} = 0 \qquad (3.2.7)$$

Equation (3.2.7) is a standard eigen value problem and the natural frequencies with the corresponding modal matrix $[X]$ with the jth mode shape in the jth column can be obtained, by any standard computer program, using say, the Jacobian method of orthogonalisation by successive rotation.

The modal matrix $[X]$ obtained for a given problem can be normalised with respect to a specified basic rotor and normalised matrix, thus obtained is called twist modal matrix.

To determine the dynamic stresses, we will later use the modal expansion method, and for this purpose we determine the stress modal matrix as follows. The resisting torque in ith shaft is given by

$$T_i = K_i(\phi_{i+1} - \phi_i) \qquad (3.2.8)$$

Adding the first i equations in (3.2.1), we notice that this resisting torque can also be written as

$$T_i = \sum_{k=1}^{i} I_k \ddot{\phi}_k \qquad (3.2.9)$$

If the system is vibrating in the jth mode, then $\ddot{\phi}_k = -p_j^2 x_{kj}$, where p_j is the jth mode natural frequency and x_{kj} is the angular twist of kth rotor in jth mode. Hence the corresponding internal torque developed in the ith shaft and jth mode is

$$T_{ij} = \sum_{k=1}^{i} \bar{I}_k p_j^2 x_{kj} \qquad (3.2.10)$$

If d_i is the minimum diameter of the ith shaft, then the section modulus for the minimum diameter is

$$Z_i = \frac{\pi d_i^3}{16} \qquad (3.2.11)$$

The stress value corresponding to the minimum diameter, or the maximum stress in the ith shaft and jth mode of vibration is

$$f_{ij} = \frac{T_{ij}}{Z_i} \qquad (3.2.12)$$

Wilson [2] called this value as specific stress per unit twist at the basic rotor and Lloyd's [7] define this as nominal shaft stress factor. The matrix $[F] = [f_{ij}]$ is called the stress modal matrix, which will be used later in the modal analysis for forced vibration studies.

Example 3.1

The arrangement of rotors of a diesel locomotive engine is given in Fig. 3.4, along with the polar mass moments of inertia and stiffnesses.

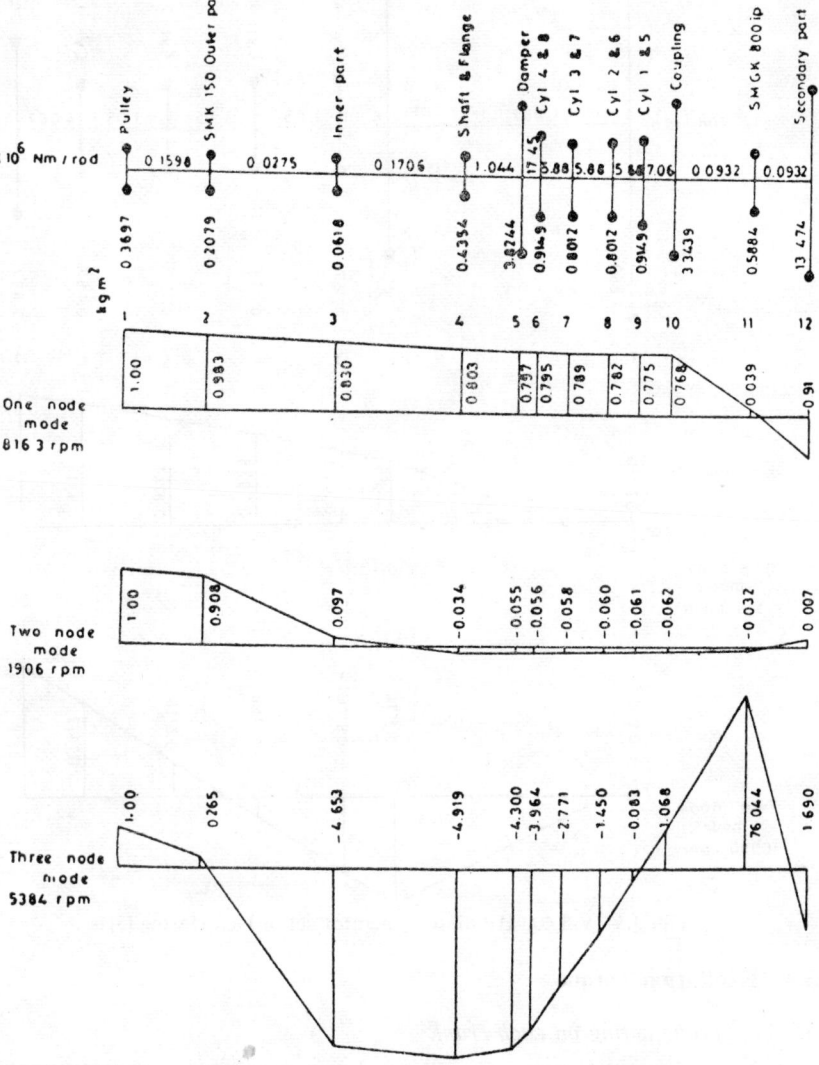

Fig. 3.4 Locomotive diesel engine used in Indian Railways [6]

The first three natural frequencies and corresponding mode shapes obtained by a computer program [4] are given in Fig. 3.4.

Example 3.2

Figure 3.5 gives the data for a 400 HP diesel generator set. The first two natural frequencies obtained are given in the same figure.

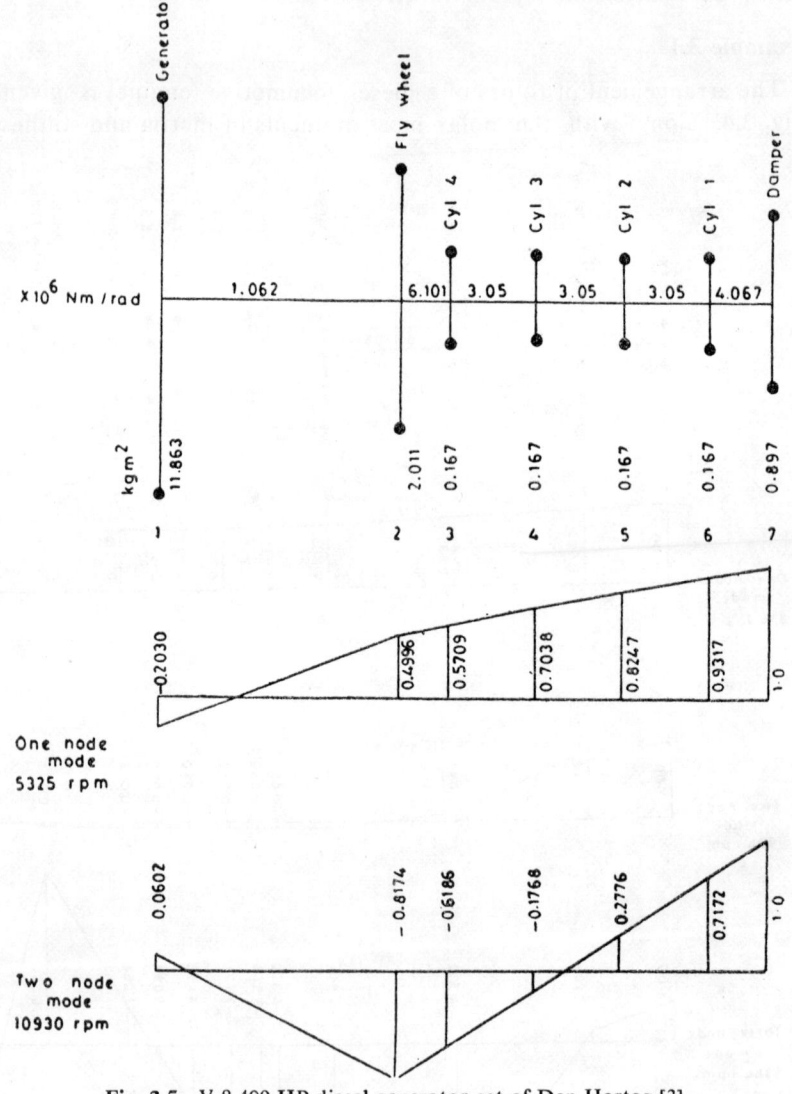

Fig. 3.5 V-8 400 HP diesel generator set of Den Hartog [3]

3.3 Excitation Torque

3.3.1 *Torque acting on each crank*

The gas pressure $p(\theta)$ on the piston varies with the crank angle $\theta = \omega t$

(ω is the angular velocity of the crank), because of firing-exhaust-intake-compression cycles. Knowing $p(\theta)$ from test measurements, the gas torque can be computed by, see Wilson [2], using the following:

$$T_g(\theta) = ARp(\theta)(\sin\theta + a\sin 2\theta - b\sin 4\theta + c\sin 6\theta \ldots) \quad (3.3.1)$$

where A is area of piston, R is crank radius and

$$a = \frac{1}{2r} + \frac{1}{8r^3} + \frac{15}{256r^5}$$

$$b = \frac{1}{16r^3} + \frac{3}{64r^5}$$

$$c = \frac{3}{256r^5} \quad (3.3.2)$$

with $r = L/R$, connecting rod length to crank radius ratio.

It is usual to express T acting on the crank, in terms of the tangential effort t, which is

$$t = \frac{T}{AR} \quad (3.3.3)$$

A reciprocating engine can either be single acting or double acting. We consider only single acting type with either 2 or 4 strokes. For such an engine, the torque cycle for each cylinder is completed for every one or two revolutions of the crankshaft, depending on 2 stroke or 4 stroke cycle of the engine. Hence the tangential effort due to gas pressure, t_g, can be expressed in Fourier series as

$$t_g = t_{g0} + \sum_{j=1}^{\infty}\left[a'_j \cos\frac{j\theta}{l} + b'_j \sin\frac{j\theta}{l}\right] \quad (3.3.4)$$

where $l = 1$ for 2 stroke engines; $l = 2$ for 4 stroke engines; t_{g0} is the mean tangential effort per cylinder; and a'_j and b'_j are jth cosine and sine harmonic components of the tangential effort respectively.

In the above

$$n = \frac{j}{l} \quad (3.3.5)$$

is called the order of the jth harmonic component. For 2 stroke engines, $n = 1, 2, 3, \ldots$ and for 4 stroke engines $n = \frac{1}{2}, 1, 1\frac{1}{2}, 2, 2\frac{1}{2}, \ldots$. Given p-θ diagram of any engine, a'_j and b'_j can be calculated by using equations (3.3.1) and (3.3.2).

In the absence of such a diagram, the harmonic components can be determined from tables and graphs given by Wilson [2], BICERA [1] and Porter [8]. These values are based on a large number of tests conducted, and for a new engine, they should be used with caution.

In addition to the gas pressure, the inertia due to reciprocating parts will also contribute to the total torque on the crank. The corresponding tangential effort due to the inertia of reciprocating parts, t_r, is given by [2],

$$t_r = \sum_{j=1}^{\infty} \left[c_j \cos \frac{j\theta}{l} + d_j \sin \frac{j\theta}{l} \right] \tag{3.3.6}$$

where

$c_j = j$th cosine harmonic component due to inertia $= 0$,

$d_j = j$th sine harmonic component due to inertia, which is given by

$d_j = 0$ for half order $n = \dfrac{j}{l} = \tfrac{1}{2}, 1\tfrac{1}{2}, 2\tfrac{1}{2} \ldots$

$$= \frac{m_{rec}\omega^2 R H_n}{A} \quad \text{for } n = \frac{j}{l} = 1, 2, 3 \ldots \tag{3.3.7}$$

In the above

$$H_1 = \frac{1}{4r} + \frac{1}{16r^3} + \frac{15}{512r^5}$$

$$H_2 = -\frac{1}{2r^2} - \frac{1}{32r^4} - \frac{1}{32r^6}$$

$$H_3 = -\frac{3}{4r} - \frac{9}{32r^3} - \frac{81}{512r^5}$$

$$H_4 = -\frac{1}{4r^2} - \frac{1}{8r^4} - \frac{1}{16r^6}$$

$$H_5 = \frac{5}{32r^3} + \frac{75}{512r^5}$$

$$H_6 = \frac{3}{32r^4} + \frac{3}{32r^6}$$

$$H_7 = H_8 = \ldots = 0 \tag{3.3.8}$$

From equations (3.3.4) and (3.3.6), the total tangential effort on the crank is given by

$$t = t_g + t_r = t_{g0} + \sum_{j=1}^{\infty} \left[a_j \cos \frac{j\theta}{l} + b_j \sin \frac{j\theta}{l} \right] \tag{3.3.9}$$

where a_j and b_j are jth cosine and sine harmonic components of total tangential effort respectively, given by $a'_j + c_j$, $b'_j + d_j$.

Equation (3.3.9) can be rewritten as

$$t = t_{g0} + \sum_{j=1}^{\infty} t_j \cos \left(\frac{j\theta}{l} - \psi_j \right) \tag{3.3.10}$$

where t_j is the amplitude of the total tangential effort, in jth harmonic

$$t_j = (a_j^2 + b_j^2)^{1/2} \tag{3.3.11}$$

and

$$\psi_j = \tan^{-1} \left(\frac{b_j}{a_j} \right) \tag{3.3.12}$$

3.3.2 Excitation torque

In general, because of the crank arrangement and the fact that the different cylinders fire at different instants as given by the firing order, the torques acting on individual cylinders will be differing from each other in phase only, assuming that each cylinder has same $p-\theta$ relation. If the ith rotor is not a cylinder, then the external torque is zero.

Let i' denote the cylinder number and i the corresponding rotor number, as for example of Fig. 3.1, the correspondence is given in Table 3.2.

Table 3.2 Cylinder number and corresponding rotor number of the installation (Fig. 3.1)

Cyl. No. i'	Rotor No. i
1	10
2	9
3	8
4	7
5	6
6	5
7	4
8	3

For i'th cylinder, measured from its own top dead center, the external torque f_i can be obtained from equation (3.3.9), with θ_i replaced by $\theta_{i'}$ and so in general

$f_i = 0$ if ith rotor is not a cylinder

$$= \left\{ t_{g0} + \sum_{j=1}^{\infty} \left[a_j \cos \frac{j\theta_{i'}}{l} + b_j \sin \frac{j\theta_{i'}}{l} \right] \right\} AR \qquad (3.3.13)$$

for ith rotor corresponding to i'th cylinder.

To take into account, the phase relationship of the torque between each cylinder, consider the crank arrangement of the engine in the example of Fig. 3.1, shown in Fig. 3.6. Let $\theta_1 = \theta =$ angle of rotation of crank 1 ($i' = 1$) measured from its top dead center. The firing order of the engine is 1-3-7-5-8-6-2-4 as shown in Fig. 3.6. At the beginning when $t = 0$, crank 1 is at its top dead center, and cylinder 1 fires. Hence the crank shaft moves counter clockwise by an angle $\theta = \omega t$. As crank 1 occupies horizontal position 0-5, in Fig. 3.6, after $\frac{1}{4}$ revolution, crank 3 ($i'=3$) comes to top dead center and cylinder 3 fires. Thus when $\theta_1 = \theta = 90°$, the angle of crank 3 after cylinder 3 fires, $\theta_3 = 0$. Hence the torque curve for cylinder 3 lags behind that of cylinder 1 by 90°. Such phase lag of torque curve of cylinder i' relative to cylinder 1 is denoted by $\beta_{i'}$

36 Rotor Dynamics

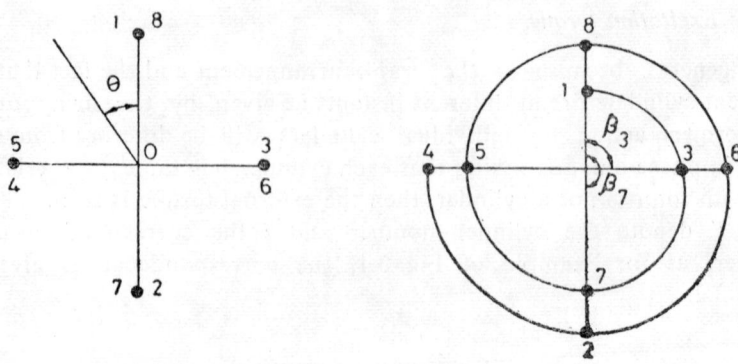

Fig. 3.6 Crank diagram and corresponding firing angle diagram of R 8 V 16/18 TL Diesel engine

and for the example under consideration, the phase differences are given in Table 3.3.

Table 3.3 Engine cylinder firing angles

Cylinder No. i'	Firing angle $\beta_{i'}$, deg.
1	0
2	540
3	90
4	630
5	270
6	450
7	180
8	360

In general
$$\theta_{i'} = \theta - \beta_{i'} \tag{3.3.14}$$
and equation (3.3.13) becomes
$$f_i = T_0 + \sum_{j=1}^{\infty} T_{ij}$$
$$= T_0 + \sum_{j=1}^{\infty} \left[A_{ij} \cos \frac{j\omega t}{l} + B_{ij} \sin \frac{j\omega t}{l} \right]$$

for ith rotor corresponding to i'th cylinder,
$$= 0, \quad \text{if } i\text{th rotor is not a cylinder} \tag{3.3.15}$$

In the above
$$T_0 = t_{g0} AR = \text{mean torque on each cylinder}$$
$$A_{ij} = \left[a_j \cos \frac{j\beta_{i'}}{l} - b_j \sin \frac{j\beta_{i'}}{l} \right] AR$$

Torsional Vibrations in Reciprocating Machinery 37

$$B_{ij} = \left[a_j \sin \frac{j\beta_{i'}}{l} + b_j \cos \frac{j\beta_{i'}}{l} \right] AR \qquad (3.3.16)$$

The force vector $\{f\}$ defined in equation (3.3.15) can be used directly for a forced vibration analysis of the engine installation. It should be noted that the relations above should be modified, if there is more than one cylinder in each bank of an engine like a V engine.

3.4 Forced Vibration and resonant stresses

3.4.1 Critical speeds

The excitation torque given in equation (3.3.15) shows that the external torque on any cylinder has several harmonic frequencies, $\frac{j\omega}{l}$, $j=1, 2, \ldots$. If the jth harmonic excitation frequency coincides with any one of the natural frequencies p_k of the system, then resonance occurs at the critical angular velocity ω_c given by

$$\frac{j\omega_c}{l} = p_k; \quad j = 1, 2, \ldots$$
$$k = 2, 3, \ldots N \qquad (3.4.1)$$

Hence the critical speeds are given by

$$N_c = \frac{lF_k}{j} = \frac{F_k}{n}; \quad j = 1, 2, \ldots$$
$$k = 2, 3, \ldots N \qquad (3.4.2)$$

where F_k = natural frequency cpm, and N_c = critical speed of the engine in rpm.

If the running speed is near any critical speed, excessive vibration stresses will be developed. These vibration stresses can be predicted by solving the equations for forced vibration of the N rotor system, by modal expansion method.

3.4.2 Solution of equations by modal expansion method

The equations of motion for forced vibration are

$$[M]\{\ddot{\phi}\} + [K]\{\phi\} = \{f\} \qquad (3.4.3)$$

The solution of the above is expressed in terms of modal components as

$$\{\phi\} = [X]\{q\} \qquad (3.4.4)$$

where $[X]$ is the twist modal matrix and $\{q\}$ is the generalised coordinate vector.

The decoupled equations of (3.4.3) will be, as in section 2.5,

$$\bar{m}_k \ddot{q}_k + \bar{k}_k q_k = \bar{f}_k; \quad k = 1, 2, \ldots N \qquad (3.4.5)$$

where

$$\bar{m}_k = \sum_{i=1}^{N} I_i x_{ik}^2 \qquad (3.4.6)$$

$$\bar{k}_k = p_k^2 \bar{m}_k \qquad (3.4.7)$$

and

$$\bar{f}_k = \sum_{i=1}^{N} x_{ik} f_i \qquad (3.4.8)$$

Since $f_i = 0$ for all rotors other than cylinders, the above equation (3.4.8), with the help of equation (3.3.15) becomes

$$\bar{f}_k = T_0 \sum' x_{ik} + \sum_{j=1}^{\infty} \sum' x_{ik} A_{ij} \cos \frac{j\omega t}{l} + \sum_{j=1}^{\infty} \sum' x_{ik} B_{ij} \sin \frac{j\omega t}{l} \qquad (3.4.9)$$

where $\sum'$ represents summation taken over rotor cylinders only.

Solution of equation (3.4.5) for the kth generalised coordinate is given by

$$q_k(t) = \frac{T_0 \sum' x_{ik}}{k_k}$$
$$+ \sum_{j=1}^{\infty} \left[\frac{1}{\bar{m}_k \left[p_k^2 - \left\{ \frac{j\omega}{l} \right\}^2 \right]} \right] \left\{ \sum' x_{ik} A_{ij} \cos \frac{j\omega t}{l} + \sum' x_{ik} B_{ij} \sin \frac{j\omega t}{l} \right\} \qquad (3.4.10)$$

The dynamic part of the solution in the above can be written as

$$q_k(t) = \sum_{j=1}^{\infty} q_{jk0} \cos \left(\frac{j\omega t}{l} - \Psi_{jk} \right) \qquad (3.4.11)$$

In the above q_{jk0} is the amplitude of kth mode generalised coordinate due to excitation by jth harmonic, which is

$$q_{jk0} = \frac{[(\sum' x_{ik} A_{ij})^2 + (\sum' x_{ik} B_{ij})^2]^{1/2}}{\bar{m}_k \left[p_k^2 - \left\{ \frac{j\omega}{l} \right\}^2 \right]} \qquad (3.4.12)$$

Using equations (3.3.11) and (3.3.16) and writing

$$T_j = t_j AR \qquad (3.4.13)$$

which is the amplitude of jth harmonic component of the total torque, equation (3.4.12) becomes

$$q_{jk0} = M q_{jk0}\}_{\omega=0} \qquad (3.4.14)$$

where M is the dynamic magnifier given by

$$M = \frac{1}{1 - \left[\frac{j\omega}{lp_k} \right]^2} \qquad (3.4.15)$$

and $q_{jk0}\}_{\omega=0}$ is the equilibrium amplitude, which is

$$q_{jk0}\}_{\omega=0} = \frac{T_j}{\bar{m}_k p_k^2} \left[\left\{ \sum' x_{ki} \cos \frac{j\beta_{i'}}{l} \right\}^2 + \left\{ \sum' x_{ik} \sin \frac{j\beta_{i'}}{l} \right\}^2 \right]^{1/2} \qquad (3.4.16)$$

At resonance, $\frac{j\omega}{l} = p_k$ and the dynamic magnifier given by equation (3.4.15) will be infinite. Actually there is always some internal damping and hence M is limited to a finite value. It is extremely difficult to assess analytically the factors which influence the dynamic magnifier at resonance. Hence empirical formulae are developed for the dynamic magnifier at resonance, based on extensive experience and experimental observations, which are listed below.

1. Wilson's formula [2]

$$M_{e1} = \frac{500}{(16 + 0.00015 f_{jk0})^{1/2}} \qquad (3.4.17)$$

where $f_{jk0} = \max f_{ik} q_{jk0}\}_{\omega=0}$, is the maximum equilibrium stress in the kth mode of vibration, N/m^2.

2. Archer's formula [7]

$$M_{e2} = \frac{3.8}{[q_{jk0}\}_{\omega=0}]^{1/4}} \qquad (3.4.18)$$

3. MAN formula [2]

$$M_{e3} = \frac{p_k \bar{m}_k}{S_e \sum' x_{ik}^2} \qquad (3.4.19)$$

where S_e denotes the viscous damping per cylinder per unit vibratory velocity, Nm sec/rad, and is given by

$$S_e = K_e A R^2 \qquad (3.4.20)$$

MAN recommends $K_e = 190$ N sec/m^2 rad, for 4 stroke medium speed oil engines.

These empirical formulae give fairly good estimates for the dynamic magnifier. However, they should be used with caution, for an engine with unknown experimental data.

Since the free vibration mode shapes are normalised with respect to the amplitude of vibration of the basic rotor, the equilibrium amplitude $q_{jk0}\}_{\omega=0}$ given by equation (3.4.16) represents the angle of twist in the basic rotor in kth mode of vibration due te jth harmonic torque. The corresponding maximum dynamic resonance stress developed in the main shafting of the installation in kth mode of vibration, due to jth harmonic, τ_{jk} is obtained by

$$\tau_{jk} = \max f_{ik} M q_{jk0}\}_{\omega=0} \qquad (3.4.21)$$

Based on the analysis given above, a computer program can be developed [4], to determine the resonant stresses of a reciprocating machine installation.

Example 3.3

The data of a 400 HP six cylinder diesel engine [2], is given in Fig. 3.7. The first two natural frequencies and the corresponding mode shapes obtained are also given in this figure. The critical speeds corresponding to different harmonics of the one node mode, that fall in the speed range of the engine are given in Fig. 3.7. The resonance stresses at the critical speeds are evaluated, by using the computer program [4], using an average value for the dynamic magnifier evaluated from equations (3.4.17) to (3.4.19). It can be seen from Fig. 3.7, that the performance of this engine is not satisfactory, with the dynamic stress in the operating speed range being more than the permissible stress.

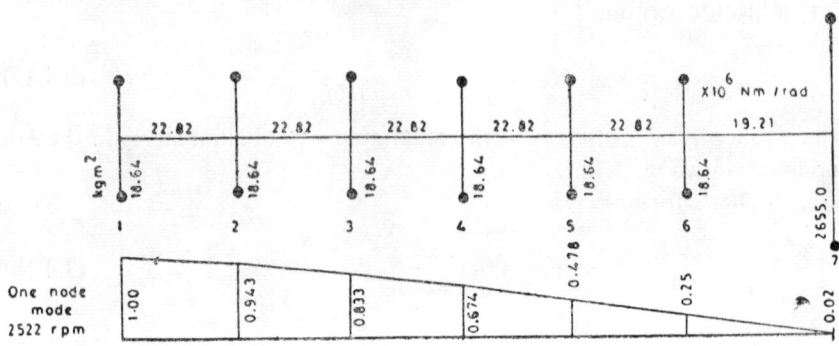

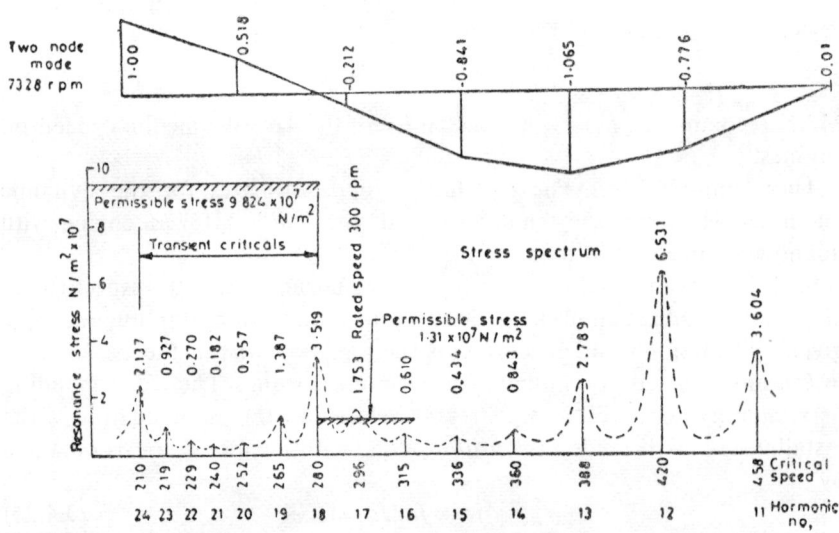

Fig. 3.7 6-cylinder 400 HP diesel engine [2]

In the analysis discussed in this section, the stresses are evaluated at the critical speeds only, with the assumption that these stress values are lower at other speeds. This assumption may not be valid, particularly if two major criticals are closer and then the stress response due to different harmonics, called flank effects, should be vectorially summed to determine the dynamic response at speeds other than criticals.

3.5 Cyclic Irregularity

For the rigid body mode, $p_1 = 0$ and $x_{i1} = 1$, for all rotors. Hence the generalised mass in the rigid body mode is

$$\bar{m}_1 = \sum_{i=1}^{N} I_i \qquad (3.5.1)$$

The generalised stiffness for the rigid body mode is

$$\bar{k}_1 = 0 \qquad (3.5.2)$$

Hence the equilibrium amplitude given by equation (3.4.12) for the rigid body mode is

$$q_{j10} = -\sum_{j=1}^{\infty} \frac{T_j}{\sum I_i \left\{\frac{j\omega}{l}\right\}^2} \left[\left\{\sum{}' \cos \frac{j\beta_{i'}}{l}\right\}^2 + \left\{\sum{}' \sin \frac{j\beta_{i'}}{l}\right\}^2 \right]^{1/2} \qquad (3.5.3)$$

We find that

$$\sum{}' \sin \frac{j\beta_{i'}}{l} = 0 \qquad (3.5.4)$$

For an m cylinder engine, we also find the following

$$\sum{}' \cos \frac{j\beta_{i'}}{l} = 0; \quad j \neq m, 2m, 3m, \ldots$$

$$= m; \quad j = m, 2m, 3m \ldots \qquad (3.5.5)$$

Hence for an m cylinder engine, equation (3.4.11) for the rigid body mode becomes

$$q_1(t) = -\sum_{j=m, 2m, \ldots}^{\infty} \frac{mT_j}{\sum I_i \left\{\frac{j\omega}{l}\right\}^2} \cos\left(\frac{j\omega t}{l} - \psi_{j1}\right) \qquad (3.5.6)$$

$j = m, 2m, 3m$ etc., are defined as major harmonics and all the remaining as minor harmonics. Referring to equation (3.3.5), the major engine orders therefore are

$$n = \frac{j}{l} = \tfrac{1}{2}m, m, 1\tfrac{1}{2}m, 2m \text{ etc.,} \quad \text{for 4 stroke engines}$$

$$= m, 2m, 3m \text{ etc.,} \quad \text{for 2 stroke engines} \qquad (3.5.7)$$

Since $x_{i1} = 1$, for all rotors in the rigid body mode, equation (3.5.6) gives the rigid body rotation $\phi_{i(\text{rigid})}$ of any rotor i. The coefficient of cyclic irregularity or the coefficient of fluctuation of speed is defined by

$$c = \frac{N_{\max} - N_{\min}}{N_{\text{mean}}} \qquad (3.5.8)$$

The rigid cyclic irregularity is an important factor in the operation of the system, particularly for power generating systems. This coefficient should be kept to a minimum. For the dynamic system, equation (3.5.8) is

$$c_r = \frac{\max \dot\phi_{i(\text{rigid})} - \min \dot\phi_{i(\text{rigid})}}{\omega} \qquad (3.5.9)$$

From equation (3.5.6) we get c_r as

$$c_r = \sum_{j=m, 2m, \ldots} c_{rj} \qquad (3.5.10)$$

where the contribution of jth major harmonic to the rigid cyclic irregularity c_{rj} is

$$c_{rj} = - \frac{2mT_j \frac{j}{l}}{\sum I_i \left\{\frac{j\omega}{l}\right\}^2} \qquad (3.5.11)$$

A general computer program can be developed to determine the natural frequencies, mode shapes, resonant stresses at the critical speeds and the rigid cyclic irregularity of diesel engine driven installations [4].

Example 3.4

Consider the diesel generator set given in Fig. 3.1, with its data in Table 3.1. The twist and stress mode shapes obtained for the one and two node modes are given in Fig. 3.8. The critical speeds with their corres-

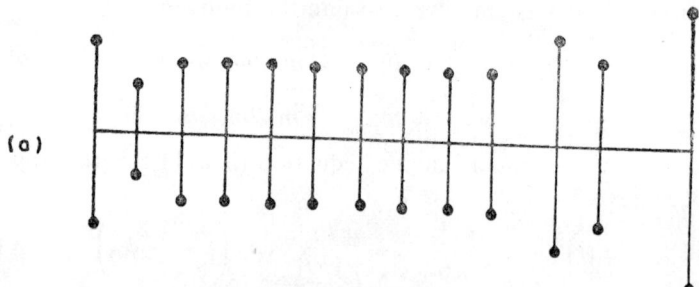

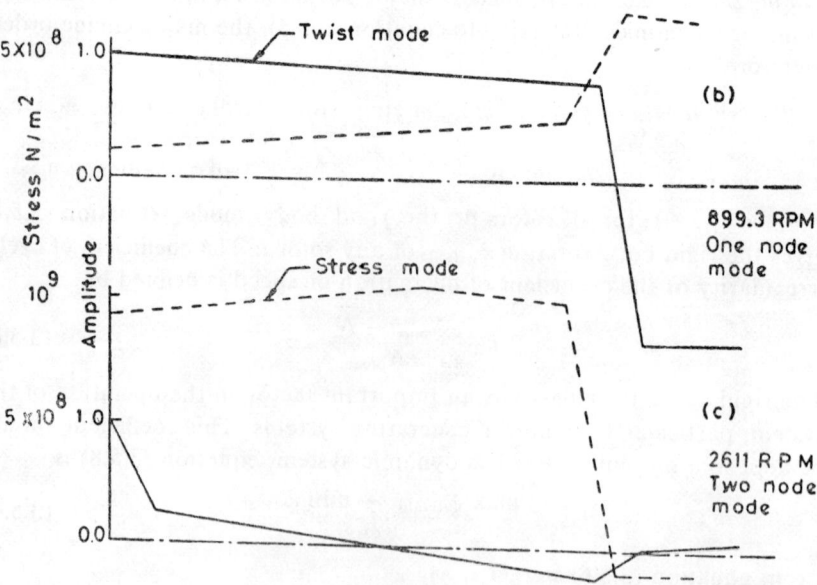

Fig. 3.8 Twist and stress modes R8V 16/18 TL diesel generator set (Fig. 3.1)

Torsional Vibrations in Reciprocating Machinery 43

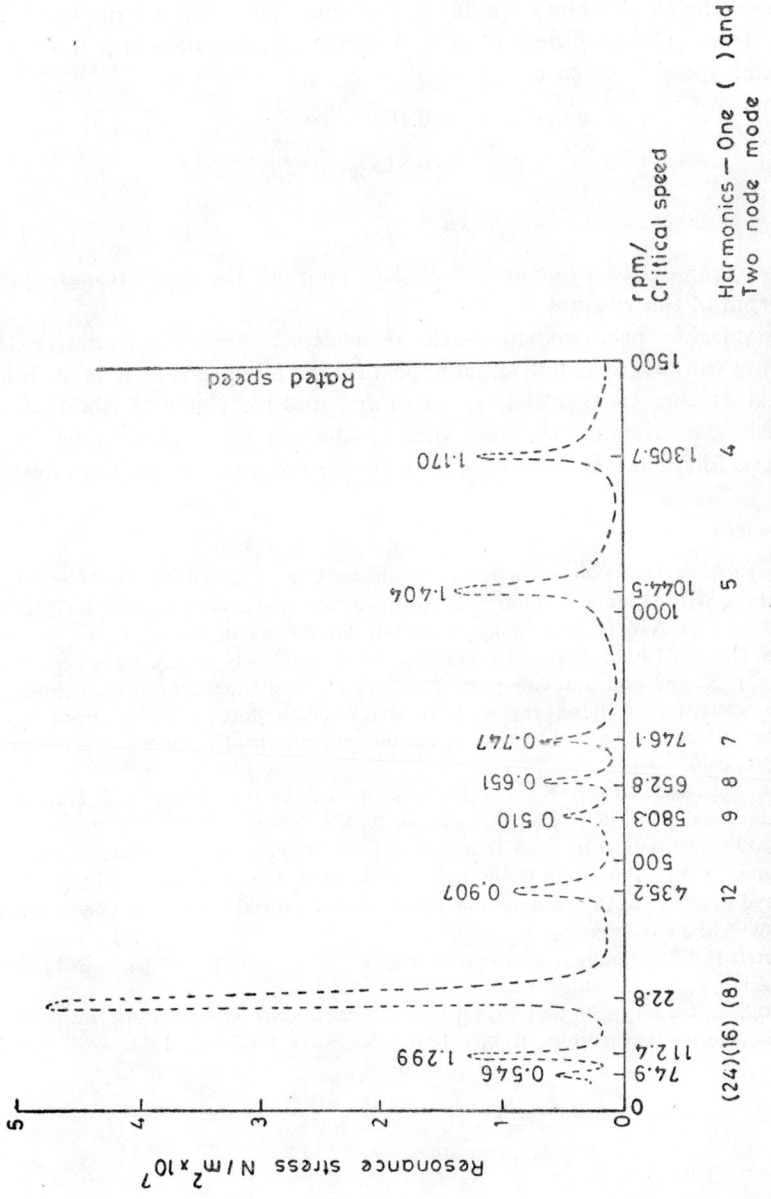

Fig. 3.9 Resonant stress spectrum of diesel generator set (Fig. 3.1)
Harmonics — One ( ) and Two node mode

ponding orders are given in Fig. 3.9. The resonant stresses at these critical speeds of operation are also given in Fig. 3.9. Critical speeds which give rise to stresses more than 5×10^6 N/m² are only shown in the figure. It may be noted that the major orders as defined in equation (3.5.7) based on the rigid cyclic irregularity need not have the maximum resonant stress. The coefficient of cyclic irregularity, calculated for this set at the rated speed is given by

$$c_{r1} = -0.1020 \times 10^{-2}$$
$$c_{r2} = -0.9137 \times 10^{-4}$$

and

$$c_r = 0.0011$$

which works out to a fluctuation of 1.65 rpm at the operational speed 1500 rpm of the engine.

For variable speed systems such as a diesel electric locomotive the coupling stiffness does not remain constant for all speeds as it is a function of torque transmitted, temperature and frequency of vibration. In such cases, transfer matrix method as in Chapter 2 may have to be used. Analysis for such a system is given in [9] for a diesel electric locomotive.

References

1. BICERA, A handbook of torsional vibrations, Cambridge University Press, 1958.
2. Wilson, W.K., Practical Solution of torsional vibration problems, Vol. I, Chapman & Hall, 1956; Vol. II, John Wiley, 1963; Vol. III, Chapman & Hall, 1965.
3. Den Hartog, J.P. Mechanical Vibrations, McGraw-Hill Book Co., 1956.
4. Rao, D.K. and Rao, J.S. Computer Programs for predicting the torsional vibration characteristics of diesel engine drive sets, Project report to Garden Reach Workshops, Calcutta, Dept. of Mech. Engng., Indian Institute of Technology, Kharagpur, India, 1974.
5. Rao, J. S. and Gupta, K., An Introductory course on Theory and Practice of Mechanical Vibration, Wiley Publication, 1984.
6. Rao, J.S., Banerji, J.R. and Bhattacharya, B. Vibrations in ZDM2 diesel loco engine, Part II, Proc. 16th Cong. Ind. Soc. Theo. & Appld. Mechs., 1972, p. 75.
7. Lloyd's register of shipping, Guidance notes on torsional vibration characteristics of main and auxiliary engines, 1968.
8. Porter, F.P., Harmonic coefficients of engine torque curves, Trans ASME, 1943, p. A33.
9. Rao, J.S., Gupta, K.N. and Shyam Lal, Critical Speeds and Vibratory response of diesel-electric locomotives, RDSO, Indian Railways, Lucknow, 1981.

A 110 MW Generator Rotor, (Courtesy BHEL, Hyderabad)

Chapter 4

Bending Critical Speeds of Simple Shafts

The phenomenon of bending vibrations and critical speeds of rotating shafts is perhaps the most common problem that is discussed by a vibration engineer, as it is a vexing day-to-day problem in design and maintenance of the machinery. Some of the rotors have a mass as much as 100 tonnes as in the case of big steam turbines and obviously they deserve utmost attention in this regard. The rotors have always some amount of residual unbalance however well they are balanced, and will get into resonance when they rotate at speeds equal to bending natural frequency. These speeds are called critical speeds by Rankine[1] and as far as possible they should be avoided. Even while taking the rotor through a critical speed to an operational speed, special precaution should be taken.

While the calculations of bending natural frequency of a simple shaft in rigid bearings is somewhat an easy matter, the problem in practice becomes complex because of:

1. Gyroscopic effects of disks,
2. Dissimilar moments of area of the shaft,
3. Stiffness and damping properties of oil film bearings, and
4. Coupling between two rotors.

To avoid failures of shafting, the general practice in the design of rotors is to determine the bending critical speeds, check the out-of-balance response and adopting a suitable balancing procedure. In this chapter, we will study the bending critical speeds of simple shafts.

4.1 Whirling of an Unbalanced Simple Elastic Rotor

Consider a heavy disk of mass M mounted at mid-span of a massless elastic shaft in radially rigid bearings, as shown in Fig. 4.1. The geometric center of the disk is E and its center of gravity is at a distance a, as shown. Since the system is symmetric, the whirling speed and rotational speed will be same and we have a synchronous whirl. In the fixed OYZ axis system, the geometry of unbalance whirl at the disk is shown in Fig. 4.2. K is lateral stiffness of the shaft and C is equivalent viscous damping of the system. Jeffcott[2] analysed such a rotor, to determine the unbalance response. The equations of motion for the rotor are

$$M\frac{d^2}{dt^2}(z + a\cos\omega t) + C\frac{dz}{dt} + Kz = 0$$

$$M\frac{d^2}{dt^2}(y + a\sin\omega t) + C\frac{dy}{dt} + Ky = 0 \qquad (4.1.1)$$

Equations (4.1.1) can be rewritten as

$$M\ddot{z} + C\dot{z} + Kz = M\omega^2 a\cos\omega t$$

$$M\ddot{y} + C\dot{y} + Ky = M\omega^2 a\sin\omega t \qquad (4.1.2)$$

The whirl radius r can be expressed as a complex quantity,

$$r = z + iy \qquad (4.1.3)$$

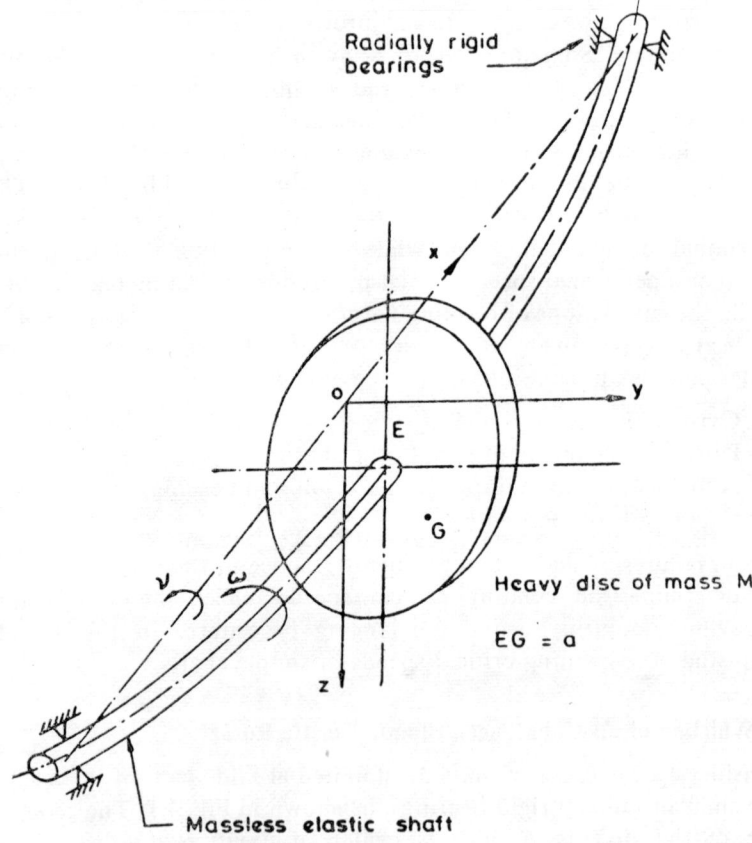

Fig. 4.1 Single mass flexible rotor with residual unbalance. Residual unbalance = Ma, $\nu = \omega$ synchronous whirl

and equations (4.1.2) can be combined to give

$$M\ddot{r} + C\dot{r} + Kr = M\omega^2 a e^{i\omega t} \qquad (4.1.4)$$

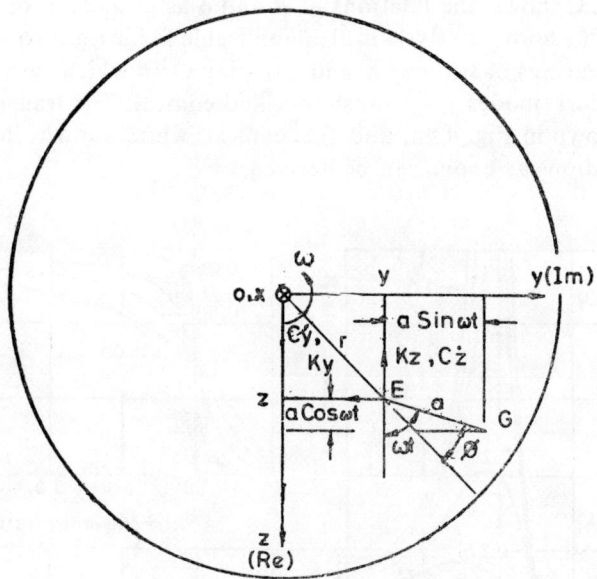

Fig. 4.2 Geometry of shaft whirl at the disk location

In the above equations, we note that the magnitude of the exciting force is the centrifugal force of the eccentric disk rotating about the shaft elastic axis, but not about the bearing axis.

The steady state solution of equation (4.1.4) is

$$r = Re^{i(\omega t - \phi)} \tag{4.1.5}$$

$$R = \frac{M\omega^2 a}{\sqrt{(K - M\omega^2)^2 + C^2\omega^2}} \tag{4.1.6}$$

Nondimensionalising, we get

$$\bar{R} = \frac{R}{a} = \frac{\Omega^2}{\sqrt{(1 - \Omega^2)^2 + (2\zeta\Omega)^2}} \tag{4.1.7}$$

$$\phi = \tan^{-1} \frac{2\zeta\Omega}{1 - \Omega^2} \tag{4.1.8}$$

where

$$\Omega = \frac{\omega}{p}; \quad p = \sqrt{\frac{K}{M}};$$

$$\zeta = \frac{C}{C_c}; \quad C_c = 2\sqrt{KM} \tag{4.1.9}$$

For

$$\Omega = \frac{\omega}{p} = 1$$

$$\bar{R} = \frac{1}{2\zeta} \tag{4.1.10}$$

Figure 4.3 shows the relations of $\bar{R}$ and ϕ as a function of shaft speed $\Omega = \omega/p$. The above analysis is also applicable for a rigid rotor in simple isotropic bearings of stiffness K and damping C, in which case there will be two whirl modes viz., translatory and conical. The translatory whirl mode is shown in Fig. 4.4a, and for conical whirl shown in Fig. 4.4b, similar relations as above can be derived.

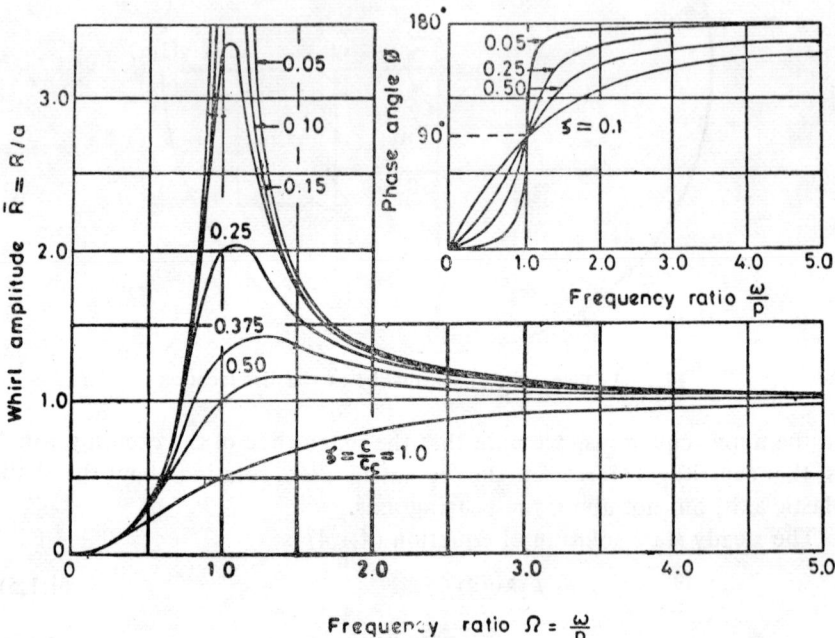

Fig. 4.3 Dynamic magnification of rotor whirl amplitude as a function of speed of shaft

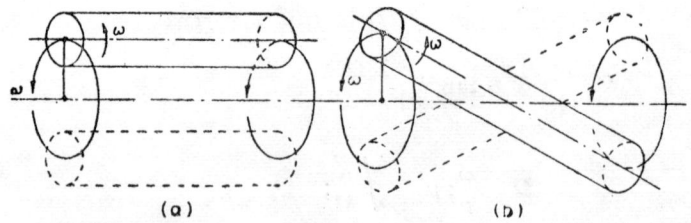

Fig. 4.4 Whirl modes of a rigid rotor in flexible end bearings (a) translatory whirl; (b) conical whirl

In writing the solution for the differential equation (4.1.4), we implicitly assumed forward synchronous whirl, that is the whirl frequency ν is the same as rotational frequency. We could have written the solution

$$r = R \exp \{i(\nu t - \phi)\} \tag{4.1.11}$$

and show that
$$\nu = \omega$$
or
$$\nu = -\omega \qquad (4.1.12)$$

The second case gives rise to backward synchronous whirl. The forward synchronous whirl occurs most readily in practice, since the frequency of disturbance due to out-of-balance coincides with the shaft rotational frequency. The backward whirl motion occurs, under certain conditions with oil film bearings and we shall discuss this motion in Chapter 8.

For an undamped rotor, we notice from Fig. 4.3, that resonance occurs when $\omega = p$. Hence the rotor whirls with large amplitudes at resonance with the natural frequency of the stationary shaft in lateral vibration and hence this speed is called critical speed. At resonance, we also observe that $\phi = 90°$, and from Fig. 4.2, we can write

$$M\omega^2 a = C\omega R \qquad (4.1.13)$$

and

$$\bar{R} = \frac{R}{a} = \frac{M\omega}{C} = \frac{1}{2\zeta} \qquad (4.1.14)$$

This result is same as obtained before in equation (4.1.10), since at $\phi = 90°$, $\omega = p$. For $\zeta < 0.25$, the maximum whirl amplitude can be predicted by the above equation (4.1.14), as it occurs at a slightly higher frequency than p.

4.2 Simple Shafts with Several Disks

In the previous section, we used a simple rotor to illustrate the fact that the critical speed of a shaft is same as the natural frequency of the rotor in lateral bending. In practice the rotor carries several components such as gears, disks, flywheels etc., and it has several critical speeds corresponding to the bending natural frequencies. For most of the rotors, it is the fundamental mode which falls in the running speed zone and there are several methods of calculation of the first critical speed, two of which will be discussed here, because of their simplicity.

4.2.1 Dunkerley's Method [3]

This is a very convenient method, proposed by Dunkerley [3] to determine the fundamental critical speed of a shaft carrying a number of components. The method is quite simple and consists of reducing the actual system into a number of simple subsystems, calculating the critical speeds of each by a direct formula, and combining these critical speeds according to the following equation

$$\frac{1}{\omega^2} = \frac{1}{\omega_1^2} + \frac{1}{\omega_2^2} + \ldots + \frac{1}{\omega_n^2} \qquad (4.2.1)$$

to obtain actual critical speed ω of the system. For a formal proof to the above equation and the approximation involved, reference may be made to Rao and Gupta [4]. The value of ω obtained is always a lower bound approximation.

4.2.2 Rayleigh's Method [5]

This is another simple method proposed by Rayleigh [5] based on the fact that the maximum kinetic energy must be equal to maximum potential energy for a conservative system under free vibration conditions. For a shaft carrying several components, we can use static deflection or any other suitable function to represent the fundamental mode of the shaft.

The fundamental frequency can be obtained from

$$\omega^2 = \frac{g \sum My}{\sum My^2} \tag{4.2.2}$$

where $M_1, M_2 \ldots$ are masses of different components and $y_1, y_2 \ldots$ are deflections of the shaft at the locations of these components. Since the frequency is a minimum, it is always an upper bound value.

The main advantages of Dunkerley's and Rayleigh's methods are that they use simple strength of material formulae for beams (either analytical or graphical methods can be chosen) and that Dunkerley's method gives a lower bound value, while Rayleigh's method gives an upper bound value. Therefore the exact range where the critical speed lies is well established by these two methods taken together and thus they are very popular. There are several other methods that can be used to determine the critical speeds, such as Stodola's method, Ritz method etc., see Jacobsen and Ayre [6], but we will follow the Myklestad method [7] in the form of transfer matrices, as it is most amenable for the rotor bending vibration problems.

4.3 Transfer Matrix Analysis for Bending Critical Speeds

In a manner similar to Holzer method [6] for torsional vibrations, Myklestad [7] and Prohl [8] developed highly successful methods of computation for the bending critical speeds of shafts. We will use this method in transfer matrix form, see Pestel and Leckie [9], for bending critical speed and out-of-balance response calculations. A polynomial frequency equation for Myklestad equations is given by Rao and Reddy [10].

Consider an n mass system, each mass representing either a gear, a disk or a flywheel etc. (Fig. 4.5). All these masses are taken as lumped with their gyroscopic inertia neglected. The ith shaft element of length l_i and mass m_i are separately shown in Fig. 4.5, and $\{S\}$ represents the state vector containing the deflection w, slope θ, bending moment M_y and shear force V_z, for bending in the x-z plane. The state vector can be defined to the left or right of station i, with a superscript as shown in Fig. 4.5. w and θ are positive in positive z and y directions. V_z and M_y represent the internal forces and their positive signs are along z and

y directions for a positive face. A positive face is defined as one which has its outward normal in the x direction. On a negative face V_z and M_y are positive in the negative directions of z and y axes respectively.

We set up the transfer matrices for each element as follows.

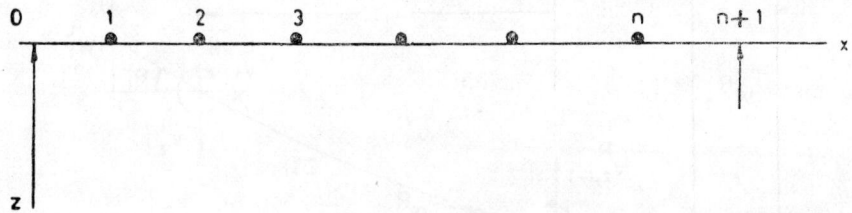

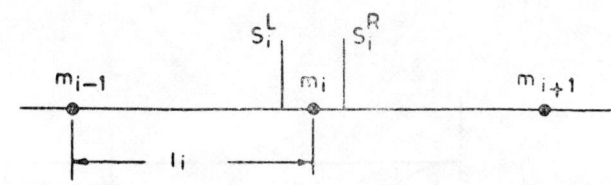

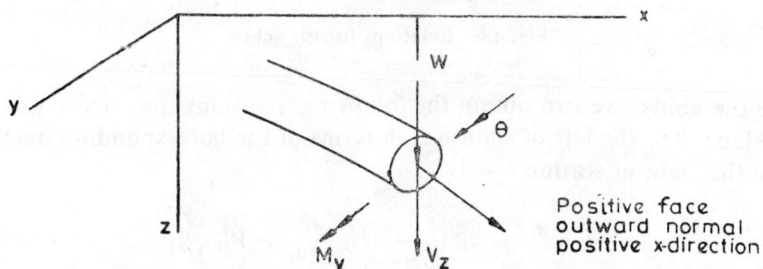

Fig. 4.5 n mass sytsem with notations

4.3.1 Field Matrix

Figure 4.6 gives the equilibrium relations for the ith field, from which we have

$$V_{zi}^L = V_{z,i-1}^R \tag{4.3.1}$$

$$M_{yi}^L = M_{y,i-1}^R + V_{zi}^L l_i \tag{4.3.2}$$

To derive the transfer relations for deflection w and slope θ, for the element shown in Fig. 4.6, we use the relations for a cantilever beam,

$$w = -\frac{Ml^2}{2EI} + \frac{Vl^3}{3EI}$$

$$\theta = \frac{Ml}{EI} - \frac{Vl^2}{2EI} \qquad (4.3.3)$$

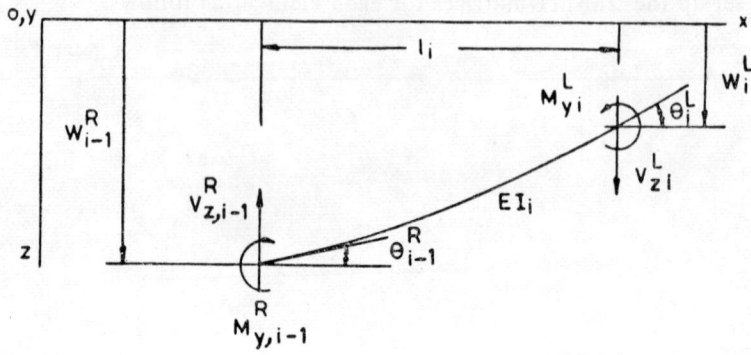

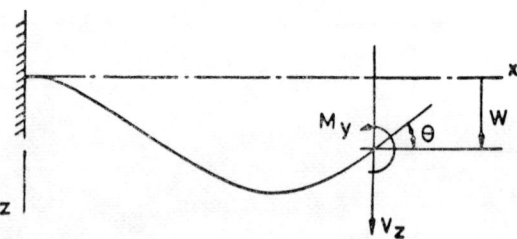

Fig. 4.6 Relations for ith field

Using the above, we can obtain the following relations for deflection w and slope θ to the left of station i, in terms of the corresponding quantities to the right of station $i-1$.

$$w_i^L = w_{i-1}^R - \theta_{i-1}^R l_i - M_{yi}^L \frac{l_i^2}{2EI_i} + V_{zi}^L \frac{l_i^3}{3EI_i} \qquad (4.3.4)$$

$$\theta_i^L = \theta_{i-1}^R + M_{yi}^L \frac{l_i}{EI_i} - V_{zi}^L \frac{l_i^2}{2EI_i} \qquad (4.3.5)$$

Using equations (4.3.1) and (4.3.2) in the above and simplifying, we get

$$-w_i^L = -w_{i-1}^R + \theta_{i-1}^R l_i + M_{y,i-1}^R \frac{l_i^2}{2EI_i} + V_{z,i-1}^R \frac{l_i^3}{6EI_i} \qquad (4.3.6)$$

$$\theta_i^L = \theta_{i-1}^R + M_{y,i-1}^R \frac{l_i}{EI_i} + V_{z,i-1}^R \frac{l_i^2}{2EI_i} \qquad (4.3.7)$$

We combine equations (4.3.6), (4.3.7), (4.3.2) and (4.3.1) to form the following transfer matrix relation.

$$\begin{Bmatrix} -w \\ \theta \\ M_y \\ V_z \end{Bmatrix}_i^L = \begin{bmatrix} 1 & l & \dfrac{l^2}{2EI} & \dfrac{l^3}{6EI} \\ 0 & 1 & \dfrac{l}{EI} & \dfrac{l^2}{2EI} \\ 0 & 0 & 1 & l \\ 0 & 0 & 0 & 1 \end{bmatrix}_i \begin{Bmatrix} -w \\ \theta \\ M_y \\ V_z \end{Bmatrix}_{i-1}^R \qquad (4.3.8)$$

Symbolically the above equation can be written as

$$\{S\}_i^L = [F]_i \{S\}_{i-1}^R \qquad (4.3.9)$$

where $\{S\}$ is the state vector and $[F]_i$ is the transfer matrix for the ith field.

4.3.2 Point Matrix

Figure 4.7 shows the equilibrium relations for the mass at station i, and we can directly write the following transfer matrix

$$\begin{Bmatrix} -w \\ \theta \\ M_y \\ V_z \end{Bmatrix}_i^R = \begin{bmatrix} 1 & 0 & 0 & 0 \\ 0 & 1 & 0 & 0 \\ 0 & 0 & 1 & 0 \\ mp^2 & 0 & 0 & 1 \end{bmatrix}_i \begin{Bmatrix} -w \\ \theta \\ M_y \\ V_z \end{Bmatrix}_i^L \qquad (4.5.10)$$

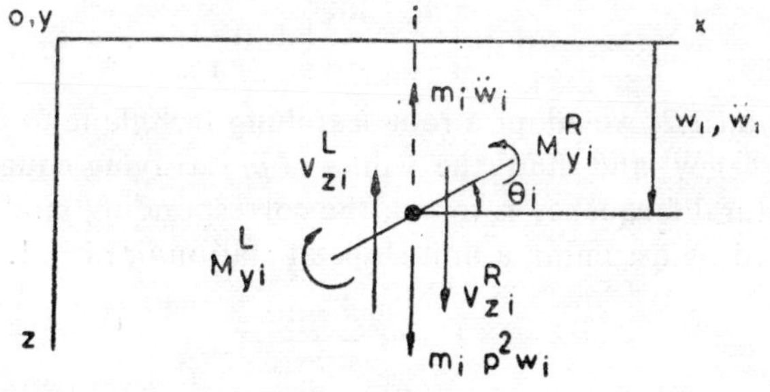

Fig. 4.7 Relations for ith mass

i.e.,

$$\{S\}_i^R = [P]_i \{S\}_i^L \qquad (4.3.11)$$

In the above $[P]_i$ is the point matrix for ith mass vibrating in a normal mode with frequency p.

4.3.3 Overall Transfer Matrix and Frequency Equation

Starting from station 0 of the shaft in Fig. 4.5, we can write the following, with the help of equations (4.3.9) and (4.3.11).

$$\{S\}_1^L = [F]_1\{S\}_0$$
$$\{S\}_1^R = [P]_1\{S\}_1^L = [P]_1[F]_1\{S\}_0$$
$$\{S\}_2^L = [F]_2\{S\}_1^R = [F]_2[P]_1[F]_1\{S\}_0$$
$$\cdots \quad \cdots \quad \cdots \quad \cdots \quad \cdots \quad \cdots \quad \cdots$$
$$\{S\}_{n+1} = [F]_{n+1}[P]_n[F]_n[P]_{n-1}\ldots[F_1]\{S\}_0 \quad (4.3.12)$$

Defining the product of all field matrices and point matrices in the order given in the above equation as overall transfer matrix $[U]$, we get

$$\{S\}_{n+1} = [U]\{S\}_0 \quad (4.3.13)$$

Since $[F]$ and $[P]$ are all 4×4 matrices, $[U]$ is also 4×4 in size. We write equation (4.3.13) in expanded form

$$\begin{Bmatrix} -w \\ \theta \\ M_y \\ V_z \end{Bmatrix}_{n+1} = \begin{bmatrix} u_{11} & u_{12} & u_{13} & u_{14} \\ u_{21} & u_{22} & u_{23} & u_{24} \\ u_{31} & u_{32} & u_{33} & u_{34} \\ u_{41} & u_{42} & u_{43} & u_{44} \end{bmatrix} \begin{Bmatrix} -w \\ \theta \\ M_y \\ V_z \end{Bmatrix}_0 \quad (4.3.14)$$

For simply supported ends

$$w = 0$$
$$M_y = 0 \quad \text{at stations 0 and } n+1 \quad (4.3.15)$$

Since θ and V_z are not zero at the stations 0 and $n+1$, we get from equation (4.3.14)

$$\Delta = \begin{vmatrix} u_{12} & u_{14} \\ u_{32} & u_{34} \end{vmatrix} = 0 \quad (4.3.16)$$

As in section 2.2, we adopt a root searching technique to determine the natural frequency and find the values of p_i satisfying equation (4.3.16). Once the natural frequency is found, the corresponding mode shape can be determined by assuming a unit slope at station 0, $\theta_0 = 1$. Hence

$$V_{z0} = -\frac{u_{12}}{u_{14}} \quad (4.3.17)$$

This defines the state vector at station 0 completely and we can proceed from station 0 to 1, then to 2..., until n, to obtain the values of w from equations (4.3.9) and (4.3.11), thus obtaining the mode shape.

4.4 Shaft with Overhangs

In Fig. 4.8, an n mass shaft with overhangs on both sides is shown with rigid intermediate supports at stations a and b. From the previous section, we can write the following equations.

$$\{S\}_a^L = [A]\{S\}_0 \quad (4.4.1)$$
$$\{S\}_b^L = [B]\{S\}_a^R \quad (4.4.2)$$

Bending Critical Speeds of Simple Shafts 57

$$\{S\}_n = [C]\{S\}_b^R \qquad (4.4.3)$$

1 2   a a+1   b b+1   n

Fig. 4.8 *n* mass shaft with overhangs

where $[A]$, $[B]$ and $[C]$ are the transfer matrices for left overhang, mid-span and right overhangs respectively.

We notice that $\{S\}_0$ represents state vector to the left of station 1, which on account of boundary conditions is

$$\{S\}_0 = \begin{Bmatrix} -w_0 \\ \theta_0 \\ 0 \\ 0 \end{Bmatrix} \qquad (4.4.4)$$

At support a, we have $w_a = 0$; θ_a and M_{ya} are continuous and there is a sudden change in shear force V_{za}, because of support reaction P_a.

Consequently, we get from equation (4.4.1),

$$-A_{11}w_0 + A_{12}\theta_0 = 0$$

i.e.,

$$\theta_0 = w_0 \frac{A_{11}}{A_{12}} \qquad (4.4.5)$$

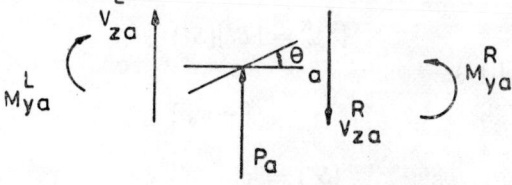

Fig. 4.9 Intermediate conditions for support *a*

Using the above equation in (4.4.1), we get

$$w_a^L = 0$$

$$\theta_a^L = \left(-A_{21} + \frac{A_{22}A_{11}}{A_{12}}\right) w_0 = A'_{21}w_0$$

$$M_{ya}^L = \left(-A_{31} + \frac{A_{32}A_{11}}{A_{12}}\right) w_0 = A'_{31}w_0$$

$$V_{za}^L = \left(-A_{41} + \frac{A_{42}A_{11}}{A_{12}}\right) w_0 = A'_{41}w_0 \qquad (4.4.6)$$

Hence

$$\begin{Bmatrix} -w \\ \theta \\ M_y \\ V_z \end{Bmatrix}_a^R = \begin{bmatrix} 0 & 0 \\ -A'_{21} & 0 \\ -A'_{31} & 0 \\ -A'_{41} & 1 \end{bmatrix} \begin{Bmatrix} -w_0 \\ P_a \end{Bmatrix} \qquad (4.4.7)$$

The above can be written as

$$\{S\}_a^R = [A']\{S'\} \qquad (4.4.8)$$

Equation (4.4.8) is

$$\begin{Bmatrix} -w \\ \theta \\ M_y \\ V_z \end{Bmatrix}_a^R = \begin{bmatrix} 0 & 0 & 0 & 0 \\ -A'_{21} & 0 & 0 & 0 \\ -A'_{31} & 0 & 0 & 0 \\ -A'_{41} & 0 & 0 & 1 \end{bmatrix} \begin{Bmatrix} -w_0 \\ 0 \\ 0 \\ P_a \end{Bmatrix} \qquad (4.4.9)$$

which defines $[A']$ and $\{S'\}$

Now we can rewrite equation (4.4.2) as

$$\{S\}_b^L = [B]\{S\}_a^R$$
$$= [B][A']\{S'\}$$

i.e.,

$$\{S\}_b^L = [B']\{S'\} \qquad (4.4.10)$$

where $[B']$ is product of $[B]$ and $[A']$ matrices.

At support b, we again have $w_b = 0$, θ_b and M_{yb} are continuous and there is a sudden change in V_{zb} because of support reaction P_b. Following similar steps as before, we get

$$\{S\}_b^R = [B'']\{S''\} \qquad (4.4.11)$$

In equation (4.4.11)

$$\{S''\} = \begin{Bmatrix} -w_0 \\ 0 \\ 0 \\ P_b \end{Bmatrix} \qquad (4.4.12)$$

$$\{B''\} = \begin{bmatrix} 0 & 0 & 0 & 0 \\ -B''_{21} & 0 & 0 & 0 \\ -B''_{31} & 0 & 0 & 0 \\ -B''_{41} & 0 & 0 & 1 \end{bmatrix} \qquad (4.4.13)$$

$$B''_{21} = -B'_{21} + \frac{B'_{24}B'_{11}}{B'_{14}}$$

$$B''_{31} = -B'_{31} + \frac{B'_{34}B'_{11}}{B'_{14}}$$

$$B''_{41} = -B'_{41} + \frac{B'_{44}B'_{11}}{B'_{14}} \qquad (4.4.14)$$

In view of equation (4.4.11), (4.4.3) becomes

$$\{S_n\} = [C][B'']\{S''\} \qquad (4.4.15)$$
$$= [U]\{S''\} \qquad (4.4.16)$$

We can find the appropriate determinant as before and determine the frequencies and mode shapes. The frequency equation is

$$\Delta = u_{31}u_{44} - u_{41}u_{34} = 0 \qquad (4.4.17)$$

Setting $w_0 = 1$, we get

$$P_b = u_{31}/u_{34} \qquad (4.4.18)$$

From here, we can work backwards to define $\{S\}_0$, as follows

$$\{S\}_b^R = [B'']\{S''\} \qquad (4.4.11)$$

$$\{S\}_b^L = \{S\}_b^R + \begin{Bmatrix} 0 \\ 0 \\ 0 \\ -P_b \end{Bmatrix} \qquad (4.4.19)$$

$$\{S'\} = [B']^{-1}\{S\}_b^L \qquad (4.4.20)$$

$$\{S\}_a^R = [A']\{S'\} \qquad (4.4.8)$$

$$\{S\}_a^L = \{S\}_a^R + \begin{Bmatrix} 0 \\ 0 \\ 0 \\ -P_a \end{Bmatrix}; P_a = S'(4) \qquad (4.4.21)$$

$$\{S\}_0 = [A]^{-1}\{S\}_a^L \qquad (4.4.22)$$

Once $\{S_0\}$ is defined, the mode shape can be determined by working forwards again. A computer program can be developed for overhung rotors, based on the analysis given in the section.

Example 4.1

Consider Kikuchi's rotor given in Fig. 4.10. The shaft is made of steel with Young's modulus of 2.0682×10^{11} N/m². The total mass of the rotor including the shaft is 51.8 kg. We can lump the shaft mass equally at the three disk locations and apply Dunkerley's formula to obtain the lower bound for the fundamental frequency, which is

$$\omega_{\text{Dunkerley}} = 40.9 \text{ Hz}$$

By using Rayleigh's formula, we get the upper bound, which is

$$\omega_{\text{Rayleigh}} = 43.5 \text{ Hz}$$

The natural frequencies and mode shapes can be obtained from the computer program, following the analysis of this section, and they are given in Fig. 4.11.

60 Rotor Dynamics

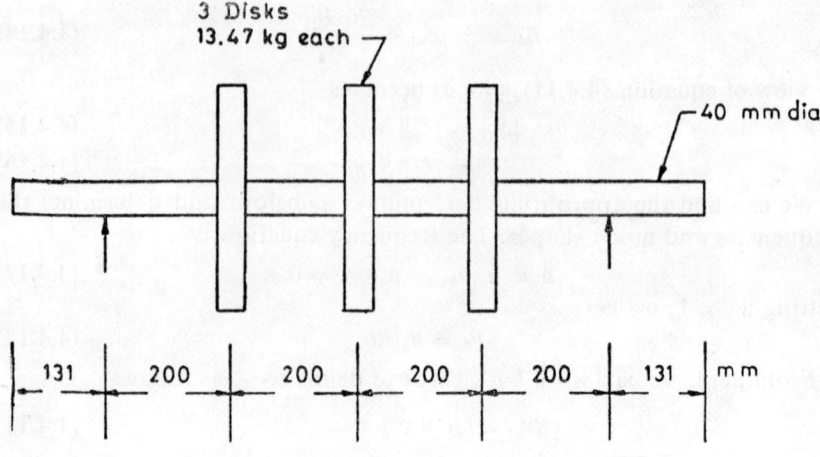

Fig. 4.10 Kikuchi's rotor [11] on rigid bearings

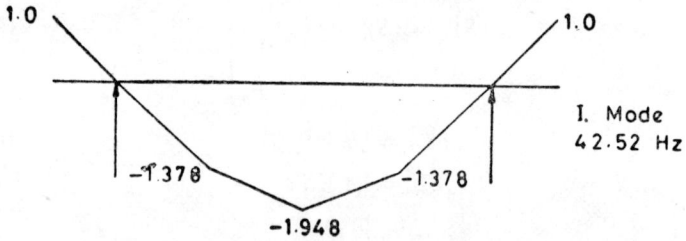

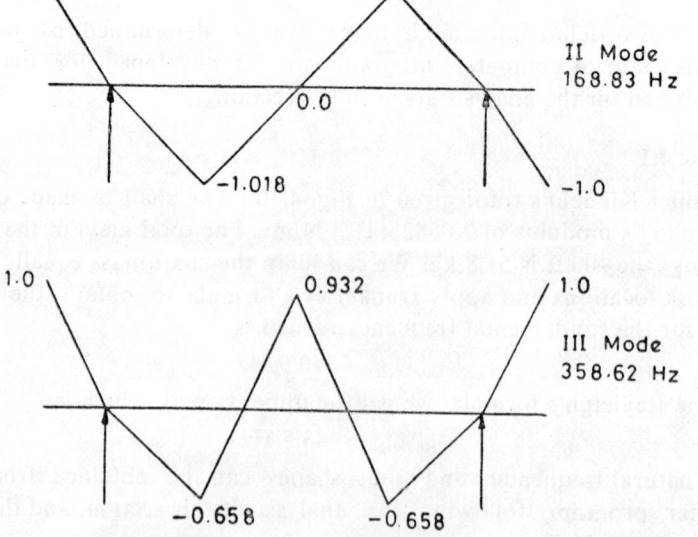

Fig. 4.11 Natural frequencies and mode shapes of Kikuchi's rotor [11]

References

1. Rankine, W.J. On the centrifugal force of rotating shafts, Engineer, Vol. 27, 1869, p. 249.
2. Jeffcott, H.H. The lateral vibration of loaded shafts in the neighbourhood of a whirling speed—The effect of want of balance, Phil. Mag., series 6, Vol. 37, 1919, p. 304.
3. Dunkerley, S. On the whirling of vibration of shafts, Phil. Trans. Roy. Soc., series A, Vol. 185, 1894, p. 279.
4. Rao, J.S. and Gupta, K., An introductory course on theory and practice of mechanical vibrations, Wiley Publication, 1984.
5. Rayleigh, J.W.S. The theory of sound, Dover Publication, 1945.
6. Jacobsen, L.S. and Ayre, R.S. Engineering Vibrations, McGraw-Hill Book Co., 1958.
7. Myklestad, N.O. A new method of calculating natural modes of uncoupled bending vibrations of airplane wings and other types of beams, J. Aero. Sci., 1944, p. 153.
8. Prohl, M.A. A general method for calculating critical speeds of flexible rotors, Trans. ASME, 1945, p. A-142.
9. Pestel, E.C. and Leckie, F.A. Matrix methods in elastomechanics, McGraw-Hill Book Co., 1963.
10. Rao, J.S. and Reddy, K.B.V. Flexural vibrations of simply supported beams, Proc. 11th Cong. Ind. Soc. Theo. and Appld. Mechs., 1966, p. 128.
11. Kikuchi, K. Analysis of unbalance vibration of rotating shaft system with many bearings and disks, Bull. JSME, Vol. 13, No. 61, 1970, p. 864.

Chapter 5

Out-of-Balance Response of Rotors in Rigid Supports

The major cause of excessive vibrations in rotating shafts is the residual unbalance. The unbalance in the rotor due to material inhomogeneities, manufacturing processes, keyways, slots etc., can be removed by a proper balancing procedure. We will discuss more about the balancing procedure in Chapter 12. However, during the operation the rotor deteriorates in its balance condition, due to wear, thermal bending, process dirt collection etc., and gradually develops more vibratory response due to this unbalance. Naturally, it is important to determine the response of a rotor due to a specified unbalance, to study its dynamic behaviour so as to determine, whether a rebalance is necessary during the rotor life.

In Chapter 8, we will find that the dynamic behaviour of the rotors mounted in fluid film bearings are significantly effected by the stiffness and damping properties of the oil film. In fact the rotor may not exhibit a critical speed under certain conditions and it is necessary to determine the out-of-balance response to assess the performance of the rotor. The transfer matrix method discussed before, is extended in this chapter, to determine the response of rotors due to unbalance.

5.1 Field Matrix

The discrete rotor shown in Fig. 4.5 would bend in x-z plane as well as in x-y plane, while it is whirling due to unbalance, which can also be considered as discretised at the lumped mass stations. In Fig. 5.1 the ith element of the shaft is shown. It is obvious that we have to extend the state vector $\{S\}$ to include the bending quantities in x-y plane also, in addition to those in x-z plane we considered while determining the natural frequencies in the previous chapter.

$$\{S\} = \begin{Bmatrix} \{S_z\} \\ \{S_y\} \end{Bmatrix} \qquad (5.1.1)$$

The projection of the elastic line of the shaft in the x-z plane is shown in Fig. 5.2 and the projection of the same in x-y plane is shown in Fig. 5.3. Actually Fig. 5.2 is the same as Fig. 4.6 and the field matrix relation in equation (4.3.9) for this plane is valid here also. $\{S_z\}$ would

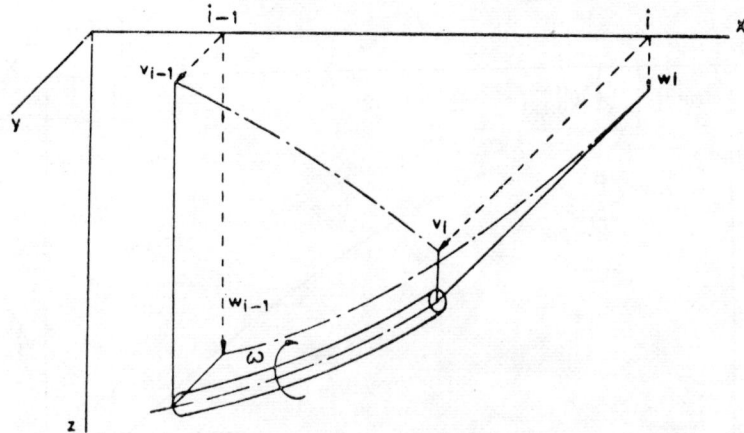

Fig. 5.1 *i*th element of the rotor

represent the state vector defined in equation (4.3.9). Defining $\{S_y\}$, the state vector for *x-y* plane of bending as

$$\{S_y\} = \begin{Bmatrix} v \\ \phi \\ M_z \\ -V_y \end{Bmatrix} \qquad (5.1.2)$$

we can show that the following relation is valid, in accordance with the equilibrium relations given in Fig. 5.3.

$$\{S_y\}_i^L = [F]_i \{S_y\}_{i-1}^R \qquad (5.1.3)$$

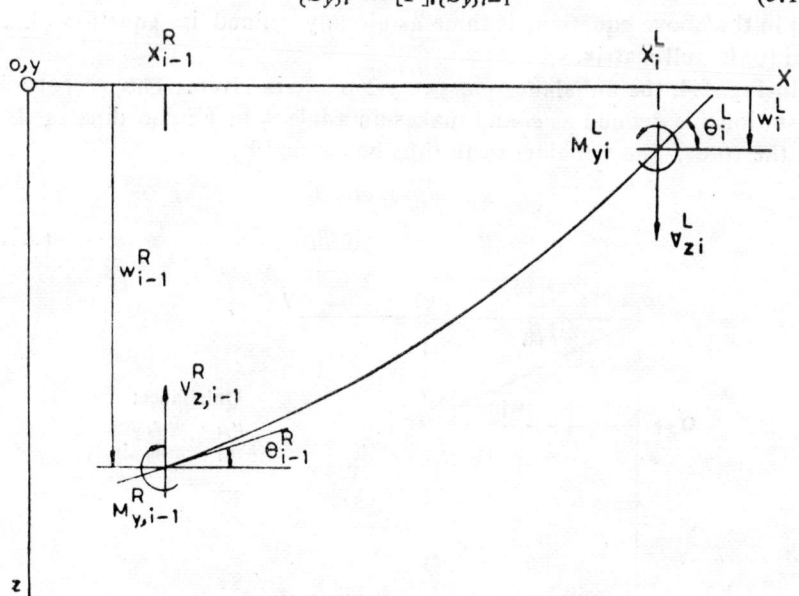

Fig. 5.2 Elastic field of *j*th element in *x-z* plane

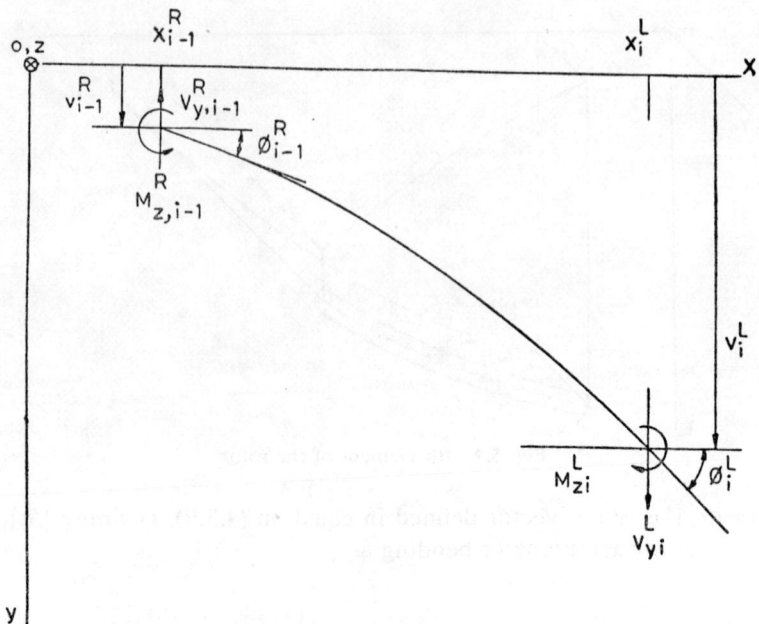

Fig. 5.3 Elastic field of ith element in x-y plane

We can combine the relations (4.3.9) and (5.1.3) to give the overall state vector relation as follows:

$$\begin{Bmatrix}\{S_z\}\\\{S_y\}\end{Bmatrix}_i^L = \begin{bmatrix}[F] & [0]\\ [0] & [F]\end{bmatrix}_i \begin{Bmatrix}\{S_z\}\\\{S_y\}\end{Bmatrix}_{i-1}^R \quad (5.1.4)$$

$[F]$ in the above equations is same as already defined in equation (4.3.9) and $[0]$ is null matrix.

In Fig. 5.4, the unbalance mass in y-z plane is given. The eccentricity at station i is defined as e_i and makes an angle β_i in $\bar{y}\bar{z}$ coordinates fixed on the rotor. The unbalance can thus be defined by

$$u_{yi} = m_i e_i \cos \beta_i$$
$$u_{zi} = m_i e_i \sin \beta_i \quad (5.1.5)$$

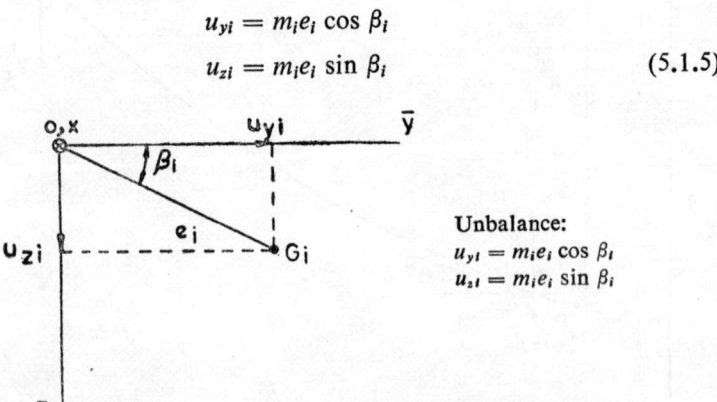

Unbalance:
$u_{yi} = m_i e_i \cos \beta_i$
$u_{zi} = m_i e_i \sin \beta_i$

Fig. 5.4a The ith unbalanced mass in rotor coordinates

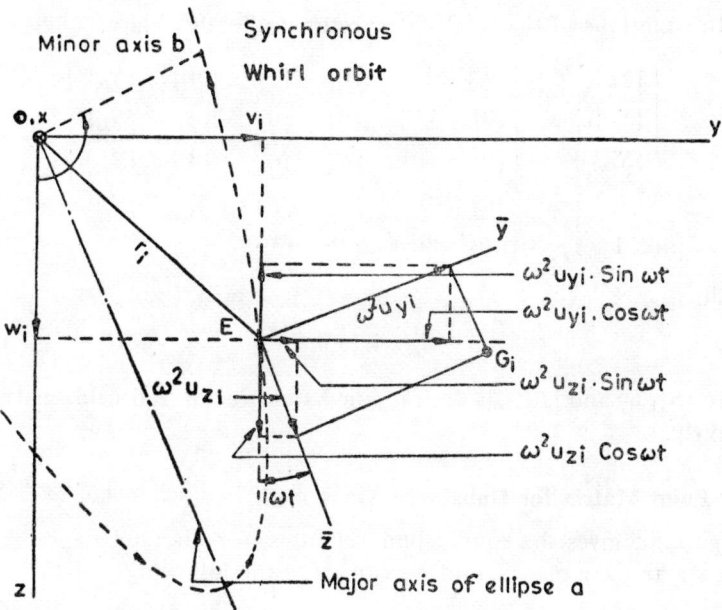

Fig. 5.4b Equilibrium relations of ith unbalance mass

In general, the response in the y direction will be different from that in the z direction. Further for anisotropic bearings the response in both the x-y and x-z planes will consist of cosine and sine components (Also see (5.2.3). Hence we split $\{S_z\}$ and $\{S_y\}$ further

$$\{S_z\} = \begin{Bmatrix} \{S_{zc}\} \\ \{S_{zs}\} \end{Bmatrix}$$

$$\{S_y\} = \begin{Bmatrix} \{S_{yc}\} \\ \{S_{ys}\} \end{Bmatrix} \tag{5.1.6}$$

$$\{S_{zc}\} = \begin{Bmatrix} -w_c \\ \theta_c \\ M_{yc} \\ V_{zc} \end{Bmatrix}; \quad \{S_{zs}\} = \begin{Bmatrix} -w_s \\ \theta_s \\ M_{ys} \\ V_{zs} \end{Bmatrix} \tag{5.1.7}$$

$$\{S_{yc}\} = \begin{Bmatrix} v_c \\ \phi_c \\ M_{zc} \\ -V_{yc} \end{Bmatrix}; \quad \{S_{ys}\} = \begin{Bmatrix} v_s \\ \phi_s \\ M_{zs} \\ -V_{ys} \end{Bmatrix} \tag{5.1.8}$$

For the purpose of response calculations, we add the identity $1 \equiv 1$, to the 16 equations of the total state vector, to finally obtain

$$\begin{Bmatrix} \{S_{zc}\} \\ \{S_{zs}\} \\ \{S_{yc}\} \\ \{S_{ys}\} \\ 1 \end{Bmatrix}_i^L = \begin{bmatrix} [F] & [0] & [0] & [0] & \{0\} \\ [0] & [F] & [0] & [0] & \{0\} \\ [0] & [0] & [F] & [0] & \{0\} \\ [0] & [0] & [0] & [F] & \{0\} \\ \{0\}^T & \{0\}^T & \{0\}^T & \{0\}^T & 1 \end{bmatrix}_i \begin{Bmatrix} \{S_{zc}\} \\ \{S_{zs}\} \\ \{S_{yc}\} \\ \{S_{ys}\} \\ 1 \end{Bmatrix}_{i-1}^R \quad (5.1.9)$$

which is

$$\{\bar{S}\}_i^L = [\bar{F}]_i \{\bar{S}\}_{i-1}^R \quad (5.1.10)$$

where $\{\bar{S}\}_{1 \times 17}$ and $[\bar{F}]_{17 \times 17}$ are modified state vector and field matrix respectively.

5.2 Point Matrix for Unbalance Mass

Figure 5.5 gives the equilibrium relations for the unbalance mass in both x-z and x-y planes, and we can write the following

$$V_{zi}^R = V_{zi}^L - m_i\omega^2 w_i - \omega^2 u_{zi} \cos \omega t + \omega^2 u_{yi} \sin \omega t \quad (5.2.1)$$

$$V_{yi}^R = V_{yi}^L - m_i\omega^2 v_i - \omega^2 u_{yi} \cos \omega t - \omega^2 u_{zi} \sin \omega t \quad (5.2.2)$$

The deflection, slope and moment are continuous and are same to the right or left of the station i. Because of the excitation forces which have both cosine and sine terms in the above, the state vector quantities w, θ, M_y, V_z, v, ϕ, M_z and V_y will also have cosine and sine response terms as follows:

$$w = w_c \cos \omega t + w_s \sin \omega t$$

$$\theta = \theta_c \cos \omega t + \theta_s \sin \omega t \ldots, \text{etc.} \quad (5.2.3)$$

Hence equations (5.2.1) and (5.2.2) can now be written as

$$V_{zic}^R = V_{zic}^L - m_i\omega^2 w_{ic} - u_{zi}\omega^2$$

$$V_{zis}^R = V_{zis}^L - m_i\omega^2 w_{is} + u_{yi}\omega^2 \quad (5.2.4)$$

$$V_{yic}^R = V_{yic}^L - m_i\omega^2 v_{ic} - u_{yi}\omega^2$$

$$V_{yis}^R = V_{yis}^L - m_i\omega^2 v_{is} - u_{zi}\omega^2 \quad (5.2.5)$$

Hence

$$\begin{Bmatrix} \{S_{zc}\} \\ \{S_{zs}\} \\ \{S_{yc}\} \\ \{S_{ys}\} \\ 1 \end{Bmatrix}_i^R = \begin{bmatrix} [P] & [0] & [0] & [0] & \{m_{zc}\} \\ [0] & [P] & [0] & [0] & \{m_{zs}\} \\ [0] & [0] & [P] & [0] & \{m_{yc}\} \\ [0] & [0] & [0] & [P] & \{m_{ys}\} \\ \{0\}^T & \{0\}^T & \{0\}^T & \{0\}^T & 1 \end{bmatrix} \begin{Bmatrix} \{S_{zc}\} \\ \{S_{zs}\} \\ \{S_{yc}\} \\ \{S_{ys}\} \\ 1 \end{Bmatrix}_i^L \quad (5.2.6)$$

where $[P]$ is the point matrix defined in equation (4.3.10) with p replaced

Out-of-Balance Response 67

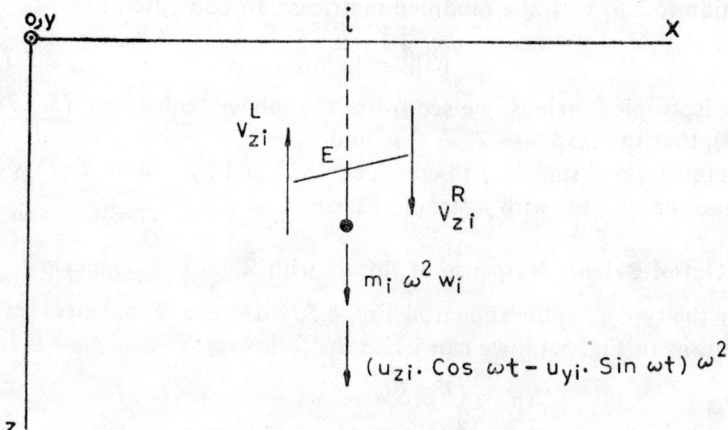

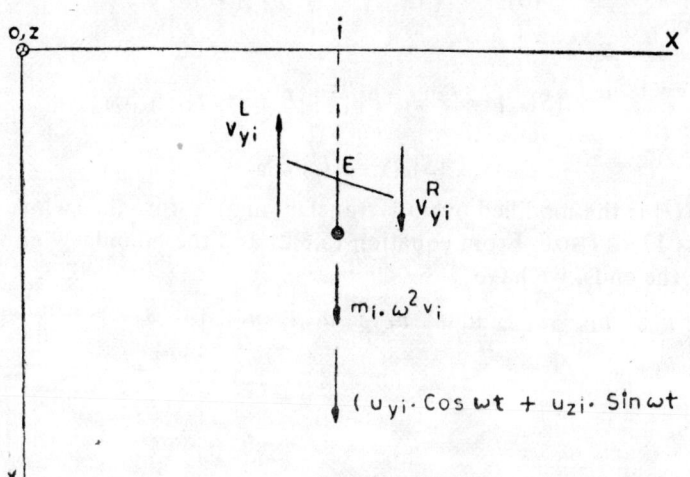

Fig. 5.5 Equilibrium relations of unbalance mass in xz and xy planes.

by ω, and

$$\{m_{zc}\} = \begin{Bmatrix} 0 \\ 0 \\ 0 \\ -u_z\omega^2 \end{Bmatrix}_i ; \{m_{zs}\} = \begin{Bmatrix} 0 \\ 0 \\ 0 \\ u_y\omega^2 \end{Bmatrix}_i \quad (5.2.7)$$

and

$$\{m_{yc}\} = \begin{Bmatrix} 0 \\ 0 \\ 0 \\ u_y\omega^2 \end{Bmatrix}_i ; \quad \{m_{ys}\} = \begin{Bmatrix} 0 \\ 0 \\ 0 \\ u_z\omega^2 \end{Bmatrix}_i \quad (5.2.8)$$

Equation (5.2.6) with the modified matrices can be written as

$$\{\bar{S}\}_i^R = [\bar{P}]_i\{\bar{S}\}_i^L \tag{5.2.9}$$

For isotropic bearings, we see from the above equations (5.2.7) and (5.2.8), that the response $w_c = -v_s$ and $w_s = v_e$.

Therefore, it is sufficient to consider $\{S_{zc}\}$ and $\{S_{yc}\}$ state vectors in the analysis for systems with isotropic bearings.

5.3 Out-of-Balance Response of Rotors with Rigid End Supports

For the type of rotor shown in Fig. 4.5, with the unbalances at each station, as in Fig. 5.4a, we can write the following:

$$\{\bar{S}\}_1^L = [\bar{F}]_1\{\bar{S}\}_0$$
$$\{\bar{S}\}_1^R = [\bar{P}]_1\{\bar{S}\}_1^L = [\bar{P}]_1[\bar{F}]_1\{\bar{S}\}_0$$
$$\{\bar{S}\}_2^L = [\bar{F}]_2\{\bar{S}\}_1^R = [\bar{F}]_2[\bar{P}]_1[\bar{F}]_1\{\bar{S}\}_0$$
$$\vdots$$
$$\{\bar{S}\}_{n+1} = [\bar{F}]_{n+1}[\bar{P}]_n[\bar{F}]_n[\bar{P}]_{n-1}\ldots[\bar{F}]_1\{\bar{S}\}_0 \tag{5.3.1}$$

Hence
$$\{\bar{S}\}_{n+1} = [\bar{U}]\{\bar{S}\}_0 \tag{5.3.2}$$

where $[\bar{U}]$ is the modified overall transfer matrix for the whole rotor, which is 17×17 size. From equation (5.3.2) and the boundary conditions at both the ends, we have

$$\begin{bmatrix} u_{1,2} & u_{1,4} & u_{1,6} & u_{1,8} & u_{1,10} & u_{1,12} & u_{1,14} & u_{1,16} \\ u_{3,2} & u_{3,4} & & \cdots & & & & u_{3,16} \\ u_{5,2} & & & & & & & \\ u_{7,2} & & & & & & & \\ u_{9,2} & & & & & & & \\ u_{11,2} & & & & & & & \\ u_{13,2} & & & & & & & \\ u_{15,2} & & & \cdots & & & & u_{15,16} \end{bmatrix} \begin{Bmatrix} \theta_c \\ V_{zc} \\ \theta_s \\ V_{zs} \\ \phi_c \\ -V_{yc} \\ \phi_s \\ -V_{ys} \end{Bmatrix} = -\begin{Bmatrix} u_{1,17} \\ u_{3,17} \\ - \\ - \\ - \\ u_{15,17} \end{Bmatrix}_0 \tag{5.3.3}$$

Solving the system of simultaneous equations in (5.3.3), we obtain the state vector at left support station 0. From there, the state vector quantities at any station can be determined using the appropriate transfer matrices, to give w_c, w_s, v_c and v_s at all stations.

Referring to Fig. 5.4b, the radius vector of the whirl orbit is

$$r = \sqrt{w^2 + v^2} \tag{5.3.4}$$

From equation (5.2.3), we get

$$r = [(w_c \cos \omega t + w_s \sin \omega t)^2 + (v_c \cos \omega t + v_s \sin \omega t)^2]^{1/2} \tag{5.3.5}$$

The above can be written as

$$r = [\tfrac{1}{2}\{w_c^2 + w_s^2 + v_c^2 + v_s^2 + (w_c^2 - w_s^2 + v_c^2 - v_s^2)\cos 2\omega t \\ + 2(w_c w_s + v_c v_s)\sin 2\omega t\}]^{1/2} \tag{5.3.6}$$

The radius vector is maximum or minimum when

$$\frac{d}{d(\omega t)}[(w_c^2 - w_s^2 + v_c^2 + v_s^2)\cos 2\omega t + 2(w_c w_s + v_c v_s)\sin 2\omega t] = 0$$

i.e.,

$$\tan 2\omega t = \frac{2(w_c w_s + v_c v_s)}{w_c^2 - w_s^2 + v_c^2 - v_s^2} \tag{5.3.8}$$

Using equation (5.3.8) in (5.3.6), we get the major and minor axes of the elliptical orbit shown in Fig. 5.4b.

$$a, b = [\tfrac{1}{2}\{w_c^2 + w_s^2 + v_c^2 + v_s^2 \\ \pm [(w_c^2 - w_s^2 + v_c^2 - v_s^2)^2 + 4(w_c w_s + v_c v_s)^2]^{1/2}\}]^{1/2} \tag{5.3.9}$$

For systems with isotropic bearings, the elliptical orbit becomes circular, since $w_c = -v_s$ and $w_s = v_c$ and the above equation becomes

$$a = b = \sqrt{w_c^2 + v_c^2} \tag{5.3.9a}$$

The radius vector r in Fig. 5.4b, can also be considered as two rotating vectors with angular velocity ω, one in positive ω direction and the other in the opposite direction. Figure 5.6 shows these two vectors at $t = 0$.

$$r(t) = z(t) + iy(t) \tag{5.3.10}$$

with the real quantities measured along the z axis and imaginary quantities measured along the y axis respectively. Therefore

$$r(t) = (w_c \cos \omega t + w_s \sin \omega t) + i(v_c \cos \omega t + v_s \sin \omega t) \tag{5.3.11}$$

Using

$$2 \cos \omega t = \exp(i\omega t) + \exp(-i\omega t)$$
$$2 \sin \omega t = -i\{\exp(i\omega t) - \exp(-i\omega t)\} \tag{5.3.12}$$

we can express equation (5.3.11) as follows:

$$r(t) = \exp(i\omega t)[\tfrac{1}{2}(w_c + v_s) + \tfrac{1}{2}i(v_c - w_s)] + \exp(-i\omega t)[\tfrac{1}{2}(w_c - v_s) + \tfrac{1}{2}i(v_c + w_s)] \tag{5.3.13}$$

$$= r^+ \exp(i\omega t) + r^- \exp(-i\omega t) \tag{5.3.14}$$

The angle α^+ made by r^+ vector from the z axis is

$$\tan \alpha^+ = \frac{v_c - w_s}{w_c + v_s} \tag{5.3.15}$$

Similarly the angle α^- made by r^- vector from the z axis is

$$\tan \alpha^- = \frac{v_c + w_s}{w_c - v_s} \tag{5.3.16}$$

The two vectors should rotate by an angle ωt, which is

$$\omega t = \tfrac{1}{2}(\alpha^- - \alpha^+) \tag{5.3.17}$$

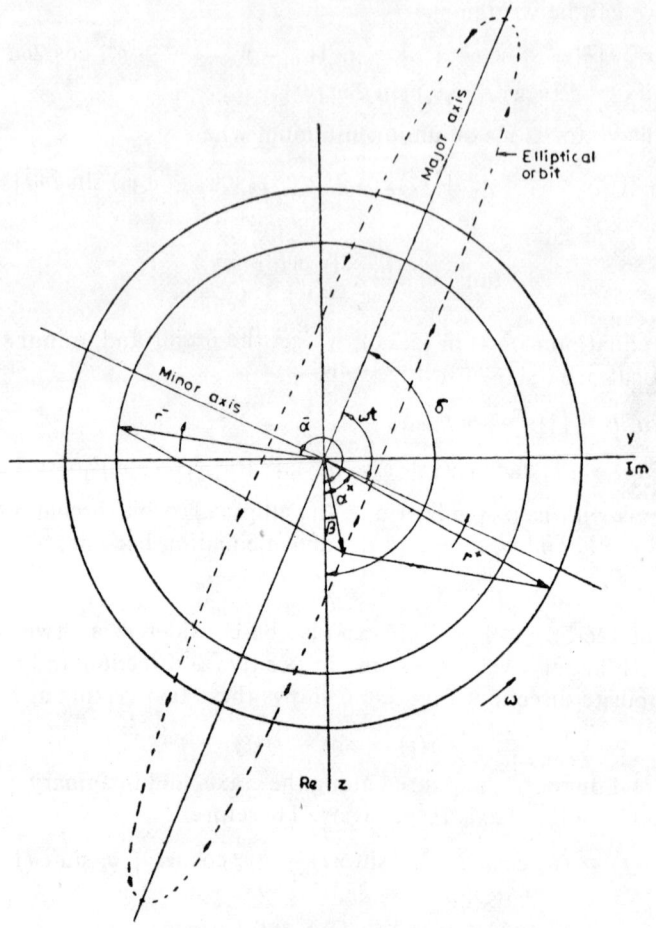

Fig. 5.6 Orbital relations

to be in line and give the location of the major axis of the ellipse. Hence for this location,

$$\tan 2\omega t = \tan(\alpha^- - \alpha^+) \tag{5.3.18}$$

which gives the same value as given in equation (5.3.8)

$$\tan 2\omega t = \frac{2(v_c v_s + w_c w_s)}{w_c^2 - v_s^2 + v_c^2 - w_s^2} \tag{5.3.8}$$

The angle β made by r, at the reference time $t = 0$, is given by

$$\tan \beta = \frac{v_c}{w_c}$$

i.e.,

$$\beta = \tan^{-1} \frac{v_c}{w_c} \tag{5.3.19}$$

The angle δ made by the major axis of the ellipse with the real z axis is

$$\delta = \omega t + \alpha^+ \quad (5.3.20)$$

Using equation (5.3.17) the above becomes

$$\delta = \tfrac{1}{2}(\alpha^- + \alpha^+) \quad (5.3.21)$$

i.e.,

$$\delta = \tfrac{1}{2} \tan^{-1} \frac{2(w_c v_c + w_s v_s)}{w_c^2 - v_c^2 + w_s^2 - v_s^2} \quad (5.3.22)$$

Since $\tan(\pi + \theta) = \tan \theta$, the above formula can also give the position of minor axis of the ellipse, from the real z axis and therefore caution should be exercised in evaluating the angle δ.

It may also be noted that

$$|r^+| > |r^-| \quad (5.3.23)$$

gives forward whirl, that is the direction of whirl is same as the direction of rotation of the shaft, while

$$|r^+| < |r^-| \quad (5.3.24)$$

gives a backward whirl for the rotor

5.4 Simply Supported Rotor with Overhangs

For the rotor shown in Fig. 4.8, we can write

$$\{S\}_a^L = [A]\{S\}_0 \quad (5.4.1)$$

$$\{S\}_b^L = [\underline{B}]\{S\}_a^R \quad (5.4.2)$$

$$\{S\}_n = [\underline{C}]\{S\}_b^R \quad (5.4.3)$$

where $\{S\}$ is the state vector 17×1 in size, $[A]$, $[\underline{B}]$ and $[\underline{C}]$ are overall transfer matrices for the left overhang, mid-span and right overhang respectively. $\{S\}_0$ for the free left hand end is given by

$$\{S\}_0^T = \{-w_{c0},\ \theta_{c0},\ 0,\ 0,\ -w_{s0},\ \theta_{s0},\ 0,\ 0,\ v_{c0},\ \phi_{c0},\ 0,\ 0,\ v_{s0},\ \phi_{s0},\ 0,\ 0,\ 1\} \quad (5.4.4)$$

At the support a we have the following conditions:

$$w_{ca} = w_{sa} = v_{ca} = v_{sa} = 0$$

$$\left.\begin{array}{c} \theta_{ca},\ \theta_{sa},\ \phi_{ca},\ \phi_{sa} \\ M_{yca},\ M_{ysa},\ M_{zca},\ M_{zsa} \end{array}\right\} \text{ are continuous}$$

Jump in the shear forces

$V_{zca},\ V_{zsa},\ V_{yca},\ V_{ysa}$ by the support reactions

$$P_{zca},\ P_{zsa},\ P_{yca},\ P_{ysa} \quad (5.4.5)$$

From the conditions of deflections in the above equation, equation (5.4.1) gives

$$[A_d]\{d_0\} + [A_s]\{s_0\} + \{f_a\} = 0 \quad (5.4.6)$$

In expanded form the above equation is

$$\begin{bmatrix} A_{1,1} & A_{1,5} & A_{1,9} & A_{1,13} \\ A_{5,1} & A_{5,5} & A_{5,9} & A_{5,13} \\ A_{9,1} & A_{9,5} & A_{9,9} & A_{9,13} \\ A_{13,1} & A_{13,5} & A_{13,9} & A_{13,13} \end{bmatrix} \begin{Bmatrix} -w_{c0} \\ -w_{s0} \\ v_{c0} \\ v_{s0} \end{Bmatrix}$$

$$+ \begin{bmatrix} A_{1,2} & A_{1,6} & A_{1,10} & A_{1,14} \\ A_{5,2} & A_{5,6} & A_{5,10} & A_{5,14} \\ A_{9,2} & A_{9,6} & A_{9,10} & A_{9,14} \\ A_{13,2} & A_{13,6} & A_{13,10} & A_{13,14} \end{bmatrix} \begin{Bmatrix} \theta_{c0} \\ \theta_{s0} \\ \phi_{c0} \\ \phi_{s0} \end{Bmatrix} + \begin{Bmatrix} A_{1,17} \\ A_{5,17} \\ A_{9,17} \\ A_{13,17} \end{Bmatrix} = 0 \quad (5.4.6a)$$

Multiplying by $[A_s]^{-1}$, the above equation gives

$$\{s_0\} = [A']\{d_0\} + \{f'_a\} \quad (5.4.7)$$

where

$$[A'] = -[A_s]^{-1}[A_d]$$
$$\{f'_a\} = -[A_s]^{-1}\{f_a\} \quad (5.4.8)$$

Now we can write the state vector $\{S\}_{ca}^L$, in terms of the deflections of station 0 only. For example, from equation (5.4.1), the first line is

$$-w_{ca}^L = -A_{1,1}w_{c0} + A_{1,2}\theta_{c0} - A_{1,5}w_{s0} + A_{1,6}\theta_{s0} + A_{1,9}v_{c0} + A_{1,10}\phi_{c0}$$
$$+ A_{1,13}v_{s0} + A_{1,14}\phi_{s0} + A_{1,17} \quad (5.4.9)$$

Substituting for the slopes θ_{c0}, θ_{s0}, ϕ_{c0} and ϕ_{s0} in the above equation (5.4.9) from (5.4.7) we get

$$-w_{ca}^L = -A_{1,1}w_{c0} + A_{1,2}(-A'_{1,1}w_{c0} - A'_{1,2}w_{s0} + A'_{1,3}v_{c0} + A'_{1,4}v_{s0} + f'_{a1})$$
$$- A_{1,5}w_{s0} + A_{1,6}(-A'_{2,1}w_{c0} - A'_{2,2}w_{s0} + A'_{2,3}v_{c0} + A'_{2,4}v_{s0} + f'_{a2})$$
$$+ A_{1,9}v_{c0} + A_{1,10}(-A'_{3,1}w_{c0} - A'_{3,2}w_{s0} + A'_{3,3}v_{c0} + A'_{3,4}v_{s0} + f'_{a3})$$
$$+ A_{1,13}v_{s0} + A_{1,14}(-A'_{4,1}w_{c0} - A'_{4,2}w_{s0} + A'_{4,3}v_{c0} + A'_{4,4}v_{s0} + f'_{a4}) + A_{1,17}$$
$$(5.4.10)$$

Hence

$$-w_{ca}^L = -\bar{A}_{1,1}w_{c0} - \bar{A}_{1,5}w_{s0} + \bar{A}_{1,9}v_{c0} + \bar{A}_{1,13}v_{s0} + \bar{A}_{1,17} \quad (5.4.11)$$

where

$$\bar{A}_{1,1} = A_{1,1} + A_{1,2}A'_{1,1} + A_{1,6}A'_{2,1} + A_{1,10}A'_{3,1} + A_{1,14}A'_{4,1}$$
$$\bar{A}_{1,5} = A_{1,5} + A_{1,2}A'_{1,2} + A_{1,6}A'_{2,2} + A_{1,10}A'_{3,2} + A_{1,14}A'_{4,2}$$
$$\bar{A}_{1,9} = A_{1,9} + A_{1,2}A'_{1,3} + A_{1,6}A'_{2,3} + A_{1,10}A'_{3,3} + A_{1,14}A'_{4,3}$$
$$\bar{A}_{1,13} = A_{1,13} + A_{1,2}A'_{1,4} + A_{1,6}A'_{2,4} + A_{1,10}A'_{3,4} + A_{1,14}A'_{4,4}$$
$$\bar{A}_{1,17} = A_{1,17} + A_{1,2}f'_{a1} + A_{1,6}f'_{a2} + A_{1,10}f'_{a3} + A_{1,14}f'_{a4} \quad (5.4.12)$$

In a similar way the rth element of state vector $\{S\}_a^L$ is

$$S_{ra}^L = -\bar{A}_{r,1}w_{c0} - \bar{A}_{r,5}w_{s0} + \bar{A}_{r,9}v_{c0} + \bar{A}_{r,13}v_{s0} + A_{r,17} \qquad (5.4.13)$$

In the above equation $\bar{A}_{r,1}$, $\bar{A}_{r,5}$, $\bar{A}_{r,9}$, $\bar{A}_{r,13}$ and $\bar{A}_{r,17}$ are given by equation (5.4.12) with the first subscript of $\bar{A}$ and A elements changed to r, A' elements remaining the same. Now we can introduce the support reactions to obtain $\{S\}_a^R$ as follows:

$$\{S\}_a^R = \begin{Bmatrix} -w_c \\ \theta_c \\ M_{yc} \\ V_{zc} \\ -w_s \\ \theta_s \\ M_{ys} \\ V_{zs} \\ v_c \\ \phi_c \\ M_{zc} \\ -V_{yc} \\ v_s \\ \phi_s \\ M_{zs} \\ -V_{ys} \\ 1 \end{Bmatrix}_a^R = \begin{Bmatrix} -w_c \\ \theta_c \\ M_{yc} \\ V_{zc} \\ -w_s \\ \theta_s \\ M_{ys} \\ V_{zs} \\ v_c \\ \phi_c \\ M_{zc} \\ -V_{yc} \\ v_s \\ \phi_s \\ M_{zs} \\ -V_{ys} \\ 1 \end{Bmatrix}_a^L + \begin{Bmatrix} 0 \\ 0 \\ 0 \\ P_{zc} \\ 0 \\ 0 \\ 0 \\ P_{zc} \\ 0 \\ 0 \\ 0 \\ -P_{yc} \\ 0 \\ 0 \\ 0 \\ -P_{ys} \\ 1 \end{Bmatrix}_a \qquad (5.4.14)$$

With the help of equation (5.4.13), the above can be written as

$$\{S\}_a^R = [\bar{A}]\{S'\}_0 \qquad (5.4.15)$$

$[\bar{A}]$ matrix of 17×17 size in the above, contains all the elements defined in equation (5.4.13) and in addition $\bar{A}_{i,i} = 1$, $i = 4, 8, 12, 16$, with the rest of the elements being zero. In $\{S'\}_0$ vector, $S'(1) = -w_{c0}$, $S'(4) = P_{zca}$, $S'(5) = -w_{s0}$, $S'(8) = P_{zsa}$, $S'(9) = v_{c0}$, $S'(12) = -P_{yca}$, $S'(13) = v_{s0}$, $S'(16) = -P_{ysa}$ and $S'(17) = 1$, with the rest of the elements being zero.

From equation (5.4.2) we now have

$$\{S\}_b^L = [B][\bar{A}]\{S'\}_0 \qquad (5.4.16)$$
$$= [B]\{S'\}_0 \qquad (5.4.16a)$$

At support b, we have

$$w_{cb} = w_{sb} = v_{cb} = v_{sb} = 0$$

$$\left.\begin{matrix} \theta_{cb}, & \theta_{sb}, & \phi_{cb}, & \phi_{sb} \\ M_{ycb}, & M_{ysb}, & M_{zcb}, & M_{zsb} \end{matrix}\right\} \text{continuous}$$

Jump in the shear forces

V_{zcb}, V_{zsb}, V_{ycb}, V_{ysb} by the support reactions

P_{zcb}, P_{zsb}, P_{ycb}, P_{ysb} (5.4.17)

For the conditions of support deflections at b in the above, we obtain from equation (5.4.16a) the relations between support reactions at a and the deflections at the left free end.

$$[B_d]\{d_0\} + [B_p]\{P_a\} + \{f_b\} = 0 \qquad (5.4.18)$$

In expanded form the above equation is

$$\begin{bmatrix} B_{1,1} & B_{1,5} & B_{1,9} & B_{1,13} \\ B_{5,1} & B_{5,5} & B_{5,9} & B_{5,13} \\ B_{9,1} & B_{9,5} & B_{9,9} & B_{9,13} \\ B_{13,1} & B_{13,5} & B_{13,9} & B_{13,13} \end{bmatrix} \begin{Bmatrix} -w_{c0} \\ -w_{s0} \\ v_{c0} \\ v_{s0} \end{Bmatrix}$$

$$+ \begin{bmatrix} B_{1,4} & B_{1,8} & B_{1,12} & B_{1,16} \\ B_{5,4} & B_{5,8} & B_{5,12} & B_{5,16} \\ B_{9,4} & B_{9,8} & B_{9,12} & B_{9,16} \\ B_{13,4} & B_{13,8} & B_{13,12} & B_{13,16} \end{bmatrix} \begin{Bmatrix} P_{zca} \\ P_{zsa} \\ -P_{yca} \\ -P_{ysa} \end{Bmatrix} + \begin{Bmatrix} B_{1,17} \\ B_{5,17} \\ B_{9,17} \\ B_{13,17} \end{Bmatrix} = 0$$

(5.4.19)

Multiplying the above equation by $[B_p]^{-1}$, we obtain

$$\{P_a\} = [B']\{d_0\} + \{f'_b\} \qquad (5.4.20)$$

where

$$[B'] = -[B_p]^{-1}[B_d]$$
$$\{f'_b\} = -[B_p]^{-1}\{f_b\} \qquad (5.4.21)$$

As in equation (5.4.9), we have for support b

$$-w_{cb}^L = -B_{1,1}w_{c0} + B_{1,4}P_{zca} - B_{1,5}w_{s0} + B_{1,8}P_{zsa} + B_{1,9}v_{c0} - B_{1,12}P_{yca}$$
$$+ B_{1,13}v_{s0} - B_{1,16}P_{ysa} + B_{1,17} \qquad (5.4.22)$$

Using equation (5.4.20) in the above we get

$$-w_{cb}^L = -B_{1,1}w_{c0} + B_{1,4}(-B'_{1,1}w_{c0} - B'_{1,2}w_{s0} + B'_{1,3}v_{c0} + B'_{1,4}v_{s0} + f'_{b1})$$
$$- B_{1,5}w_{s0} + B_{1,8}(-B'_{2,1}w_{c0} - B'_{2,2}w_{s0} + B'_{2,3}v_{c0} + B'_{2,4}v_{s0} + f'_{b2})$$
$$+ B_{1,9}v_{c0} + B_{1,12}(-B'_{3,1}w_{c0} - B'_{3,2}w_{s0} + B'_{3,3}v_{c0} + B'_{3,4}v_{s0} + f'_{b3})$$
$$+ B_{1,13}v_{s0} + B_{1,16}(-B'_{4,1}w_{c0} - B'_{4,2}w_{s0} + B'_{4,3}v_{c0} + B'_{4,4}v_{s0} + f'_{b4})$$
$$+ B_{1,17} \qquad (5.4.23)$$

As in equation (5.4.11), we have for support b,

$$-w_{cb}^L = -\bar{B}_{1,1},w_{c0} - \bar{B}_{1,5}w_{s0} + \bar{B}_{1,9}v_{c0} + \bar{B}_{1,13}v_{s0} + \bar{B}_{1,17} \qquad (5.4.24)$$

and for the rth element in $\{S\}_b^L$, we have

$$S_{rb}^L = -\bar{B}_{r,1}w_{c0} - \bar{B}_{r,5}w_{s0} + \bar{B}_{r,9}v_{c0} + \bar{B}_{r,13}v_{s0} + \bar{B}_{r,17} \quad (5.4.25)$$

The elements, $\bar{B}_{r,1}, \bar{B}_{r,5}, \bar{B}_{r,9}, \bar{B}_{r,13}$ and $\bar{B}_{r,17}$ for $r = 1, 2, \ldots, 17$ are given by

$$\bar{B}_{r,1} = B_{r,1} + B_{r,4}B'_{1,1} + B_{r,8}B'_{2,1} + B_{r,12}B'_{3,1} + B_{r,16}B'_{4,1}$$

$$\bar{B}_{r,5} = B_{r,5} + B_{r,4}B'_{1,2} + B_{r,8}B'_{2,2} + B_{r,12}B'_{3,2} + B_{r,16}B'_{4,2}$$

$$\bar{B}_{r,9} = B_{r,9} + B_{r,4}B'_{1,3} + B_{r,8}B'_{2,3} + B_{r,12}B'_{3,3} + B_{r,16}B'_{4,3}$$

$$\bar{B}_{r,13} = B_{r,13} + B_{r,4}B'_{1,4} + B_{r,8}B'_{2,4} + B_{r,12}B'_{3,4} + B_{r,16}B'_{4,4}$$

$$\bar{B}_{r,17} = B_{r,17} + B_{r,4}f'_{b1} + B_{r,8}f'_{b2} + B_{r,12}f'_{b3} + B_{r,16}f'_{b4} \quad (5.4.26)$$

Working further, in a similar way to that of support a, we arrive at the following equation:

$$\{S\}_b^R = [\bar{B}]\{S''\}_0 \quad (5.4.27)$$

where $[\bar{B}]$ has all elements defined in equation (5.4.26), and in addition $\bar{B}_{i,i} = 1$, $i = 4, 8, 12, 16$, with the rest of the elements being zero, just like in equation (5.4.15). $\{S''\}_0$ is defined by

$$\{S''\}_0^T = (-w_{c0}, 0, 0, P_{zcb}, -w_{s0}, 0, 0, P_{zsb}, v_{c0}, 0, 0, -P_{ycb}, v_{s0}, 0, 0, -P_{y,sb}, 1) \quad (5.4.28)$$

Using equation (5.4.27) in (5.4.3), we get

$$\{S\}_n = [\underline{C}][\bar{B}]\{S''\}_0 = [C]\{S''\}_0 \quad (5.4.29)$$

At station n, shear force and bending moments are zero, hence

$$\begin{bmatrix} c_{3,1} & c_{3,4} & c_{3,5} & c_{3,8} & c_{3,9} & c_{3,12} & c_{3,13} & c_{3,16} \\ c_{4,1} & & & & & & & \\ c_{7,1} & & & & & & & \\ c_{8,1} & & & & & & & \\ c_{11,1} & & & & & & & \\ c_{12,1} & & & & & & & \\ c_{15,1} & & & & & & & \\ c_{16,1} & c_{16,4} & c_{16,5} & c_{16,8} & c_{16,9} & c_{16,12} & c_{16,13} & c_{16,16} \end{bmatrix} \begin{Bmatrix} -w_{c0} \\ P_{zcb} \\ -w_{s0} \\ P_{zsb} \\ v_{c0} \\ -P_{ycb} \\ v_{s0} \\ -P_{ysb} \end{Bmatrix} = - \begin{Bmatrix} c_{3,17} \\ c_{4,17} \\ c_{7,17} \\ c_{8,17} \\ c_{11,17} \\ c_{12,17} \\ c_{15,17} \\ c_{16,17} \end{Bmatrix}$$

$$(5.4.30)$$

Solving the above equations, we obtain the deflections at station 0, and support reactions at b. Then we use equation (5.4.7) to determine the slopes at station 0, which defines $\{S\}_0$ completely. Hence we can determine the deflections at all stations in the left overhang. Next we use equation (5.4.20) to evaluate the support reactions at a, which defines $\{S'\}_0$ completely. With the help of equation (5.4.15), we can now determine the deflections

76 Rotor Dynamics

at all stations in the mid-span. $\{S''\}_0$ is defined completely from the solution of equations (5.4.30) already obtained. Hence using equation (5.4.27) we determine the deflections in the right overhang. From these we have the deflections of the total rotor defining the unbalance response for a given speed of rotation. A computer program can be developed, based on this analysis to determine the unbalance response of such rotors.

Example 5.1

As an example consider Kikuchi's rotor [5] given in Fig. 4.10. Since this rotor is isotropic in character, at all stations, it will be sufficient to consider 9×9 size matrices in the calculations. The response due to an unbalance of 120 kg micron at the middle disk is shown in Fig. 5.7, around

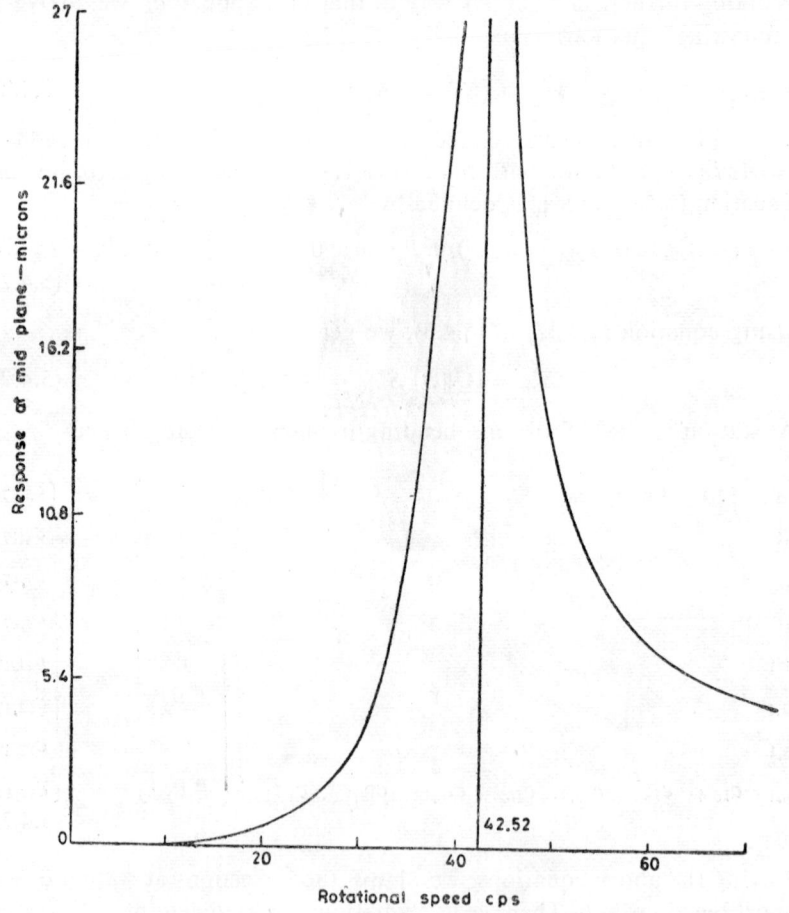

Fig. 5.7 Unbalance response of Kikuchi's rotor

the first mode region. The response of rotors mounted in anisotropic bearings is considered later in Chapter 8.

References

1. Pestel, E.C. and Leckie, F.A. Matrix methods in elastomechanics, McGraw-Hill Book Co. 1963.
2. Piikey, W.D. and Chang, P.Y. Modern formulas for statics and dynamics, McGraw-Hill Book Co., 1978.
3. Bigret, R. Vibrations des machines tournantes et des structures, Tome I to IV, Technique et Documentation, Paris, 1980.
4. Lalanne, M. and Queau, J.P. Calcul par elements finis du comportment dynamique des chaines cinematiques de reducteur, Societe Nationale des Industries Aerospatiales, May 1979.
5. Kikuchi, K., Analysis of unbalance vibration of rotating shaft system with many bearings and disks, Bull. JSME, Vol. 13, No. 61, 1970, p. 864.

A 1.5 tonne impeller on a balancing machine
(Courtesy: ABRO balancing machines, Delhi)

Chapter 6

Gyroscopic Effects

Rotors of turbines, compressors and pumps carry one or more disks along their length. Besides contributing to a lumped mass, as considered in the previous chapters, the disk also contributes to the gyroscopic couple. This couple changes the bending moment equations across the mass, which we have not considered earlier. If the disks are located at nodal points, their gyroscopic effects are predominant, because there is a precession of the disk corresponding to the changes in slope, which is maximum at this location. At antinodal points, there is a pure translation of the disk and there is no gyroscopic effect, because of zero slope. Hence for overhung rotors, with a disk at the free end, the gyroscopic couple has a considerable influence on the dynamic behaviour of the shaft. The effect of gyroscopic couple on the bending critical speeds of a rotor in radially rigid bearings is considered in this chapter.

6.1 Gyroscopics of a Spinning Disk

Using the same notations as in previous chapters, we consider a spinning disk, corresponding to the ith disk of the rotor, which is rotating at an angular velocity ω and precessing about y and z axes with amplitudes θ and ϕ respectively as shown in Fig. 6.1. θ and ϕ correspond to slopes of bending in the x-z and x-y planes respectively. The precessional frequency will thus correspond to whirling frequency ν rad/sec. The transverse and polar mass moments of inertia of the disk are I_T and I_p respectively. The disk is shown projected in x-z plane, which is spinning at angular velocity ω about the 0-x axis as shown in Fig. 6.2. Because of precessional motion about 0-y axis, there is a gyroscopic couple, which is

$$T_z = I_p \omega \dot{\theta} \qquad (6.1.1)$$

The direction of the couple is as shown in Fig. 6.2, with the spin vector closing onto the torque vector in the direction of precession [1].

The projection of the disk in x-y plane is shown in Fig. 6.3, with the gyroscopic couple T_y, which is

$$T_y = I_p \omega \dot{\phi} \qquad (6.1.2)$$

For free vibrations we have

$$I_T \ddot{\theta} + I_p \omega \dot{\phi} = 0 \qquad (6.1.3)$$

$$I_T\ddot{\phi} - I_p\omega\dot{\theta} = 0 \qquad (6.1.4)$$

Also
$$\theta = \theta_0 \exp(i\nu t)$$
$$\phi = \phi_0 \exp\{i(\nu t - \tfrac{1}{2}\pi)\} = -i\phi_0 \exp(i\nu t) \qquad (6.1.5)$$

For a circular whirl, $\theta_0 = \phi_0$ and
$$\phi = -i\theta \qquad (6.1.6)$$

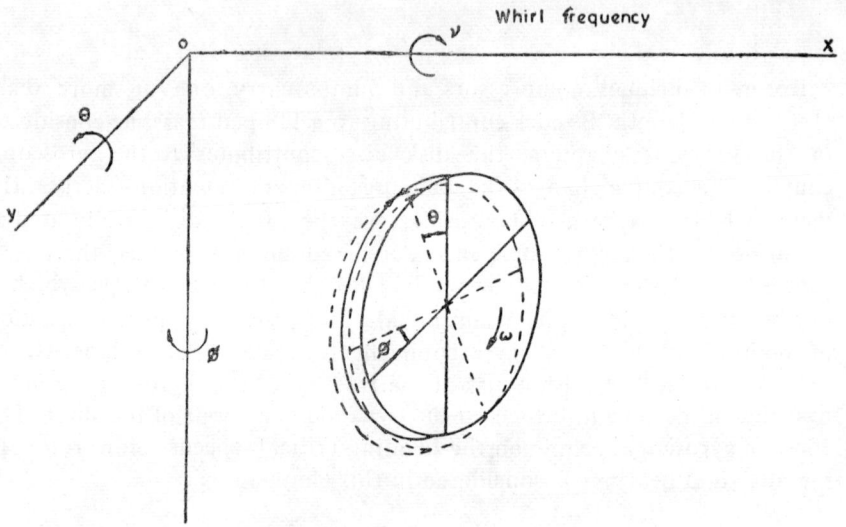

Fig. 6.1 A spinning disk with precessions about y and z axes
ω-rotational frequency; θ, ϕ-precession frequency $= \nu$

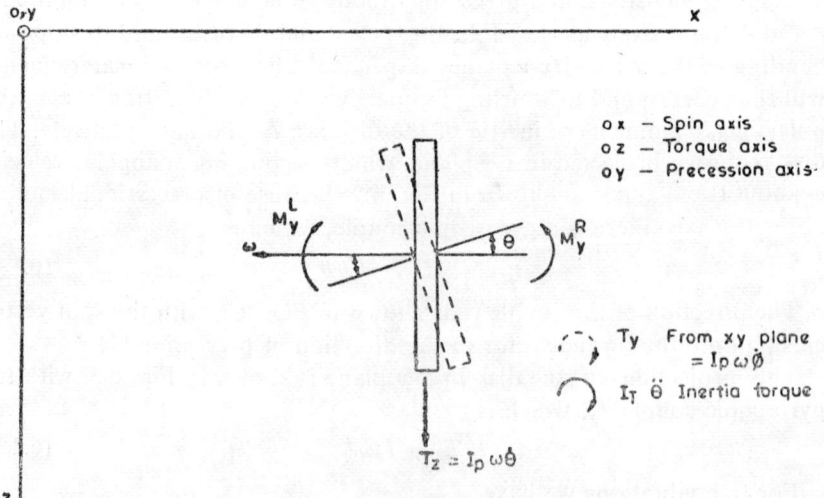

Fig. 6.2 Motion of spinning disk in xz plane

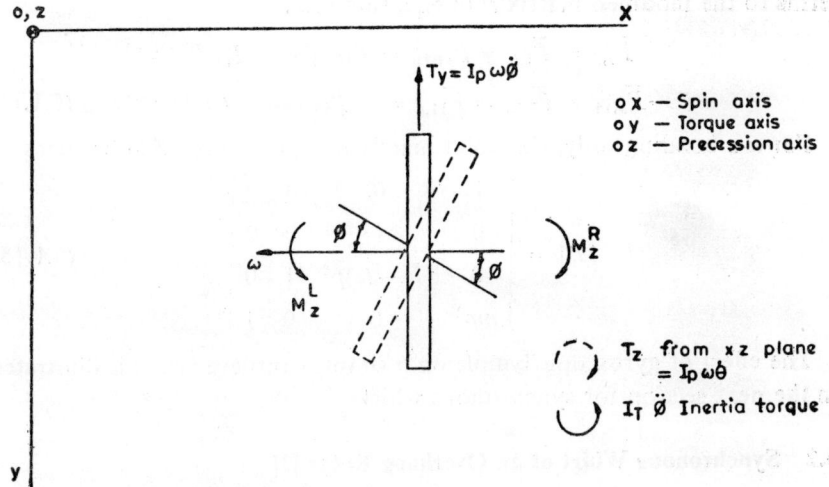

Fig. 6.3 Motion of spinning disk in xy plane

Hence from equation (6.1.3)

$$\nu = \frac{I_p}{I_T} \omega \qquad (6.1.7)$$

This is the precessional frequency of a freely spinning disk, following a disturbance θ and ϕ from the mean equilibrium position.

From Figs. 6.2 and 6.3, we can write

$$M_{yi}^R = M_{yi}^L + I_p\omega\dot{\phi}_i + I_T\ddot{\theta}_i \qquad (6.1.8)$$

$$M_{zi}^R = M_{zi}^L - I_p\omega\dot{\theta}_i + I_T\ddot{\phi}_i \qquad (6.1.9)$$

From equations (5.2.3), we have, for synchronous whirl

$$M_{yic}^R \cos \omega t + M_{yis}^R \sin \omega t = M_{yic}^L \cos \omega t + M_{yis}^L \sin \omega t$$
$$+ I_p\omega^2(-\phi_{ic} \sin \omega t + \phi_{is} \cos \omega t) - I_T\omega^2(\theta_{ic} \cos \omega t + \theta_{is} \sin \omega t) \qquad (6.1.10)$$

$$M_{zic}^R \cos \omega t + M_{zis}^R \sin \omega t = M_{zic}^L \cos \omega t + M_{zis}^L \sin \omega t$$
$$- I_p\omega^2(-\theta_{ic} \sin \omega t + \theta_{is} \cos \omega t) - I_T\omega^2(\phi_{ic} \cos \omega t + \phi_{is} \sin \omega t) \qquad (6.1.11)$$

Hence

$$M_{yic}^R = M_{yic}^L + I_p\omega^2\phi_{is}^L - I_T\omega^2\theta_{ic}^L$$
$$M_{yis}^R = M_{yis}^L - I_p\omega^2\phi_{ic}^L - I_T\omega^2\theta_{is}^L$$
$$M_{zic}^R = M_{zic}^L - I_p\omega^2\theta_{is}^L - I_T\omega^2\phi_{ic}^L$$
$$M_{zis}^R = M_{zis}^L + I_p\omega^2\theta_{ic}^L - I_T\omega^2\phi_{is}^L \qquad (6.1.13)$$

For synchronous whirl all that we have to do is to add the following

82 Rotor Dynamics

terms to the modified matrix $\bar{P}$ in equation (5.2.9)

$$\bar{P}_{3,2} = \bar{P}_{7,6} = \bar{P}_{11,10} = \bar{P}_{15,14} = -I_T\omega^2$$
$$-\bar{P}_{3,14} = \bar{P}_{7,10} = \bar{P}_{11,6} = -\bar{P}_{15,2} = -I_p\omega^2 \qquad (6.1.14)$$

For x-z bending only, the point matrix in equation (4.3.11) becomes

$$[P]_l = \begin{bmatrix} 1 & 0 & 0 & 0 \\ 0 & 1 & 0 & 0 \\ 0 & (\bar{I}_p - I_T)p^2 & 1 & 0 \\ mp^2 & 0 & 0 & 1 \end{bmatrix} \qquad (6.1.15)$$

The effect of gyroscopic couple on a simple cantilever shaft is illustrated in the next section for synchronous whirl.

6.2 Synchronous Whirl of an Overhung Rotor [2]

Consider the overhung rotor shown in Fig. 6.4. The two bearings of the rotor are very close and we can consider it as a fixed end, i.e., the

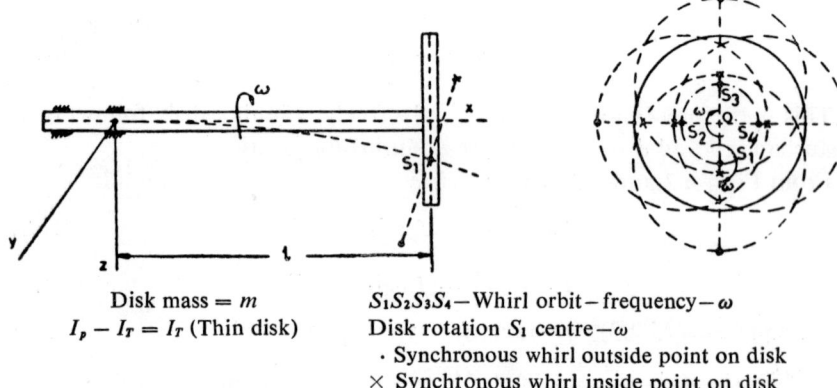

Disk mass = m
$I_p - I_T = I_T$ (Thin disk)

$S_1 S_2 S_3 S_4$ — Whirl orbit – frequency – ω
Disk rotation S_1 centre – ω
· Synchronous whirl outside point on disk
× Synchronous whirl inside point on disk

Fig. 6.4 Overhung cantilever shaft gyroscopics.

deflection and slope can be taken as zero at this station. The disk is shown whirling synchronously in the forward direction. Four different positions of the disk are shown, which indicate that a point on outside the rotor, marked •, remains always outside, while an inside point, marked ×, remains always inside. The disk is considered thin, so that $I_p - I_T = I_T$. Then

$$\begin{Bmatrix} -w \\ \theta \\ 0 \\ 0 \end{Bmatrix}_d^R = \begin{bmatrix} 1 & 0 & 0 & 0 \\ 0 & 1 & 0 & 0 \\ 0 & I_T p^2 & 1 & 0 \\ mp^2 & 0 & 0 & 1 \end{bmatrix} \begin{bmatrix} 1 & l & \frac{l^2}{2EI} & \frac{l^3}{6EI} \\ 0 & 1 & \frac{l}{EI} & \frac{l^2}{2EI} \\ 0 & 0 & 1 & l \\ 0 & 0 & 0 & 1 \end{bmatrix} \begin{Bmatrix} 0 \\ 0 \\ M_y \\ V_z \end{Bmatrix}_0 \qquad (6.2.1)$$

or

$$\begin{Bmatrix} -w \\ \theta \\ 0 \\ 0 \end{Bmatrix}_d^R = \begin{bmatrix} 1 & l & \dfrac{l^2}{2EI} & \dfrac{l^3}{6EI} \\ 0 & 1 & \dfrac{l}{EI} & \dfrac{l^2}{2EI} \\ 0 & I_T p^2 & \dfrac{I_T p^2 l}{EI}+1 & \dfrac{I_T p^2 l^2}{2EI}+l \\ mp^2 & mp^2 l & \dfrac{mp^2 l^2}{2EI} & \dfrac{mp^2 l^3}{6EI}+1 \end{bmatrix} \begin{Bmatrix} 0 \\ 0 \\ M_y \\ V_z \end{Bmatrix}_0 \qquad (6.2.2)$$

Hence the frequency equation is

$$\begin{vmatrix} \left(\dfrac{I_T p^2 l}{EI}+1\right) & \left(\dfrac{I_T p^2 l^2}{2EI}+l\right) \\ \dfrac{mp^2 l^2}{2EI} & \left(\dfrac{mp^2 l^3}{6EI}+1\right) \end{vmatrix} = 0 \qquad (6.2.3)$$

Equation (6.2.3) becomes

$$p^4 + p^2 \frac{12EI}{mI_T l^3}\left[\frac{ml^2}{3} - I_T\right] - \frac{12E^2I^2}{mI_T l^4} = 0 \qquad (6.2.4)$$

Let

$$\lambda = p\sqrt{\frac{ml^3}{EI}}, \text{ frequency parameter} \qquad (6.2.5)$$

and

$$\delta = \frac{I_T}{ml^2}, \text{ disk parameter} \qquad (6.2.6)$$

Then

$$\lambda^4 + \lambda^2\left(\frac{4}{\delta} - 12\right) - \frac{12}{\delta} = 0 \qquad (6.2.7)$$

and

$$\lambda_{1,2}^2 = \left(6 - \frac{2}{\delta}\right) \pm \sqrt{\left(6 - \frac{2}{\delta}\right)^2 + \frac{12}{\delta}} \qquad (6.2.8)$$

For real values of λ, the positive sign is to be considered in the above equation and this root is plotted in Fig. 6.5.

For a lumped mass, $\delta = 0$ and the critical speed is given by

$$p^2 = \frac{3EI}{ml^3} \qquad (6.2.9)$$

If $\delta = \infty$, i.e., $I_T = \infty$, which means that the disk mass is concentrated at a large radius, no finite θ is possible and the critical speed becomes

$$p^2 = \frac{12EI}{ml^3} \qquad (6.2.10)$$

The effect of gyroscopic couple is to stiffen the rotor and raise the critical speed as shown in Fig. 6.5.

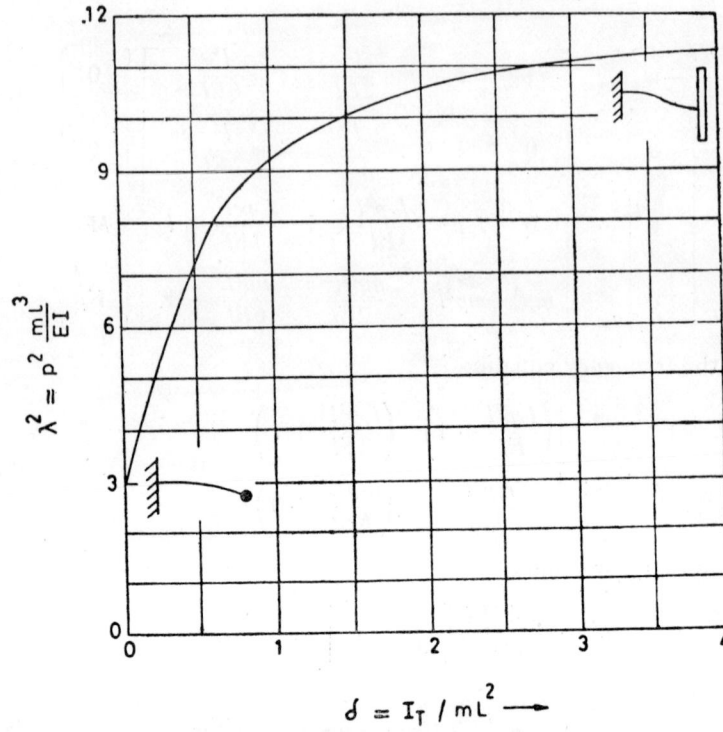

Fig. 6.5 Effect of disk inertia on critical speed of a simple overhung rotor

6.3 Nonsynchronous Whirl

For nonsynchronous whirl of the rotor at a frequency ν rad/sec, equations (5.2.3) are

$$w = w_c \cos \nu t + w_s \sin \nu t$$
$$\theta = \theta_c \cos \nu t + \theta_s \sin \nu t$$
$$\cdots \quad \cdots \quad \cdots \quad \cdots \quad \cdots \quad (6.3.1)$$

Equations (6.1.10) and (6.1.11) now become

$$M^R_{yic} \cos \nu t + M^R_{yis} \sin \nu t = M^L_{yic} \cos \nu t + M^L_{yis} \sin \nu t$$
$$+ I_p \omega \nu (-\phi_{ic} \sin \nu t + \phi_{is} \cos \nu t) - I_T \nu^2 (\theta_{ic} \cos \nu t + \theta_{is} \sin \nu t) \quad (6.3.2)$$
$$M^R_{zic} \cos \nu t + M^R_{zis} \sin \nu t = M^L_{zic} \cos \nu t + M^L_{zis} \sin \nu t$$
$$- I_p \omega \nu (-\theta_{ic} \sin \nu t + \theta_{is} \cos \nu t) - I_T \nu^2 (\phi_{tc} \cos \nu t + \phi_{is} \sin \nu t) \quad (6.3.3)$$

The $\bar{P}$ terms of equation (5.2.9) now become

$$\bar{P}_{3,2} = \bar{P}_{7,6} = \bar{P}_{11,10} = \bar{P}_{15,14} = -I_T \nu^2$$
$$\bar{P}_{3,14} = \bar{P}_{15,2} = I_p \omega \nu$$
$$\bar{P}_{7,10} = \bar{P}_{11,6} = -I_p \omega \nu$$
$$\bar{P}_{4,1} = \bar{P}_{8,5} = \bar{P}_{12,9} = \bar{P}_{16,13} = m\nu^2$$
$$\bar{P}_{i,i} = 1, \ i = 1, 2, \ldots 17 \quad (6.3.4)$$

For nonsynchronous whirl of the overhung shaft of Fig. 6.4, we get

$$\begin{Bmatrix} -w \\ \theta \\ 0 \\ 0 \end{Bmatrix}_d^R = \begin{bmatrix} 1 & 0 & 0 & 0 \\ 0 & 1 & 0 & 0 \\ 0 & I_T(2\omega - \nu)\nu & 1 & 0 \\ m\nu^2 & 0 & 0 & 1 \end{bmatrix} [F] \begin{Bmatrix} 0 \\ 0 \\ M_y \\ V_z \end{Bmatrix}_0 \quad (6.3.5)$$

The frequency equation is therefore

$$\begin{vmatrix} \left[I_T\nu(2\omega - \nu)\dfrac{l}{EI} + 1 \right] & \left[I_T\nu(2\omega - \nu)\dfrac{l^2}{2EI} + l \right] \\ \left[\dfrac{m\nu^2 l^2}{2EI} \right] & \left[\dfrac{m\nu^2 l^3}{6EI} + 1 \right] \end{vmatrix} = 0 \quad (6.3.6)$$

After expansion and simplifying, the above equation becomes

$$\nu^4 - 2\omega\nu^3 - EI\left(\frac{12}{ml^3} + \frac{4}{I_T l}\right)\nu^2 + \frac{24\omega EI}{ml^3}\nu + \frac{12E^2I^2}{I_T ml^4} = 0 \quad (6.3.7)$$

Let

$$\lambda = \nu\sqrt{\frac{ml^3}{3EI}}, \quad \text{whirl frequency parameter}$$

$$\delta = \frac{3I_T}{ml^2} \quad \text{disk parameter}$$

and

$$\Omega = \omega\sqrt{\frac{ml^3}{3EI}} \quad \text{rotational speed parameter} \quad (6.3.8)$$

Equation (6.3.7) now becomes

$$\lambda^4 - 2\Omega\lambda^3 - 4\lambda^2 \frac{\delta + 1}{\delta} + 8\Omega\lambda + \frac{4}{\delta} = 0 \quad (6.3.9)$$

For $I_T = \dfrac{ml^2}{3}$, i.e., $\delta = 1$, the above equation is

$$\lambda^4 - 2\Omega\lambda^3 - 8\lambda^2 + 8\Omega\lambda + 4 = 0 \quad (6.3.10)$$

and

$$\Omega = \frac{\lambda^4 - 8\lambda^2 + 4}{2\lambda^3 - 8\lambda} \quad (6.3.11)$$

For a given λ, determine Ω and plot λ versus Ω as in Fig. 6.6. The ---- lines in this figure represent the whirl frequency as a multiple of the rotational frequency $\nu = \omega$, $\nu = 3\omega$ for forward whirl and $\nu = -\omega$ for backward whirl. Point A represents the critical speed Ω for forward synchronous whirl, which is the most dangerous speed. The critical speed corresponding to synchronous backward whirl, point B in Fig. 6.6, is observed in the laboratory [2], but its severity is not high for practical rotors. Whirls of multiples of the rotational speed are not observed in practice because the unbalance predominantly excites the forward synchronous whirl.

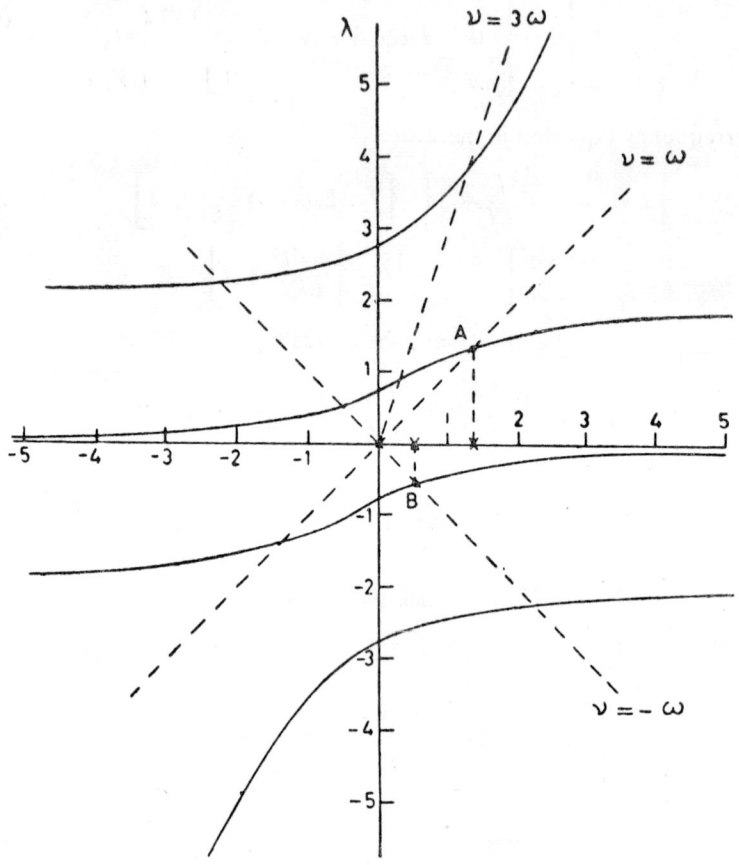

Fig. 6.6 Nonsynchronous whirl of an overhung shaft

6.4 Rotor System with a Coupling

Consider the system of two rotors connected by a spline coupling as shown in Fig. 6.7. It has six stations with three lumped masses. The disk at station 1 is thin with its diametral mass moment of inertia $I_T = .00128 \text{ kg m}^2$. The other two masses at stations 2 and 4 have negligible diametral mass moment of inertia. The left end is assumed as rigid, with deflection and slope equal to zero, as in a cantilever support, because of the two bearings which are very close to each other. At station 2, we have a coupling, where the deflection is continuous, bending moment is zero, the shear force changes because of the mass at the coupling station and the slope has a jump. Starting from station 0, we can obtain the following relation:

$$\{S\}_2^L = [A]\{S\}_0 \tag{6.4.1}$$

Gyroscopic Effects 87

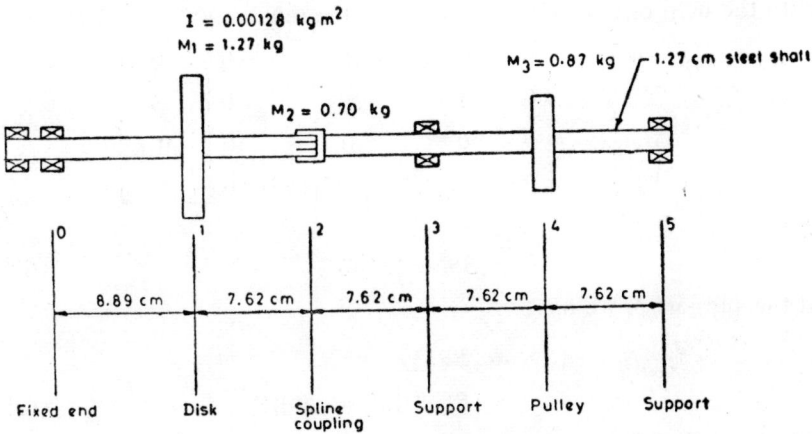

Fig. 6.7 A two rotor system coupled by a spline

where

$$\{S\}_0 = \begin{Bmatrix} 0 \\ 0 \\ M_{y0} \\ V_{z0} \end{Bmatrix} \quad (6.4.2)$$

At the coupling station 2,

$$-w_2^R = -w_2^L$$

$$\theta_2^R = \theta_2^L - \phi$$

$$M_y = 0$$

$$V_{z2}^R = V_{z2}^L - m_2 p^2 w_2^R \quad (6.4.3)$$

Using the relation $M_y = 0$ from the above, in equation (6.4.1), we get

$$V_{z0} = -\frac{A_{33}}{A_{34}} M_{y0} \quad (6.4.4)$$

Eliminating V_{z0} in equation (6.4.1), by using equation (6.4.4) we obtain

$$-w_2^L = \left(A_{13} - \frac{A_{14} A_{33}}{A_{34}} \right) M_{y0} = A'_{13} M_{y0}$$

$$\theta_2^L = \left(A_{23} - \frac{A_{24} A_{33}}{A_{34}} \right) M_{y0} = A'_{23} M_{y0}$$

$$M_{y2}^L = 0$$

$$V_{z2}^L = \left(A_{43} - \frac{A_{44} A_{33}}{A_{34}} \right) M_{y0} = A'_{43} M_{y0} \quad (6.4.5)$$

With the help of equation (6.4.3) we get the following:

$$\begin{Bmatrix} -w \\ \theta \\ M_y \\ V_z \end{Bmatrix}^R_2 = \begin{bmatrix} 0 & 0 & A'_{13} & 0 \\ 0 & -1 & A'_{23} & 0 \\ 0 & 0 & 0 & 0 \\ 0 & 0 & (A'_{43} + A'_{13}mp^2) & 0 \end{bmatrix} \begin{Bmatrix} 0 \\ \phi \\ M_{y0} \\ 0 \end{Bmatrix}$$

i.e.,
$$\{S\}^R_2 = [A']\{S'\} \tag{6.4.6}$$

At the support 3, we have
$$\{S\}^L_3 = [B]\{S\}^R_2$$
$$= [B][A']\{S'\} = [B']\{S'\} \tag{6.4.7}$$

Since $w_3 = 0$, from the above equation
$$\phi = -\frac{B'_{13}}{B'_{12}} M_{y0} \tag{6.4.8}$$

At this support, we have a jump in the shear force given by
$$V^R_{z3} = V^L_{z3} + P \tag{6.4.9}$$

where P is the support reaction.

The slope and bending moment remain continuous at this station. Using the relation in equation (6.4.8), we obtain the following from equation (6.4.7):

$$-w^R_3 = 0$$
$$\theta^R_3 = \left(B'_{23} - \frac{B'_{22}B'_{13}}{B'_{12}}\right) M_{y0} = B''_{23} M_{y0}$$
$$M^R_{y3} = \left(B'_{33} - \frac{B'_{32}B'_{13}}{B'_{12}}\right) M_{y0} = B''_{33} M_{y0}$$
$$V^R_{z3} = \left(B'_{43} - \frac{B'_{42}B'_{13}}{B'_{12}}\right) M_{y0} = B''_{43} M_{y0} \tag{6.4.10}$$

Hence
$$\begin{Bmatrix} -w \\ \theta \\ M_y \\ T_z \end{Bmatrix}^R_3 = \begin{bmatrix} 0 & 0 & 0 & 0 \\ 0 & 0 & B''_{23} & 0 \\ 0 & 0 & B''_{33} & 0 \\ 0 & 0 & B''_{43} & 1 \end{bmatrix} \begin{Bmatrix} 0 \\ 0 \\ M_{y0} \\ P \end{Bmatrix} \tag{6.4.11}$$

i.e.,
$$\{S\}^R_3 = [B'']\{S''\} \tag{6.4.12}$$

For the shaft between stations 3 and 5 we have
$$\{S\}_5 = [C]\{S\}^R_3 \tag{6.4.13}$$

Using equation (6.4.12), the above equation becomes
$$\{S\}_5 = [C][B'']\{S''\} = [U]\{S''\} \tag{6.4.14}$$

Gyroscopic Effects

Using the boundary conditions at the right support, the above equation is

$$\begin{Bmatrix} 0 \\ \theta \\ 0 \\ V_z \end{Bmatrix}_5 = \begin{bmatrix} u_{11} & u_{12} & u_{13} & u_{14} \\ u_{21} & u_{22} & u_{23} & u_{24} \\ u_{31} & u_{32} & u_{33} & u_{34} \\ u_{41} & u_{42} & u_{43} & u_{44} \end{bmatrix} \begin{Bmatrix} 0 \\ 0 \\ M_{y0} \\ P \end{Bmatrix} \quad (6.4.15)$$

Hence the frequency equation is

$$\begin{vmatrix} u_{13} & u_{14} \\ u_{33} & u_{34} \end{vmatrix} = 0 \quad (6.4.16)$$

Using the root search technique outlined in section 2.2, we find the critical speeds, and then the corresponding mode shapes are obtained by following a procedure similar to that used in section 4.4.

The critical speeds obtained by a computer program for this problem are given in Table 6.1.

Table 6.1 Critical speeds of a two rotor system with gyroscopic effect

Mode No.	Critical speed without gyroscopic effects, rpm	Critical speed with gyroscopic effects, rpm
1	8546	8851
2	22427	22427
3	26346	26430

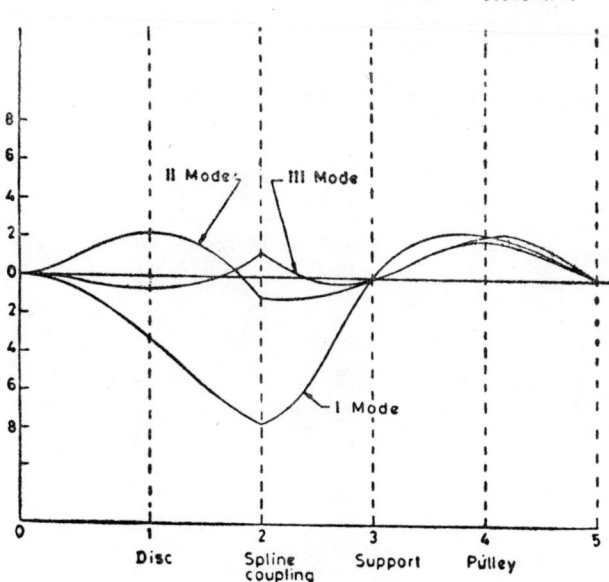

Fig. 6.8 Mode shapes of two rotor system

The critical speed is obviously affected in the fundamental mode because of nonzero slope at the disk 1 and the frequency is increased by 3.6%. The effect of gyroscopic terms on the second mode is zero, since the slope at the disk 1 is zero for this mode. In the third mode the effect of gyroscopic terms is negligible. The mode shapes are plotted in Fig. 6.8.

References

1. Shigley, J.E. Dynamic analysis of machines, McGraw-Hill Book Co., 1961.
2. Den Hartog, J.P. Mechanical Vibrations, McGraw-Hill Book Co., 1956.
3. Pilkey, W.D. and Chang, P.Y. Modern formulas for statics and dynamics, McGraw-Hill Book Co., 1978.
4. Green, R.B. Gyroscopic effects on the critical speeds of flexible rotors, J. of Appld. Mechs., ASME, Vol. 15, 1948, p. 369.
5. Muster, D. and Sternlicht, B. Proc. Int. Symp. on Lubrication and Wear, Houston, 1963.
6. Carnegie, W. Rotary inertia and gyroscopic effects in overhung shaft systems, Bull. Mech. Engng. Educ., Vol. 3, 1964, p. 191.
7. Gunter, E.J. Dynamic stability of rotor bearing system, NASA SP-113, 1966.
8. Alba, S. The effect of gyroscopic moment and distributed mass on the vibration of a rotating shaft with a rotor, Bull. JSME, Vol. 16, 1973.

Chapter 7

Fluid Film Bearings

Fluid film bearings commonly used in heavy rotating machines play a significant role in the dynamic behaviour of rotors. Because the thin film that separates the moving surfaces supports the rotor load, it acts like a spring and provides damping due to squeeze film effect. The stiffness and damping properties of the oil film significantly alter the critical speeds and out-of-balance response of a rotor. In addition, rotor instability occurs, which is a self excited vibration arising out of the bearing fluid film effects, and this is an important factor to be considered in the rotor design.

The subject of fluid film lubrication is very wide concerning hydrodynamic, hydrostatic or hybrid bearings, with the fluid medium as gas which is compressible, or liquid which can be considered as incompressible operating in laminar or turbulent regimes. Moreover, the geometry of the bearing, e.g., plain cylindrical, four axial grooved, elliptical, multilobe and tilting pad type, is to be considered while evaluating the steady state or dynamic characteristics of the bearing. It is not obviously the intention here, to consider the subject of fluid film lubrication, but certain facts of interest that concern the rotor-bearing interaction will be given, which will be used later, for the out-of-balance response and stability studies. For a full detailed study of fluid film bearings, reference may be made to the books of Pinkus and Sternlicht [1] and Smith [2].

7.1 Steady State Characteristics of Plain Cylindrical Hydrodynamic Bearings

Consider a rigid rotor operating in a hydrodynamic bearing as shown in Fig. 7.1, which also gives the notations used. Reynolds derived the governing equation for such a journal bearing, which is

$$\frac{\partial}{\partial s^*}\left[\frac{\rho h^{*3}}{\mu}\frac{\partial p^*}{\partial s^*}\right] + \frac{\partial}{\partial x^*}\left[\frac{\rho h^{*3}}{\mu}\frac{\partial p^*}{\partial x^*}\right] = 6U\frac{\partial}{\partial s^*}(\rho h^*) \qquad (7.1.1)$$

where ρ is the density of fluid, which cancels out in the above equation, if the bearing is operating under incompressible conditions; μ—viscosity of lubricant usually given in the units of centipoise, 1 $cp = 1.0054 \times$

10^{-3} N sec/m²; x^*—axial coordinate; and p^*—pressure developed in the film.

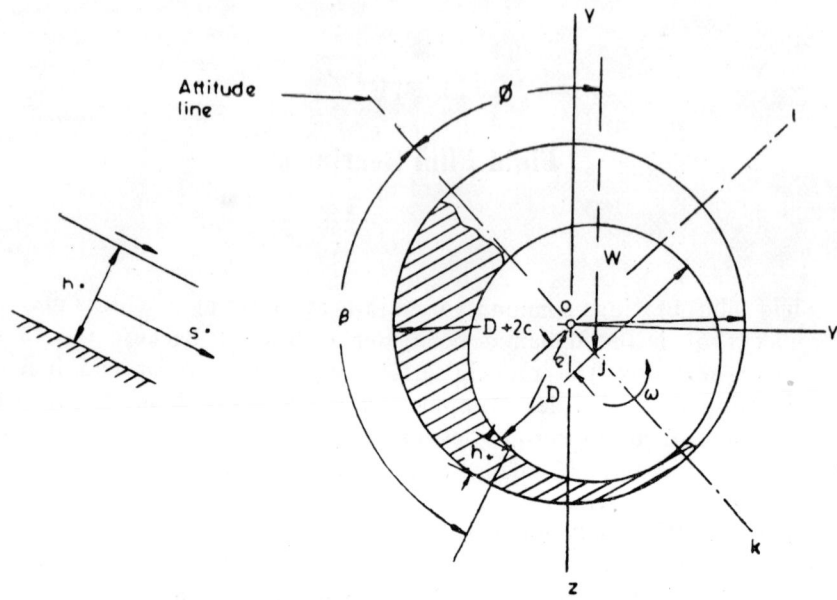

Fig. 7.1 A plain cylindrical hydrodynamic bearing:

Notation D—Journal diameter = $2R$, C—Radial clearance, e—Eccentricity, W—Radial load, h^*—Film thickness, ϕ—Attitude angle, s^*—Coordinate = $R\beta$, ω—Speed—rad/sec, N—Speed—rev/sec, U—Peripheral velocity, S—Sommerfeld number = $\dfrac{\mu DLN}{W}\left(\dfrac{R}{C}\right)^2$
μ—Coefficient of Viscosity, ϵ—Eccentricity ratio e/C

As mentioned earlier, we do not intend to obtain a solution for the above equation here, but we will discuss the solutions obtained to the extent necessary for our rotor-bearing interaction problem. The following nondimensional parameters are used:

$$h = \frac{h^*}{C}$$

$$X = \frac{x^*}{L}$$

$$p = \frac{p^*}{\mu N}\left\{\frac{C}{R}\right\}^2 \tag{7.1.2}$$

where L is the length of the bearing.

Writing $\partial s^* = R\,\partial\beta$, the Reynolds equation becomes

$$\frac{\partial}{\partial \beta}\left[h^3 \frac{\partial p}{\partial \beta}\right] + \left\{\frac{R}{L}\right\}^2 \frac{\partial}{\partial X}\left[h^3 \frac{\partial p}{\partial X}\right] = 12\pi \left[\frac{\partial h}{\partial \beta}\right] \tag{7.1.3}$$

If we restrict ourselves to a plain cylindrical bearing, the film thickness is

$$h = 1 + \epsilon \cos \beta \qquad (7.1.4)$$

where ϵ is the eccentricity ratio

$$\epsilon = \frac{e}{C} \qquad (7.1.5)$$

Reynolds equation now becomes

$$\frac{\partial}{\partial \beta}\left[(1 + \epsilon \cos \beta)^3 \frac{\partial p}{\partial \beta}\right] + \left\{\frac{R}{L}\right\}^2 \frac{\partial}{\partial X}\left[(1 + \epsilon \cos \beta)^3 \frac{\partial p}{\partial X}\right] = -12\pi\epsilon \sin \beta \qquad (7.1.6)$$

For an infinitely long bearing, i.e., $L/D \gg 1$

$$\frac{\partial p}{\partial X} \approx 0 \qquad (7.1.7)$$

For an infinitely short bearing, i.e., $L/D \ll 1$

$$\frac{\partial p}{\partial \beta} \approx 0 \qquad (7.1.8)$$

For the above two cases, solution of equation (7.1.3) in closed form is possible. For a finite bearing, however, a numerical solution using finite difference, finite element techniques, or a series solution is necessary to determine the pressure p as a function of β and/or X. Then

$$p_k = \frac{1}{2}\iint p \cos \beta \, d\beta \, dX$$

$$p_l = \frac{1}{2}\iint p \sin \beta \, d\beta \, dX$$

$$\phi = \tan^{-1} \frac{p_l}{p_k} \qquad (7.1.9)$$

The mean pressure on the projected area of the bearing is

$$p' = \frac{W}{2RL} \qquad (7.1.10)$$

In nondimensional form

$$p_m = \frac{p'}{\mu N}\left\{\frac{C}{R}\right\}^2 = \frac{W}{DL\mu N}\left\{\frac{C}{R}\right\}^2 \qquad (7.1.11)$$

The non-dimensional load of the bearing is defined commonly as duty parameter or Sommerfeld number, which is

$$S = \frac{1}{p_m} = \frac{\mu DLN}{W}\left\{\frac{R}{C}\right\}^2 \qquad (7.1.12)$$

In terms of load and speed, a heavily loaded bearing has a lower speed, high load and small S. A larger value of S indicates a lightly loaded bearing with a high speed of rotation. The steady state behaviour of a bearing is studied by finding the load carrying capacity (S), and the attitude angle (ϕ), for different eccentricity ratios (ϵ). S versus ϵ for a plain cylindrical bearing is shown in Fig. 7.2, and the journal locus in the eccentricity circle is given in Fig. 7.3.

Figure 7.2 gives the eccentricity ratio of a given bearing operating at a certain speed and radial load. For the eccentricity ratio thus obtained, by drawing an arc, we can find the attitude angle of the bearing, which defines the steady state position of the journal centre in Fig. 7.3, e.g., $S = 1$, $\phi = 54°$, $\epsilon = 0.37$, $L/D = 0.5$ is shown marked on this figure. We

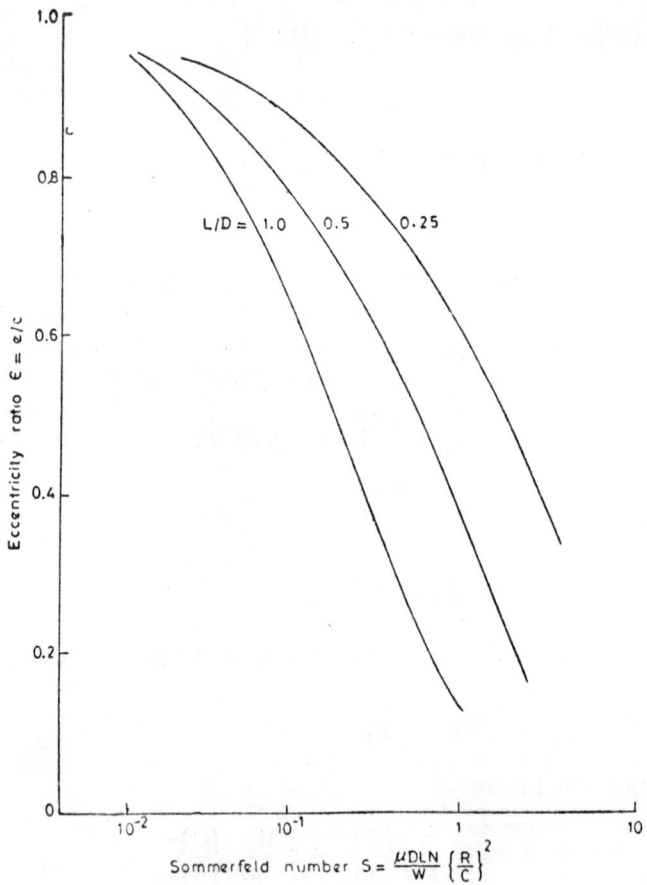

Fig. 7.2 Eccentricity ratio for plain cylindrical bearings as a function of Sommerfeld number

notice that as load increases, i.e., as S decreases, the eccentricity ratio increases and the journal centre gets far away from the bearing centre and conversely, the bearing operates more concentrically for lighter loads and high speeds.

It should also be remembered that on the steady state locus of the journal in Fig. 7.3, the load applied on the rotor is always vertically down Because of this fact, we find that the journal centre does not move verti-

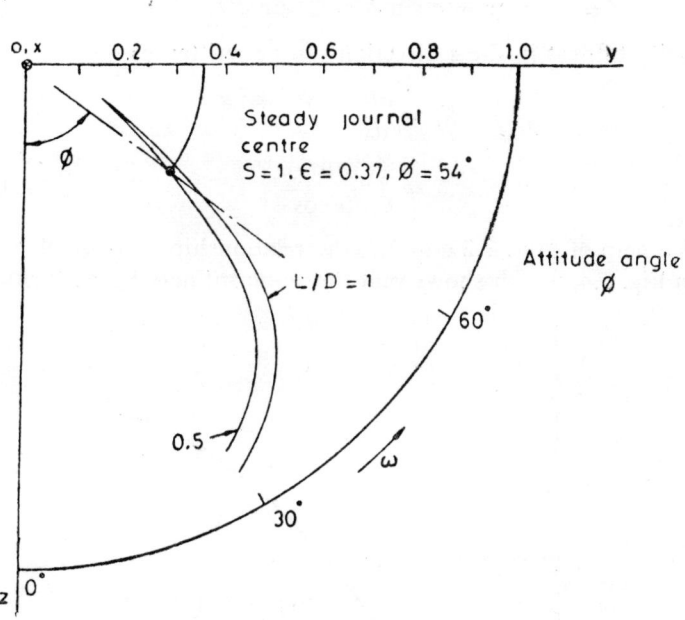

Fig. 7.3 Journal locus in eccentricity circle for plain cylindrical bearings

cally up or down, if the load is respectively released or increased, as we find in a conventional spring system. The journal bearing behaves like an asymmetric bearing, offering also a horizontal motion, in addition to a vertical motion, due to changes in vertical load and vice versa. In other words, the bearing is anisotropic in character, offering different stiffnesses in different radial directions. When we consider the rotor dynamic behaviour, we should account for these spring stiffness properties.

Furthermore, if the journal is oscillating, the velocity of the journal introduces time dependent terms on the right hand side of the Reynolds equation, which give rise to squeeze film forces, corresponding to damping forces in a viscous dashpot. These forces are also asymmetric and should be accounted for, in calculating the unbalance response for a rotor. Unlike the material damping in the rotor, which is small, the damping offered by a fluid film bearing could be substantial and it is important to consider film damping in any analysis of unbalance response of a rotor mounted on fluid film bearings. These stiffness and damping properties have also a considerable influence on the instability of a rotor, leading to dangerous oil whip phenomenon observed for such rotors in practice. The following section deals with the stiffness and damping properties of hydrodynamic bearings.

7.2 Stiffness and Damping Coefficients of a Journal Bearing

From Fig. 7.3, we have

$$z = e \cos \phi = C\epsilon \cos \phi$$
$$y = e \sin \phi = C\epsilon \sin \phi \tag{7.2.1}$$

The static stiffness in the z direction can be written as

$$K_z = \frac{dW}{dz}$$
$$= \frac{dW}{C\, d(\epsilon \cos \phi)} \tag{7.2.2}$$

With the help of Figs. 7.2 and 7.3, the relationship between W and z is plotted in Fig. 7.4, which shows that the static stiffness K_z as defined in

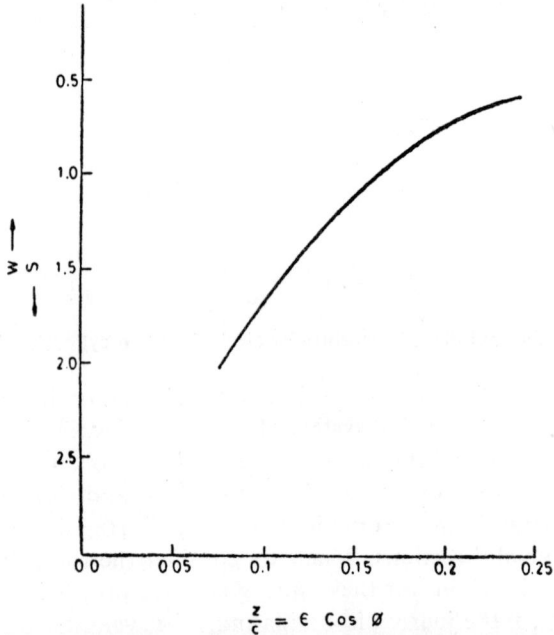

Fig. 7.4 Variation of load W with vertical displacement z

equation (7.2.2) is nonlinear and would depend upon the loading conditions. The stiffness and damping coefficients are therefore functions of the eccentricity (load), in addition to being asymmetric as we discussed before.

For a given operating condition of load W, let the journal centre coordinates be z and y. When the load changes, the journal centre assumes a new position causing changes in both z and y directions. The total differential of the load W with respect to z and y is

$$dW = \frac{\partial W}{\partial z}\, dz + \frac{\partial W}{\partial y}\, dy \tag{7.2.3}$$

which can be written as
$$dW = K_{zz}\, dz + K_{zy}\, dy \qquad (7.2.4)$$
where
$$K_{zz} = \frac{\partial W}{\partial z};\ K_{zy} = \frac{\partial W}{\partial y} \qquad (7.2.5)$$

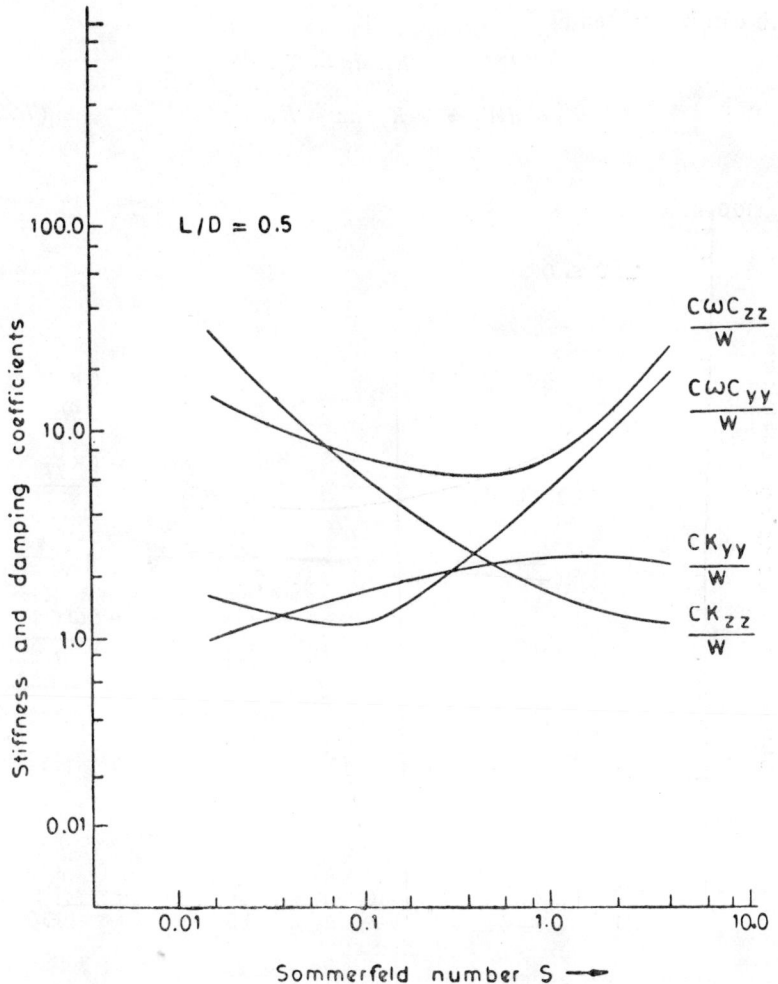

Fig. 7.5 Direct stiffness and damping coefficients of a plain cylindrical bearing

We must however consider the radial load in y direction also in making the relations more general. Under steady conditions, $W_z = -W$, $W_y = 0$. As soon as the journal is disturbed from its steady state position, a horizontal load will be generated to oppose this disturbance. Hence we must consider the radial load in y direction also, to make the above relations more general. Since both W_z and W_y are dependent on both z and y, their total differential with respect to z and y must be considered,

98 Rotor Dynamics

i.e.,

$$dW_z\bigg\}_{static} = -\frac{\partial W_z}{\partial z} dz - \frac{\partial W_z}{\partial y} dy$$

$$dW_y\bigg\}_{static} = -\frac{\partial W_y}{\partial z} dz - \frac{\partial W_y}{\partial y} dy \qquad (7.2.6)$$

which can be written as

$$dW_z = -K_{zz} dz - K_{zy} dy$$

$$dW_y = -K_{yz} dz - K_{yy} dy \qquad (7.2.7)$$

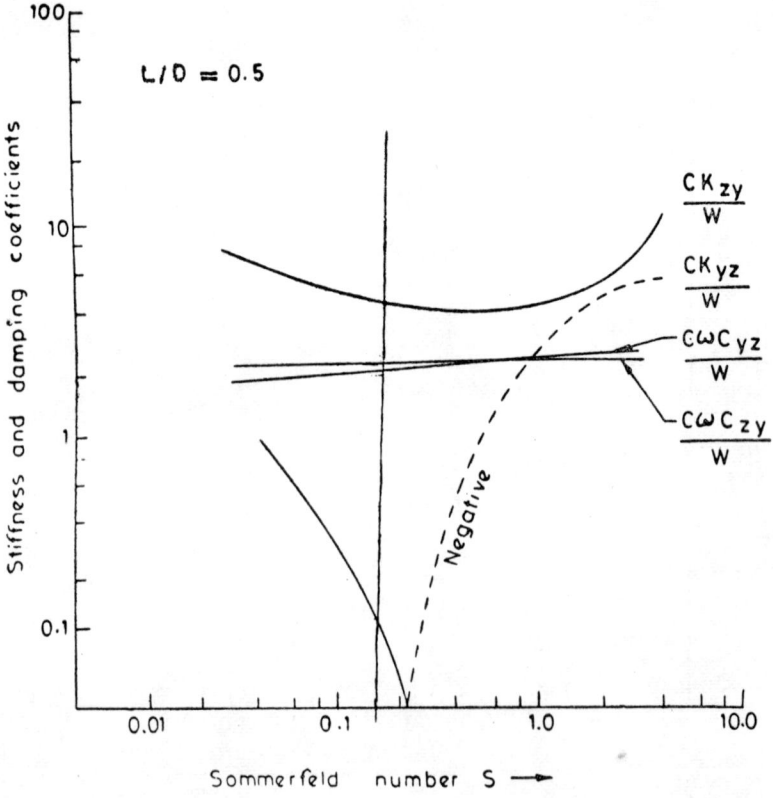

Fig. 7.6 Cross stiffness and damping coefficients of a plain cylindrical bearing

where

$$K_{zz} = \frac{\partial W_z}{\partial z}; \ K_{zy} = \frac{\partial W_z}{\partial y}$$

$$K_{yz} = \frac{\partial W_y}{\partial z}; \ K_{yy} = \frac{\partial W_y}{\partial y} \qquad (7.2.8)$$

In the above, K_{ij} are stiffness coefficients, with i representing the direction of force and j representing the direction of displacement. Under dynamic conditions, the journal centre will have velocities $\dot{z}$, $\dot{y}$ and they induce additional forces W_z and W_y due to squeeze film effect. These

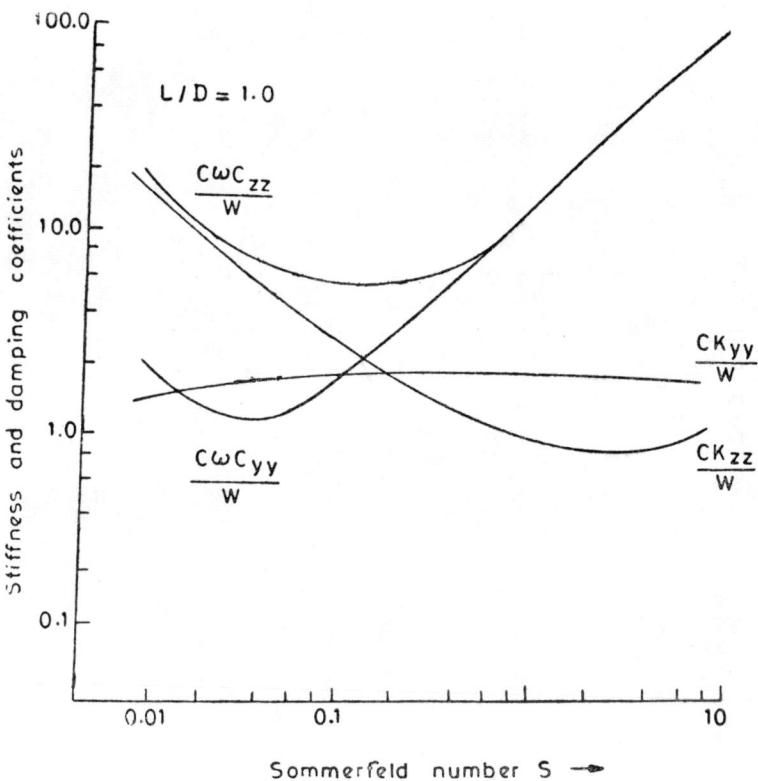

Fig. 7.7 Direct stiffness and damping coefficients of a plain cylindrical bearing

forces are denoted with coefficients C_{zz}, C_{zy}, C_{yz}, and C_{yy} to define the total changes in forces W_z and W_y. The coefficients C are called damping coefficients, since the squeeze film force is taken proportional to the velocities. Hence we can write

$$\Delta W_z = -K_{zz}\Delta z - C_{zz}\Delta \dot{z} - K_{zy}\Delta y - C_{zy}\Delta \dot{y} \qquad (7.2.9)$$

$$\Delta W_y = -K_{yz}\Delta z - C_{yz}\Delta \dot{z} - K_{yy}\Delta y - C_{yy}\Delta \dot{y} \qquad (7.2.10)$$

In the above K_{zz}, C_{zz}, K_{yy}, C_{yy} are direct stiffness and damping coefficients and K_{yz}, C_{yz}, K_{zy}, C_{zy} are cross coupling terms. It is obvious from the above discussion, that we have neglected nonlinear terms and therefore they are valid for small changes in displacement and velocities. It should also be noted that these stiffness and damping coefficients should be evaluated for the steady state position and that they are different for other steady operating conditions of the journal.

The 8 coefficients in (7.2.9) and (7.2.10) for a plain cylindrical bearing are given [3] for $L/D = 0.5$ and 1 in Figs. 7.5 to 7.8. Figures 7.9 to 7.12

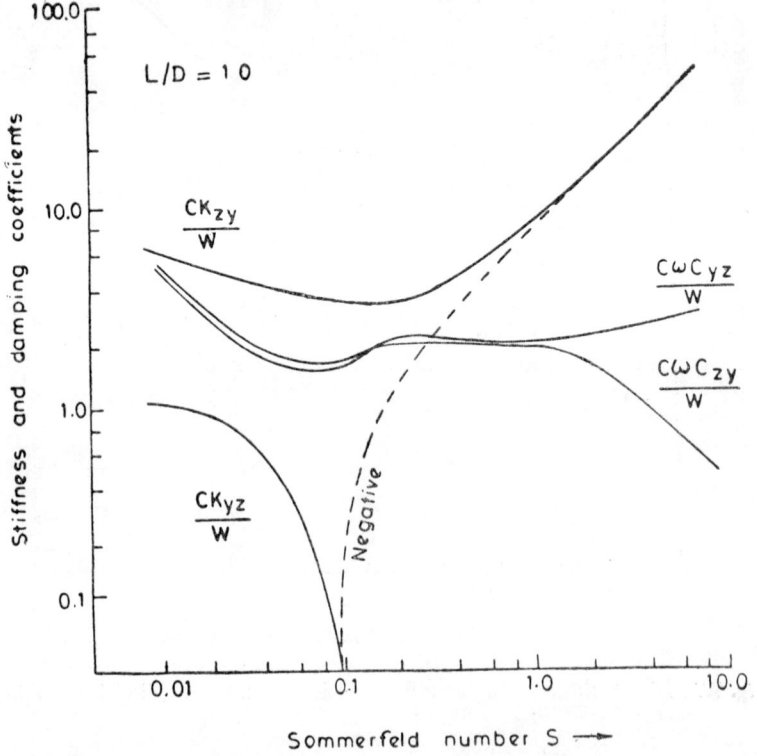

Fig. 7.8 Cross stiffness and damping coefficients of a plain cylindrical bearing

give the same for a 4 axial groove bearing [3]. The stiffness and damping coefficients are given in nondimensional form as CK_{zz}/W, $C\omega C_{zz}/W$ etc.

Though there are several references available for the data on stiffness and damping coefficients, one of the most comprehensive works on this data is due to Lund [3], which can be used to determine these values for other bearings like elliptical, multilobe and partial bearings operating under compressible and incompressible conditions and also considering turbulence in the film region.

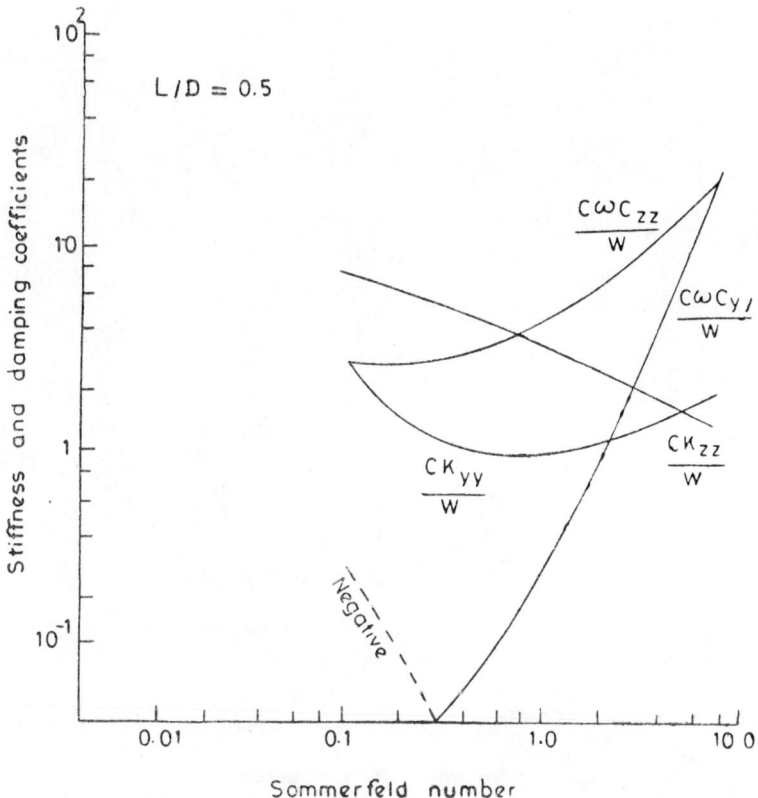

Fig. 7.9 Direct stiffness and damping coefficients of a grooved bearing

In the discussion of this chapter, it is assumed that the journal axis is parallel to the bearing axis. For flexible shafts or slightly tilted rigid journals, the relations have to be modified. In addition to the force coefficients, moment coefficients also should be calculated [4]. However these coefficients are generally insignificant and have relatively less influence on the dynamic behaviour of a flexible shaft [5]. These moment terms are not accounted in the analysis that follows in later chapters. The discussion in this chapter is also limited to the performance of bearings, based on the theory hydrodynamic lubrication. The results presented in this chapter are useful to a reasonable extent, but discrepancies have been observed in practice, particularly for heavy duty bearings. Such bearings have been found to carry heavier loads than predicted by the theory and that the speed at which oil whirl instability,

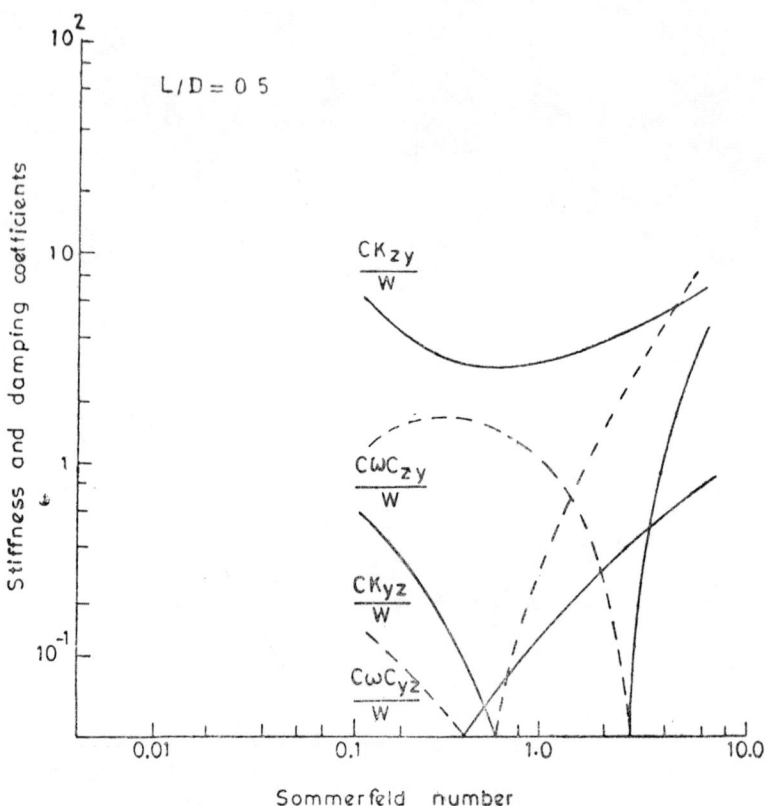

Fig. 7.10 Cross stiffness and damping coefficients of a grooved bearing

see chapter 10, occurs in heavy duty machines is also found to be significantly higher than the theoretical predictions. Morton [6] has shown that over a wide range of operating conditions, the use of experimentally derived stiffness and damping coefficients resulted in a significantly greater rotor bearing stability, than the predictions based on theories of simple hydrodynamic lubrication including thermal effects. In addition to the formation of a wedge due to hydrodynamic lubrication, two other aspects become significant for heavy duty bearings. Briefly, they are:

1. A thermal wedge which was first observed by Fogg [7] that a parallel slider thrust bearing was able to support load without the benefit of the Reynolds wedge. Dowson and March [8] used thermohydrodynamic

lubrication theory for a journal bearing allowing for variable viscosity across the film using simple thermal models for both the journal and the bearing.

2. Thermoelastic distortion that can effect the performance of the bearing, as found by Morton and Keogh [9]. Morton et al. [10] conducted an experimental investigation on large hydrodynamic bearings and

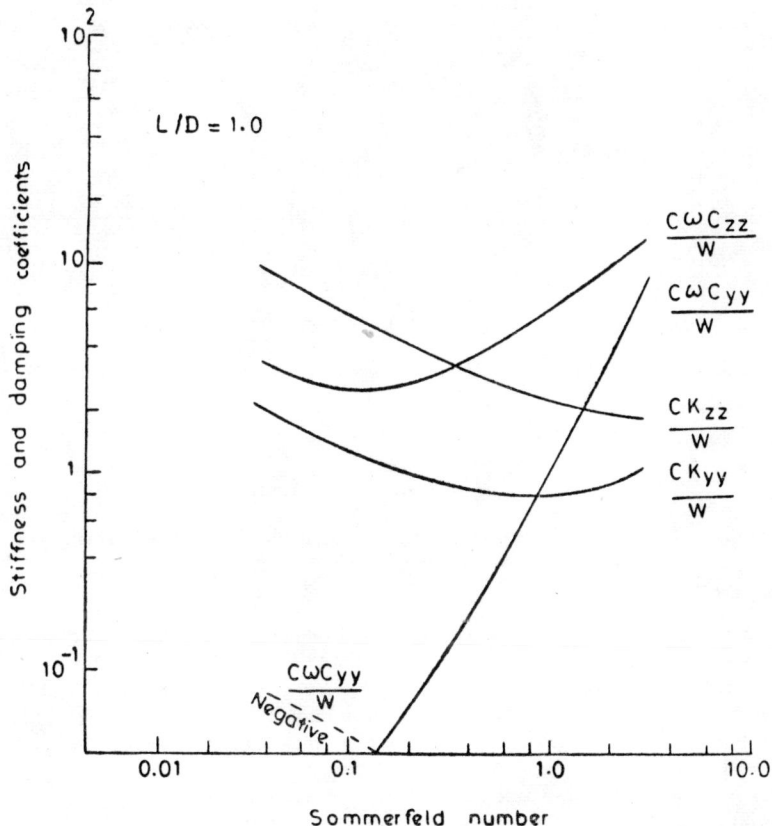

Fig. 7.11 Direct stiffness and damping coefficients of a grooved bearing

found that the bearing had significant distortion due to the thermal effects influencing the pressure distribution and the stiffness and damping coefficients. Using the experimentally observed film profile and variable viscosity of the film based on the bearing metal temperature measurements, they reported threshold instability speeds far higher than the conventional theoretical calculations.

The above may be borne in mind while applying the data of bearing stiffness and damping coefficients for rotor dynamic calculations.

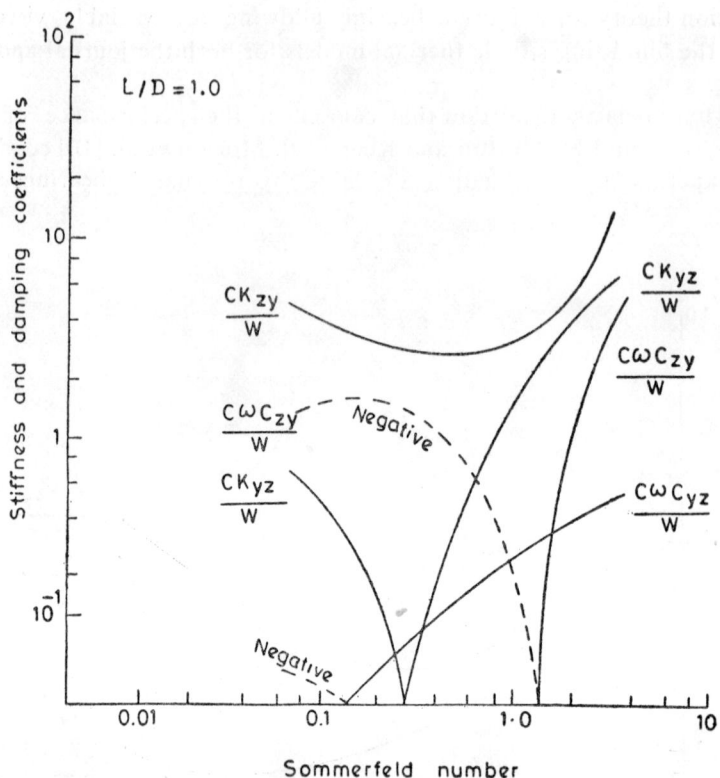

Fig. 7.12 Cross stiffness and damping coefficients of a grooved bearing

References

1. Pinkus, O. and Sternlicht, B. Theory of hydrodynamic lubrication, McGraw-Hill Book Co., 1961.
2. Smith, D.M. Journal bearings in turbomachinery, Chapman and Hall Ltd., 1969.
3. Lund, J.W. Rotor bearings dynamic design technology, part III: Design hand book for fluid film bearings, Mechanical Technology Inc., AFAPL-Tr-65-45, 1965.
4. Rao, J.S. and Mukherjee, A. Stiffness and damping coefficients of tilted journal bearing, Mechanism and Machine Theory, Vol. 12, 1977, p. 339.
5. Mukherjee, A. Effect of tilt on dynamic behaviour of flexible rotors in fluid film bearings, Ph.D. thesis, Indian Institute of Technology, Kharagpur, 1975.
6. Morton, P.G., Measurement of the dynamic characteristics of a large sleeve bearing, J. Lubrication Technology, v. 93, 1971, p. 143.
7. Fogg, A. Fluid film lubrication of parallel thrust surfaces, Proc. Instn of Mech. Engrs., v. 155, 1946, p. 49.
8. Dowson, D. and March, C.N., A thermo-hydrodynamic analysis of journal bearings, Proc. Instn. Mech. Engrs. v. 183, 1966, p. 117.
9. Morton, P.G. and Keogh, P.S., Thermoelastic influences in journal bearing lubrication, Proc. Royal Soc., v. 403, series A, 1986, p. 111.
10. Morton P.G., Johnson, J.H. and Wale, G.D., The effect of thermal distortion on the characteristics of large hydrodynamic bearing, Proc. Instn. of Mech. Engrs., v. 202, C3, 1988, p. 219.

A high speed turbine rotor (Courtesy BEHL, Hyderabad)

Chapter 8

Rotors Mounted on Fluid Film Bearings

The properties of fluid film bearings discussed in the previous chapter will be used here to determine the critical speeds and out-of-balance response of a rotor. We will consider first a simple rotor and then extend the analysis to the case of general rotor by using the transfer matrix procedure.

8.1 A Simple Rotor in Fluid Film Bearings

8.1.1 *Undamped analysis*

In order to study the effect of support stiffness and damping on the dynamic behaviour of a rotor, we can use a Jeffcott model, instead of a multimass model for the system. This simplifies the analysis and gives fairly good results for symmetrical rotors in the fundamental mode region. We first determine the rigid bearing critical speed of the rotor by the analysis outlined earlier using a transfer matrix method, and determine the equivalent stiffness of the rotor, with the total mass of the rotor lumped at the mid-span of an equivalent Jeffcott type rotor, which has the same critical speed of the system. Such a rotor is shown in Fig. 8.1, supported on two fluid film bearings at its ends. In this analysis, we consider only the stiffness coefficients of the bearings and the damping properties will be included in the next section.

For the rotor shown in Fig. 8.1, we have

$$M \frac{d^2}{dt^2}(z + a \cos \omega t) + K(z - z_0) = 0$$

$$M \frac{d^2}{dt^2}(y + a \sin \omega t) + K(y - y_0) = 0$$

(8.11)

and

$$K(z - z_0) = 2K_{zz}z_0 + 2K_{zy}y_0$$
$$K(y - y_0) = 2K_{yz}z_0 + 2K_{yy}y_0$$

(8.1.2)

From equation (8.1.2), we have

$$\begin{bmatrix} (2K_{zz} + K) & 2K_{zy} \\ 2K_{yz} & (2K_{yy} + K) \end{bmatrix} \begin{Bmatrix} z_0 \\ y_0 \end{Bmatrix} = \begin{Bmatrix} Kz \\ Ky \end{Bmatrix}$$

(8.1.3)

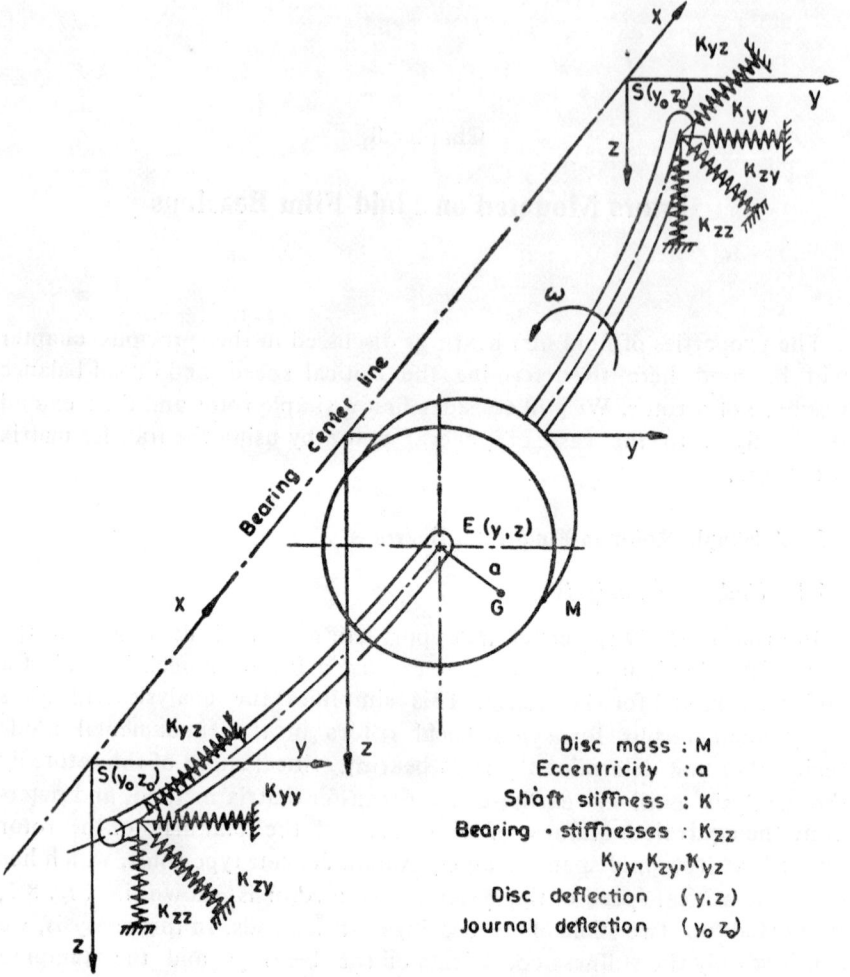

Fig. 8.1 A simple rotor on fluid film bearings

With the help of the above equation, we can evaluate

$$z_0 = K \frac{(2K_{yy} + K)z - 2K_{zy}y}{(2K_{zz} + K)(2K_{yy} + K) - 4K_{zy}K_{yz}}$$
$$y_0 = K \frac{(2K_{zz} + K)y - 2K_{yz}z}{(2K_{zz} + K)(2K_{yy} + K) - 4K_{zy}K_{yz}}$$

(8.1.4)

Equation (8.1.1) now becomes

$$M\ddot{z} + K_1 z + K_{12} y = M a \omega^2 \cos \omega t$$
$$M\ddot{y} + K_2 y + K_{21} z = M a \omega^2 \sin \omega t$$

(8.1.5)

$$K_1 = \frac{K[2K_{zz}(2K_{yy} + K) - 4K_{zy}K_{yz}]}{(2K_{zz} + K)(2K_{yy} + K) - 4K_{zy}K_{yz}}$$

$$K_2 = \frac{K[2K_{yy}(2K_{zz} + K) - 4K_{zy}K_{yz}]}{(2K_{zz} + K)(2K_{yy} + K) - 4K_{zy}K_{yz}}$$

$$K_{12} = \frac{2K_{zy}K^2}{(2K_{zz} + K)(2K_{yy} + K) - 4K_{zy}K_{yz}}$$

$$K_{21} = \frac{2K_{yz}K^2}{(2K_{zz} + K)(2K_{yy} + K) - 4K_{zy}K_{yz}} \tag{8.1.6}$$

In absence of the cross coupled stiffness terms, the above equations reduce to

$$M\ddot{z} + K_z z = Ma\omega^2 \cos \omega t$$

$$M\ddot{y} + K_y y = Ma\omega^2 \sin \omega t$$

$$K_1 \Big\}_{\substack{K_{zy}=0 \\ K_{yz}=0}} = K_z = \frac{2K_{zz}K}{(2K_{zz} + K)}$$

$$K_2 \Big\}_{\substack{K_{zy}=0 \\ K_{yz}=0}} = K_y = \frac{2K_{yy}K}{(2K_{yy} + K)} \tag{8.1.7}$$

The solution for equation (8.1.5) is

$$z = w = w_c \cos \omega t + w_s \sin \omega t$$
$$y = v = v_c \cos \omega t + v_s \sin \omega t \tag{8.1.8}$$

Substituting equations (8.1.8) in (8.1.5) and separating out cosine and sine terms, we obtain

$$-Mw_c\omega^2 + K_1 w_c + K_{12} v_c = Ma\omega^2 \tag{8.1.9}$$

$$-Mw_s\omega^2 + K_1 w_s + K_{12} v_s = 0 \tag{8.1.10}$$

$$-Mv_c\omega^2 + K_2 v_c + K_{21} w_c = 0 \tag{8.1.11}$$

$$-Mv_s\omega^2 + K_2 v_s + K_{21} w_s = Ma\omega^2 \tag{8.1.12}$$

From equations (8.1.10) and (8.1.11), we get

$$w_s = \frac{-K_{12}}{K_1 - M\omega^2} v_s; \quad v_c = \frac{-K_{21}}{K_2 - M\omega^2} w_c \tag{8.1.13}$$

Using equations (8.1.13) in (8.1.9) and (8.1.12), the following solutions can be obtained:

$$w_c = \frac{Ma\omega^2(K_2 - M\omega^2)}{(K_1 - M\omega^2)(K_2 - M\omega^2) - K_{12}K_{21}}$$

$$w_s = \frac{-K_{12}Ma\omega^2}{(K_1 - M\omega^2)(K_2 - M\omega^2) - K_{12}K_{21}}$$

$$v_c = \frac{-K_{21}Ma\omega^2}{(K_1 - M\omega^2)(K_2 - M\omega^2) - K_{12}K_{21}}$$

$$v_s = \frac{Ma\omega^2(K_1 - M\omega^2)}{(K_1 - M\omega^2)(K_2 - M\omega^2) - K_{12}K_{21}} \tag{8.1.14}$$

We introduce the following parameters:

$$\omega_1^2 = \frac{K_1}{M}; \qquad \omega_2^2 = \frac{K_2}{M}$$

$$\mu_1 = \frac{K_{12}}{K_1} \qquad \mu_2 = \frac{K_{21}}{K_2}$$

$$\bar{w}_c = \frac{w_c}{a}; \qquad \bar{w}_s = \frac{w_s}{a}$$

$$\bar{v}_c = \frac{v_c}{a}; \qquad \bar{v}_s = \frac{v_s}{a} \qquad (8.1.15)$$

Equations (8.1.14) then can be written as follows:

$$\bar{w}_c = \frac{\frac{\omega^2}{\omega_1^2}\left(1 - \frac{\omega^2}{\omega_2^2}\right)}{\left(1 - \frac{\omega^2}{\omega_1^2}\right)\left(1 - \frac{\omega^2}{\omega_2^2}\right) - \mu_1\mu_2}$$

$$\bar{w}_s = \frac{-\mu_1 \frac{\omega^2}{\omega_2^2}}{\left(1 - \frac{\omega^2}{\omega_1^2}\right)\left(1 - \frac{\omega^2}{\omega_2^2}\right) - \mu_1\mu_2}$$

$$\bar{v}_c = \frac{-\mu_2 \frac{\omega^2}{\omega_1^2}}{\left(1 - \frac{\omega^2}{\omega_1^2}\right)\left(1 - \frac{\omega^2}{\omega_2^2}\right) - \mu_1\mu_2}$$

$$\bar{v}_s = \frac{\frac{\omega^2}{\omega_2^2}\left(1 - \frac{\omega^2}{\omega_1^2}\right)}{\left(1 - \frac{\omega^2}{\omega_1^2}\right)\left(1 - \frac{\omega^2}{\omega_2^2}\right) - \mu_1\mu_2} \qquad (8.1.16)$$

The frequency equation can be written as:

$$\left(1 - \frac{\omega^2}{\omega_1^2}\right)\left(1 - \frac{\omega^2}{\omega_2^2}\right) - \mu_1\mu_2 = 0 \qquad (8.1.17)$$

which gives

$$p_{1,2}^2 = \frac{\omega_1^2 + \omega_2^2}{2} \pm \sqrt{\left[\frac{\omega_1^2 - \omega_2^2}{2}\right]^2 + \mu_1\mu_2\omega_1^2\omega_2^2} \qquad (8.1.18)$$

If $\mu_1 = \mu_2 = 0$, i.e., the cross-coupling coefficients are zero, then:

$$p_1 = \omega_1 = \sqrt{\frac{K_z}{M}}$$

$$p_2 = \omega_2 = \sqrt{\frac{K_y}{M}} \qquad (8.1.19)$$

Noting that:

$$\bar{w} = \bar{w}_c \cos \omega t + \bar{w}_s \sin \omega t$$

$$\bar{v} = \bar{v}_c \cos \omega t + \bar{v}_s \sin \omega t$$

$$\cos \omega t = \frac{\exp(i\omega t) + \exp(-i\omega t)}{2}$$

$$\sin \omega t = -i \frac{\exp(i\omega t) - \exp(-i\omega t)}{2}$$

and

$$\bar{r} = \bar{w} + i\bar{v}$$

we get

$$\bar{r} = \frac{\omega^2}{2}\left[\frac{(\omega_1^2 + \omega_2^2 - 2\omega^2) - i(\mu_2\omega_2^2 - \mu_1\omega_1^2)}{(\omega_1^2 - \omega^2)(\omega_2^2 - \omega^2) - \mu_1\mu_2\omega_1^2\omega_2^2}\right] \exp(i\omega t)$$

$$- \frac{\omega^2}{2}\left[\frac{(\omega_1^2 - \omega_2^2) + i(\mu_2\omega_2^2 + \mu_1\omega_1^2)}{(\omega_1^2 - \omega^2)(\omega_2^2 - \omega^2) - \mu_1\mu_2\omega_1^2\omega_2^2}\right] \exp(-i\omega t)$$

$$= r^+ \exp(i\omega t) + r^- \exp(-i\omega t) \qquad (8.1.20)$$

For cross coupling terms equal to zero

$$\bar{r} = \bar{w}_c \cos \omega t + i\bar{v}_s \sin \omega t$$

$$r = \frac{\omega^2}{2}\left[\frac{(\omega_1^2 + \omega_2^2 - 2\omega^2)}{(\omega_1^2 - \omega^2)(\omega_2^2 - \omega^2)}\right] \exp(i\omega t) + \frac{\omega^2}{2}\left[\frac{(\omega_2^2 - \omega_1^2)}{(\omega_1^2 - \omega^2)(\omega_2^2 - \omega^2)}\right] \exp(-i\omega t)$$

$$(8.1.21)$$

Example 8.1

Consider a rotor of mass 54.432 kg, with its rigid bearing critical speed equal to 4820 rpm, mounted on two 4 axial groove bearings 2.54 cm diameter and 1.27 cm long, with a radial clearance $C = 0.00254$ cm and viscosity at operating temperature, 0.0242 N sec/m². We will denote this rotor as rotor 1.

The shaft stiffness for the Jeffcott model is

$$K = M\omega_n^2 = 1.387 \times 10^7 \text{ N/m}$$

The stiffness coefficients for the bearing at a speed 4500 rpm ($S=0.5488$); are

$$K_{zz} = 4.16 \times 10^7 \text{ N/m}; \quad K_{yy} = 1.01 \times 10^7 \text{ N/m}$$
$$K_{zy} = 3.12 \times 10^7 \text{ N/m}; \quad K_{yz} = 4.16 \times 10^5 \text{ N/m}$$

Hence

$$K_1 = 1.185 \times 10^7 \text{ N/m}; \quad K_2 = 8.117 \times 10^6 \text{ N/m}$$
$$K_{12} = 3.700 \times 10^6 \text{ N/m}; \quad K_{21} = 4.920 \times 10^4 \text{ N/m}$$
$$\omega_1 = 466.6 \text{ rad/sec}; \quad \omega_2 = 386.2 \text{ rad/sec}$$
$$\mu_1 = 0.31; \quad \mu_2 = 0.006$$

Substituting the above quantities in equation (8.1.18), the two frequencies are:

$$p_1 = 468.74 \text{ rad/sec} = 4476 \text{ rpm}$$
$$p_2 = 385.96 \text{ rad/sec} = 3685 \text{ rpm}$$

These two frequencies can be taken as critical speeds of the rotor, as the stiffness coefficients are evaluated at a speed of 4500 rpm and we may reasonably expect the coefficients to be the same at speeds 4476 and 3685 rpm also. In reality, we cannot define critical speeds, as we do for a rigid bearing rotor, because the stiffness coefficients are functions of the speed of rotor. Consequently, it is always better to study the out-of-balance response, to locate the critical speeds.

Using the stiffness coefficients for the bearing at different speeds, the unbalance response obtained (major axis of the elliptical orbit) is plotted in Fig. 8.2. Figures 8.3 a, b and c give the elliptical orbit of the rotor at

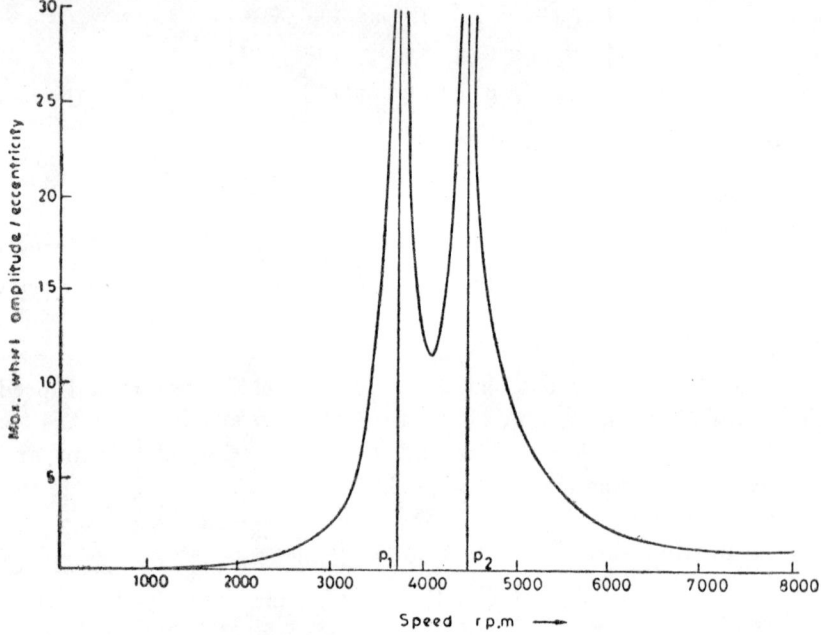

Fig. 8.2 Unbalance response of a rotor in fluid film bearing

3000 rpm, 4000 rpm and 6000 rpm respectively. In Figs. 8.3 a and c, the rotor has a forward synchronous whirl, when the speed of the rotor is either below the first critical speed or above the second critical speed. However, when the rotor is rotating with a speed between the two criticals, there is a backward synchronous whirl of the rotor. Tondl[1], Dimintberg [2], and Gasch and Pfutzner [3] have shown this property by considering the stiffness asymmetry of the bearing but not with the cross-coupled stiffness term. Hull [4] observed experimentally this backward whirl for a rotor driven by a lathe and supported on outboard bearing held between two vertical springs in tension, giving rise to asymmetric bearing stiffness.

It is not necessary that rotors in all cases will behave as in the example just considered, with two clear critical speeds corresponding to the fundamental rigid bearing critical speed, with a backward whirl between the

Rotors Mounted on Fluid Film Bearings 113

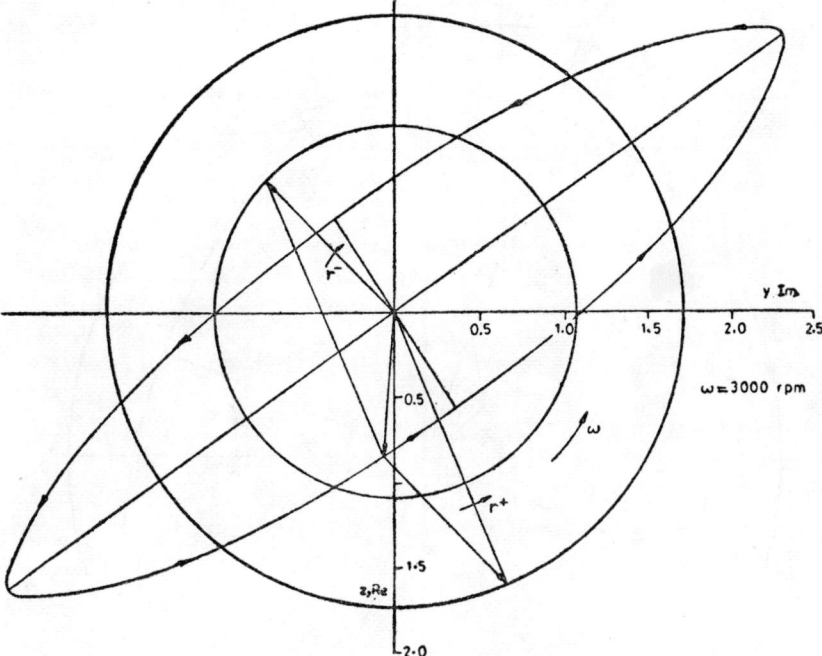

Fig. 8.3a Whirl orbit in forward direction $\omega < p_1$

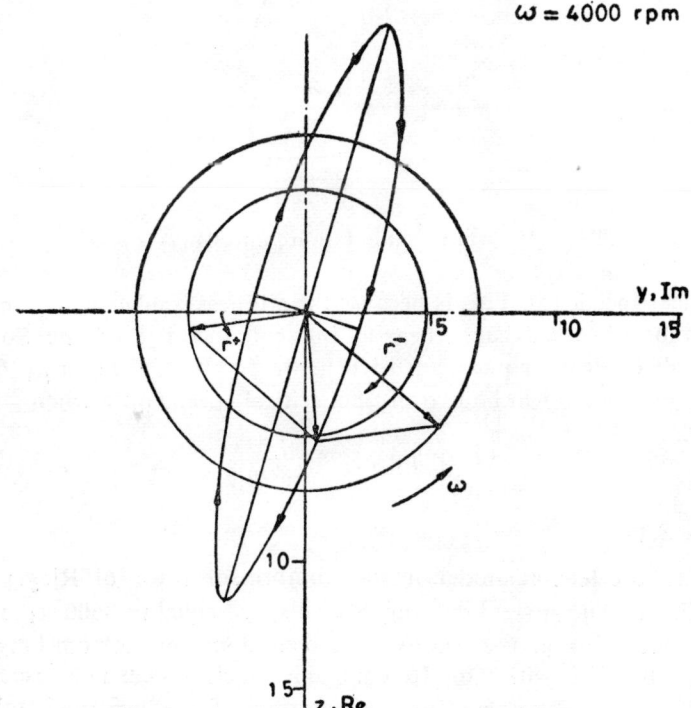

Fig. 8.3b Whirl orbit in backward direction $p_1 < \omega < p_2$

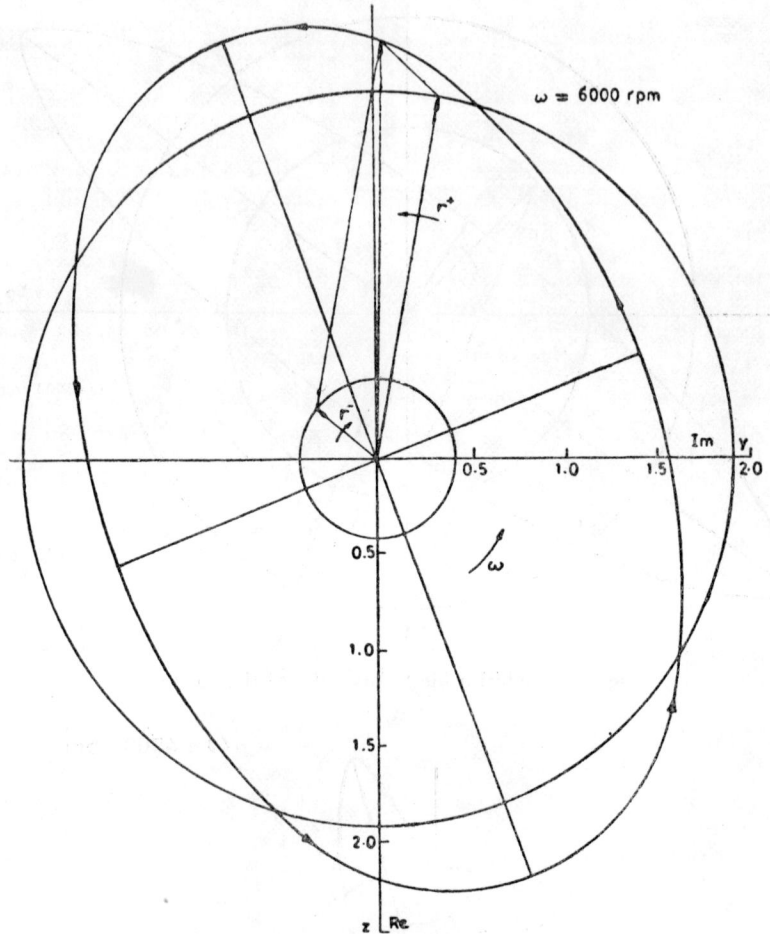

Fig. 8.3c Whirl orbit in forward direction $\omega > p_2$

two critical speeds [5]. This is because the stiffness coefficient K_{yz} can be negative for a hydro-dynamic bearing in a certain range of the Sommerfeld number. From equation (8.1.18), we find that, if either μ_1 or μ_2 is negative then there will be two distinct critical speeds only when

$$\sqrt{1 + |\mu_1||\mu_2|} < \frac{\omega_1^2 + \omega_2^2}{2\omega_1\omega_2} \tag{8.1.22}$$

Example 8.2

Consider the Jeffcott model of the compressor rotor of Rieger [6] of mass 453.6 kg with a rigid bearing critical speed equal to 8600 rpm. The rotor is mounted at the ends on 10.16 cm diameter 5.08 cm long plain cylindrical bearings with 0.01016 cm diametral clearance. The viscosity of the lubricant at the operating temperature of the bearing is taken as 5.68×10^{-3} N sec/m². We will denote this rotor as rotor 2.

The out-of-balance response obtained is given in Fig. 8.4, which does not show two distinct peaks as in the previous example. Even in the absence of damping, we find no resonance as demonstrated below.

The stiffness of the rotor is

$$K = M\omega_n^2 = 3.68 \times 10^8 \text{ N/m}$$

At

$$\omega = 7500 \text{ rpm, i.e., } S = 1.65$$

we find

$$K_1 = 1.839 \times 10^8 \text{ N/m}; \quad K_2 = 2.004 \times 10^8 \text{ N/m}$$

$$K_{12} = 1.340 \times 10^8 \text{ N/m}; \quad K_{21} = -1.026 \times 10^8 \text{ N/m}$$

$$\omega_1 = 637 \text{ rad/sec}; \quad \omega_2 = 665 \text{ rad/sec}.$$

$$\mu_1 = 0.728; \quad \mu_2 = -0.512$$

The term under the square root of equation (8.1.18) is negative, and equation (8.1.22) is not satisfied. Hence there are no two distinct critical speeds for this rotor. Further, it does not exhibit backward whirl anywhere in the region considered. It is interesting to note the following frequencies, which are recorded in Fig. 8.4. The rigid bearing critical

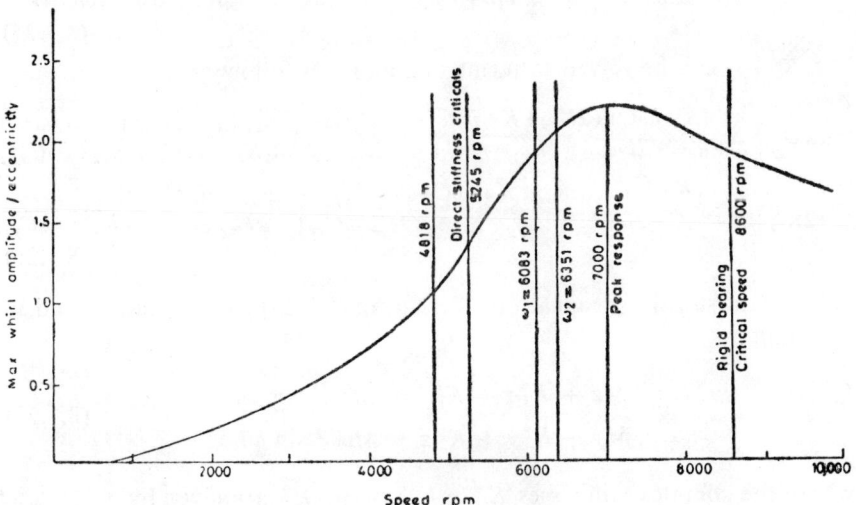

Fig. 8.4 Out of balance response of a rotor in fluid film bearings

speed, 8600 rpm modifies to two critical speeds 4818 and 5245 rpm, when the direct stiffnesses are only considered. The two frequencies in equation (8.1.15) are 6083 and 6351 rpm, which are substantially different from the direct stiffness critical speeds, unlike in the case of rotor 1. The peak unbalance occurs still at a higher speed, 7000 rpm which is however lower than the rigid bearing critical speed. The effect of bearing damping and the shaft structural damping approximated by an equivalent viscous damping is considered in the next section.

8.1.2 Analysis with damping

Considering the bearing damping coefficients and the shaft structural damping approximated by an equivalent viscous damping C, equations (8.1.1) and (8.1.2) become

$$M \frac{d^2}{dt^2}(z + a \cos \omega t) + K(z - z_0) + C(\dot{z} - \dot{z}_0) = 0$$

$$M \frac{d^2}{dt^2}(y + a \sin \omega t) + K(y - y_0) + C(\dot{y} - \dot{y}_0) = 0 \qquad (8.1.23)$$

and

$$K(z - z_0) + C(\dot{z} - \dot{z}_0) = 2K_{zz}z_0 + 2C_{zz}\dot{z}_0 + 2K_{zy}y_0 + 2C_{zy}\dot{y}_0$$

$$K(y - y_0) + C(\dot{y} - \dot{y}_0) = 2K_{yz}z_0 + 2C_{yz}\dot{z}_0 + 2K_{yy}y_0 + 2C_{yy}\dot{y}_0 \qquad (8.1.24)$$

For harmonic motion, equation (8.1.24) becomes

$$\begin{bmatrix} 2K_{zz}+K+i\omega C+2i\omega C_{zz} & 2K_{zy}+2i\omega C_{zy} \\ 2K_{yz}+2i\omega C_{yz} & 2K_{yy}+K+i\omega C+2i\omega C_{yy} \end{bmatrix} \begin{Bmatrix} z_0 \\ y_0 \end{Bmatrix} = \begin{Bmatrix} (K+i\omega C)z \\ (K+i\omega C)y \end{Bmatrix} \qquad (8.1.25)$$

z_0 and y_0 can be solved in terms of z and y as follows:

$$z_0 = \frac{(K+i\omega C)[(2K_{yy}+K+i\omega C+2i\omega C_{yy})z - (2K_{zy}+2i\omega C_{zy})y]}{(2K_{zz}+K+i\omega C+2i\omega C_{zz})(2K_{yy}+i\omega C+K+2i\omega C_{yy}) - 4(K_{zy}+i\omega C_{zy})(K_{yz}+i\omega C_{yz})}$$

$$y_0 = \frac{(K+i\omega C)[(2K_{zz}+K+i\omega C+2i\omega C_{zz})y - (2K_{yz}+2i\omega C_{yz})z]}{(2K_{zz}+K+i\omega C+2i\omega C_{zz})(2K_{yy}+K+i\omega C+2i\omega C_{yy}) - 4(K_{zy}+i\omega C_{zy})(K_{yz}+i\omega C_{yz})} \qquad (8.1.26)$$

We can substitute equations (8.1.26) in (8.1.23) to eliminate z_0 and y_0 to obtain

$$M\ddot{z} + K_1^* z + K_{12}^* y = Ma\omega^2 \cos \omega t$$
$$M\ddot{y} + K_2^* y + K_{21}^* z = Ma\omega^2 \sin \omega t \qquad (8.1.27)$$

where the complex stiffnesses K_1^*, K_2^*, K_{12}^* and K_{21}^* are given by

$$K_1^* = \frac{K^*[2K_{zz}^*(2K_{yy}^* + K^*) - 4K_{zy}^* K_{yz}^*]}{\Delta}$$

$$K_2^* = \frac{K^*[2K_{yy}^*(2K_{zz}^* + K^*) - 4K_{zy}^* \cdot K_{yz}^*]}{\Delta}$$

$$K_{12}^* = 2K_{zy}^* K^{*2}/\Delta$$

$$K_{21}^* = 2K_{yz}^* K^{*2}/\Delta$$

and

$$\Delta = (2K_{zz}^* + K^*)(2K_{yy}^* + K^*) - 4K_{zy}^* \cdot K_{yz}^*$$

$$K^* = K + i\omega C$$

$$K_{zz}^* = K_{zz} + i\omega C_{zz}$$

$$K_{yy}^* = K_{yy} + i\omega C_{yy}$$

$$K_{zy}^* = K_{zy} + i\omega C_{zy}$$

$$K_{yz}^* = K_{yz} + i\omega C_{yz} \tag{8.1.28}$$

The solution of equations (8.1.27) can be directly written as

$$z = w = w^+ \exp(i\omega t) + w^- \exp(-i\omega t)$$
$$y = v = v^+ \exp(i\omega t) + v^- \exp(-i\omega t) \tag{8.1.29}$$

Substituting these solutions in equation (8.1.27), we obtain the following:

$$w^+ = \frac{(Ma\omega^2/2)[(K_2^* - M\omega^2) + iK_{12}^*]}{(K_1^* - M\omega^2)(K_2^* - M\omega^2) - K_{12}^*K_{21}^*}$$

$$v^+ = \frac{(-iMa\omega^2/2)[(K_1^* - M\omega^2) - iK_{21}^*]}{(K_1^* - M\omega^2)(K_2^* - M\omega^2) - K_{12}^*K_{21}^*}$$

$$w^- = \frac{(Ma\omega^2/2)[K_2^* - M\omega^2) - iK_{12}^*]}{(K_1^* - M\omega^2)(K_2^* - M\omega^2) - K_{12}^*K_{21}^*}$$

$$v^- = \frac{(iMa\omega^2/2)[(K_1^* - M\omega^2) + iK_{21}^*]}{(K_1^* - M\omega^2)(K_2^* - M\omega^2) - K_{12}^*K_{21}^*} \tag{8.1.30}$$

The out-of-balance response now becomes

$$\bar{r} = \frac{(w + iv)}{a} \tag{8.1.31}$$

which can be written as

$$\bar{r} = (\omega^2/2)[\{(\omega_1^{*2} + \omega_2^{*2} - 2\omega^2) - i(\mu_2^*\omega_2^{*2} - \mu_1^*\omega_1^{*2})\} \exp(i\omega t)$$
$$- \{(\omega_1^{*2} - \omega_2^{*2}) + i(\mu_2^*\omega_2^{*2} + \mu_1^*\omega_1^{*2})\} \exp(-i\omega t)]/$$
$$[(\omega_1^{*2} - \omega^2)(\omega_2^{*2} - \omega^2) - \mu_1^*\mu_2^*\omega_1^{*2}\omega_2^{*2}] \tag{8.1.32}$$

where

$$\omega_1^{*2} = K_1^*/M; \quad \omega_2^{*2} = K_2^*/M$$
$$\mu_1^* = K_{12}^*/K_1^*; \quad \mu_2^* = K_{21}^*/K_2^* \tag{8.1.33}$$

In the absence of cross-coupled stiffness and damping coefficients, i.e., $C_{zy} = C_{yz} = K_{zy} = K_{yz} = 0$ and shaft damping, i.e., $C = 0$, the equations of motion get decoupled as follows.

$$M\ddot{z} + \bar{C}_z\dot{z} + \bar{K}_z z = Ma\omega^2 \cos \omega t$$
$$M\ddot{y} + \bar{C}_y\dot{y} + \bar{K}_y y = Ma\omega^2 \sin \omega t \tag{8.1.34}$$

where

$$\bar{C}_z = \frac{2K^2 C_{zz}}{(2K_{zz} + K)^2 + (2\omega C_{zz})^2}; \quad \bar{K}_z = \frac{K[2K_{zz}(2K_{zz} + K) + (2\omega C_{zz})^2]}{(2K_{zz} + K)^2 + (2\omega C_{zz})^2}$$

(8.1.35)

and $\bar{C}_y$ and $\bar{K}_y$ are obtained from above by replacing C_{zz} and K_{zz} with C_{yy} and K_{yy} respectively.

In the above $\bar{C}_z$ and $\bar{K}_z$ are the effective damping and stiffness for the system in the z direction and $\bar{C}_y$ and $\bar{K}_y$ are the corresponding values in the y direction. An interesting fact to be noted is that the effective stiffnesses are functions of bearing damping and rotational speed, in addition to the shaft and bearing stiffnesses. Further, the effective damping is also dependent on the shaft and bearing stiffnesses and the rotational speed. We should also remember that the stiffnesses and damping coefficients of the bearing are themselves dependent on the rotational speed. Using the analysis given above, a computer program can be written to determine the unbalance response [7] of a rotor, taking into account the bearing stiffness and damping and equivalent viscous damping of the rotor. The results obtained for the two rotors given in section 8.1.1 are discussed below.

Example 8.3 (Rotor 1 with Two Critical Speeds)

Consider rotor 1 of Example 8.1, whose undamped response is shown in Fig. 8.2. Using the stiffness and damping coefficients of the bearing at different speeds, the unbalance response obtained is given in Fig. 8.5. The presence of bearing damping reduces the unbalance response in the region of the critical speeds and the peaks are shifted to the right from these criticals. The backward whirl of the rotor in the frequency region between the two critical speeds persists in the presence of bearing damping also. From the two damped response curves in Fig. 8.5, it is observed that the cross-coupled damping enhances the response. This is because the predominant cross-coupled damping C_{zy} is negative.

Another interesting feature about the rotor bearing system is its effective damping values. Though the absolute predominant damping in the bearing, C_{zz} is of the order of 70000 N sec/m, C_{yy} being 1930 N sec/m, the effective damping values of the system from equation (8.1.35) are 2130 and 646 N sec/m respectively, in the z and y directions. The corresponding effective critical damping values are 52400 and 42400 N sec/m and the critical speeds with only effective stiffnesses are 4600 and 3725 rpm. Thus this rotor is lightly damped. It may also be observed that the response around 4600 rpm is more suppressed than around 3700 rpm, because of a higher value of effective damping ratio in the z direction than in the y direction.

The unbalance response of this rotor with both direct and cross-coupled stiffnesses and only shaft damping (damping ratios based on rigid bearing model) is given in Fig. 8.6. It is seen that when the shaft damping is 5

Rotors Mounted on Fluid Film Bearings   119

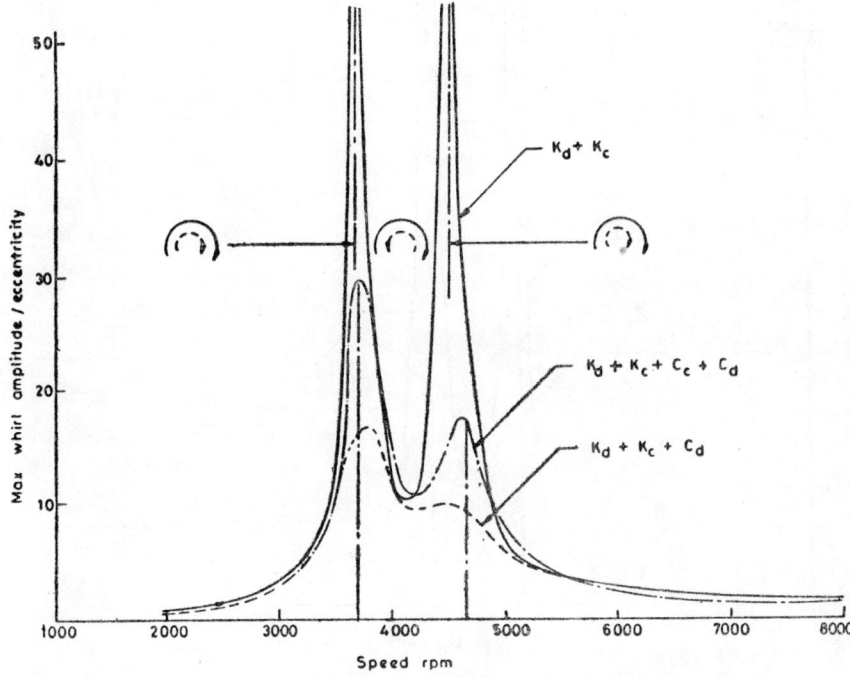

Fig. 8.5 Unbalance response of rotor 1 in hydrodynamic bearings (with no shaft damping). For legend see Fig. 8.7

percent the critical value, the response at the critical speeds is approximately equal to that when the damping in the bearings is considered.

Example 8.4 (Rotor 2 with no Critical Speeds)

The data for this rotor is given in section 8.1.1, with its undamped unbalance response in Fig. 8.4. Figure 8.7 shows the unbalance response of this rotor for the following cases:
1. With only direct and cross-coupled bearing stiffnesses.
2. With only direct and cross coupled stiffnesses and direct damping in the bearings.
3. With all stiffnesses and damping in the bearings.
4. With only direct stiffnesses and direct damping in the bearings.
5. With only direct stiffnesses, and direct and cross coupled damping in the bearings.

Considering only the direct stiffnesses, rotor 2 has two critical speeds at 4818 and 5245 rpm as shown in Fig. 8.8. However, when the different combinations of cross-coupled stiffnesses and direct and cross coupled dampings are included, the peak response occurs far away from these criticals as seen from Fig. 8.7. This is due to a large effective damping given by the bearing to the system, e.g., the effective viscous damping and

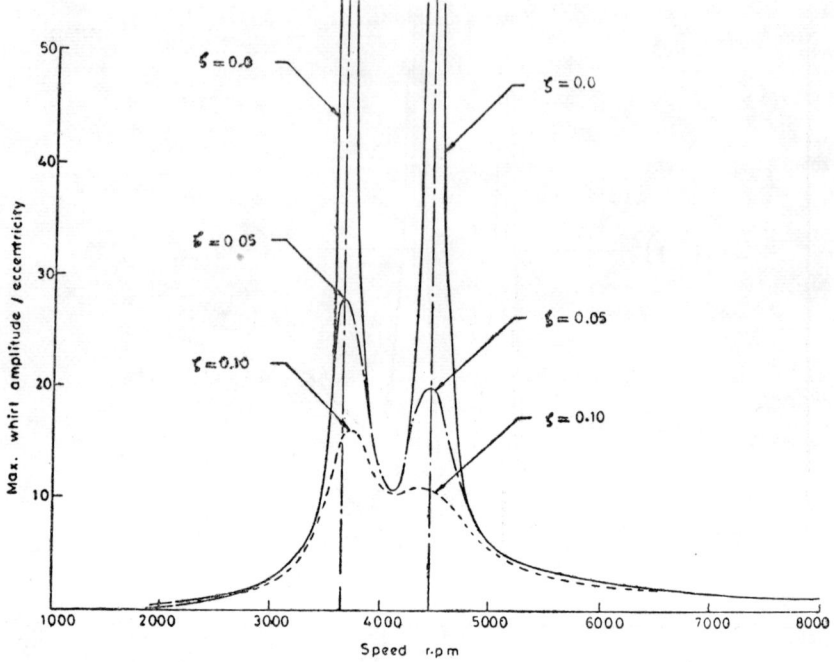

Fig. 8.6   Unbalance response of rotor 1 in hydrodynamic bearings (with only shaft damping)

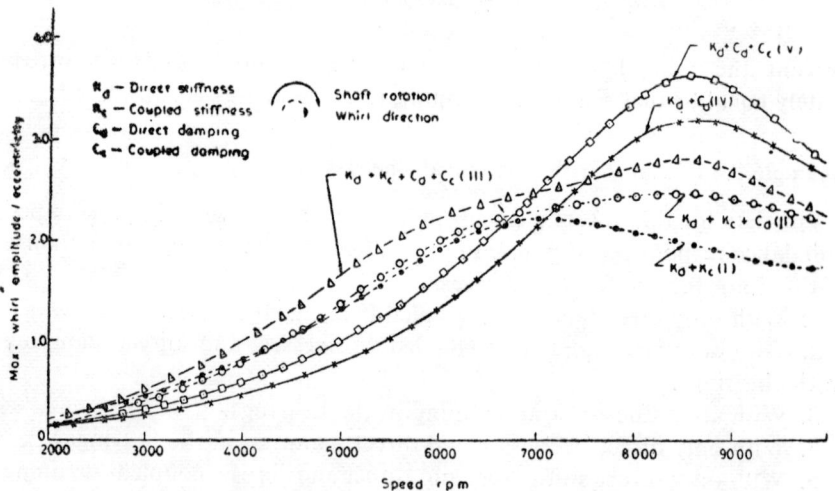

Fig. 8.7   Unbalance response of rotor 2 in hydrodynamic bearings

critical damping are 135000 and 753000 N sec/m in the z direction and 133000 and 720000 N sec/m in the y direction respectively for this system at 8600 rpm, in the absence of cross coupled stiffness and damping. In addition it may also be noted that the critical speeds with only effective stiffnesses for the y and z motions occur at 7583 and 7932 rpm respectively.

As mentioned earlier, the effective stiffness and damping of the system in the z and y directions are functions of rotational speed and hence, the net unbalance response exhibits a single peak at a frequency slightly above the two criticals.

The unbalance response for this rotor with only direct stiffnesses and shaft damping ratio based on the rigid bearing model, is shown in Fig. 8.8. The unbalance response is reduced at the critical speeds as the shaft damping is increased. However, even with 10 percent critical shaft damping, the

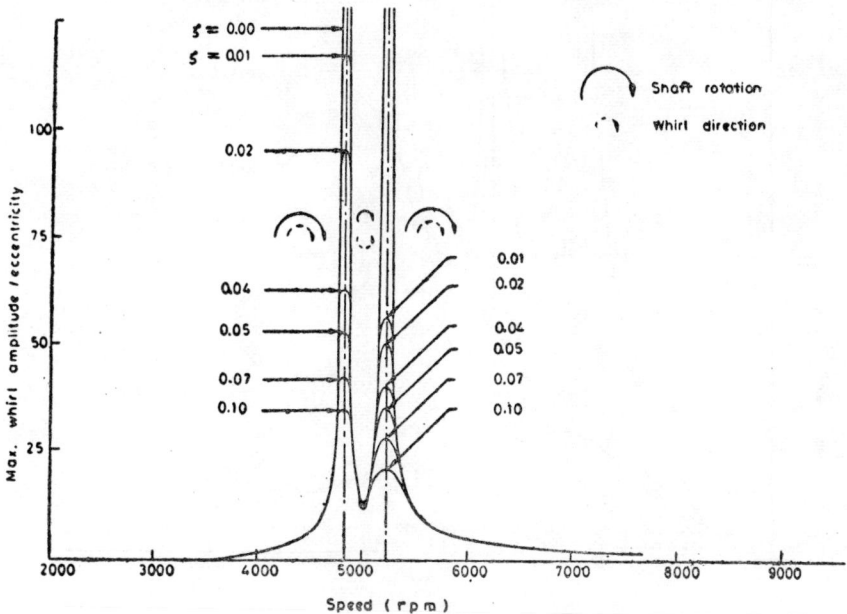

Fig. 8.8 Unbalance response of rotor 2 in hydrodynamic bearings (direct stiffness only and with shaft damping)

response near critical speeds is 7 to 10 times higher than the peak responses when damping in the bearings is included, and the effect of nominal values of shaft damping on the unbalance response is insignificant in the presence of damping in the bearings. In the absence of bearing damping and cross-coupled stiffness, the rotor exhibits backward whirl between the two criticals.

Example 8.5 (Rotor 3)

Rotors 1 and 2 discussed above exhibited two different behaviours either having two distinct critical speeds or no critical speeds. Actually a single rotor can exhibit either of these properties depending on the characteristics of the bearings, e.g., consider a 9.07 kg rotor, having a stiffness of 884000 N/m, mounted on two plain cylindrical bearings 2.54 cm diameter and 2.54 cm long with the viscosity of the lubricating oil to be 24 cp. The radial clearance in the bearings is chosen as (a) 0.00533 cm and (b) 0.0188 cm, in such a manner to exhibit the two distinct behaviours cited before.

Figure 8.9 gives the unbalance response of this rotor for both cases, out of which case (a) exhibits no critical speeds and case (b) exhibits two distinct critical speeds with backward whirl between the two. It can be observed that the peak response for case (a) is shifted to the right from the

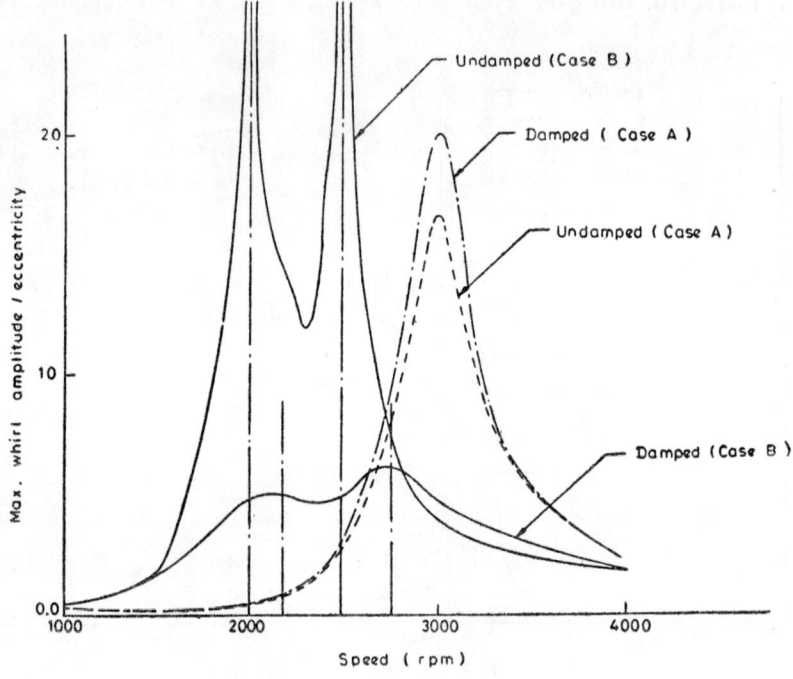

Fig. 8.9 Unbalance response of rotor 3 in hydrodynamic bearings

two distinct criticals in case (b). Also the damping reduces the response at the criticals for case (b), whereas, it enhances the response in the peak region for case (a).

Earlier it was observed that the unbalance response for rotor 2, with no critical speeds is much lower than the response of rotor 1 with two distinct critical speeds. However, Fig. 8.9 predicts an opposite trend. This is due to relative values of stiffness of the shaft and bearing stiffness coefficients, e.g., in case of rotor 2, the shaft stiffness is higher than the stiffness coefficients of the bearings, whereas for rotor 3 case (a), the shaft stiffness is smaller than the stiffness coefficients of the bearings.

Subbiah et al. [21] built the case (b) rotor to observe the backward whirl between the two critical speeds. The unbalance introduced in the rotor was 0.01084 kg cm and the oil used has a viscosity of 0.96×10^{-5} Pascal Sec at 25.5 deg C. The two bearings are mounted on cast iron pedestals at the two ends, and in turn these pedestals are rigidly fastened to the support, which is made of steel angles. The pedestal frequency was found to be much higher than the critical speeds of the rotor, during an impact test, indicating that the pedestals are rigid. The unbalance response is measured in both z and y directions by two proximity pick-ups.

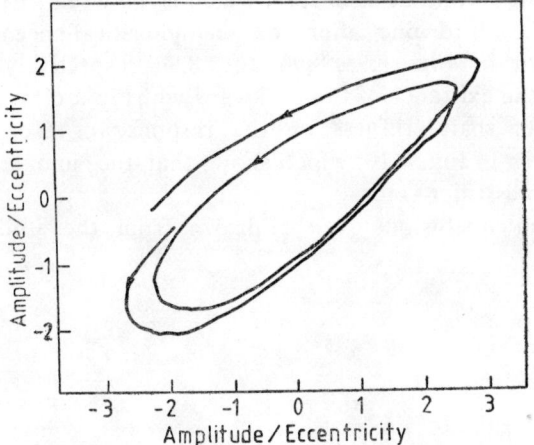

Fig. 8.9(a) Rotor whirl orbit at 2150 rpm (counterclockwise-direction corresponds to forward whirl)

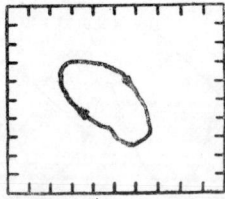

Fig. 8.9(b) Rotor whirl orbit at 2500 rpm (clockwise direction-corresponds to backward whirl)

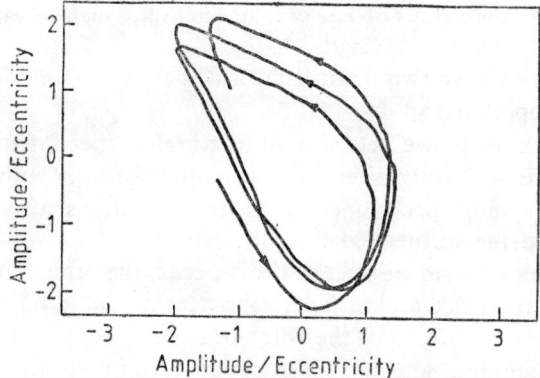

Fig. 8.9(c) Rotor whirl orbit at 3000 rpm (counter clockwise-direction corresponds to forward whirl)

The signals from the pick-ups are fed to a twin channel FFT analyzer and the orbit diagrams are obtained with a X-Y plotter. A photograph on the screen was also taken when there was a backward whirl. The direction of the plotter pen indicated the direction of the rotor whirl. The whirl orbits are obtained at three different speeds, one below the critical at 2150 rpm, the second at 2500 rpm, which is between the two critical speeds and the third one after the second critical speed at 3000 rpm. They are shown in Figs. 8.9a, 8.9b and 8.9c respectively. Figure 8.9b demonstrated the existence of the backward whirl in a clear manner.

The effect of shaft stiffness on the response of rotor 3 for case (a) bearings is given in Fig. 8.10, which shows that the unbalance response is reduced with a stiffer rotor.

The following conclusions can be drawn from the studies made in section 8.1.

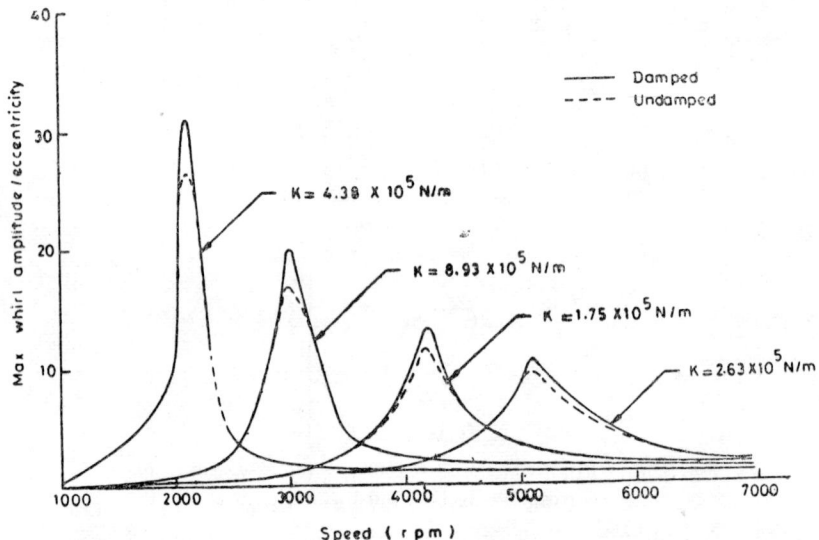

Fig. 8.10 Unbalance response of rotor 3 in hydrodynamic bearings

1. A rotor can have two distinct critical speeds or none depending on the bearing properties.
2. The peak response gets shifted to a higher speed, from the region of critical speeds with only direct stiffnesses, for a rotor with no critical speeds. This is more pronounced for rotors of large shaft stiffness compared to the bearing stiffness coefficients.
3. For rotors with no distinct critical speeds, the unbalance response is enhanced in the peak region due to the bearing damping.
4. The effective stiffness of the rotor bearing system is a function of the bearing damping and the speed of rotation, in addition to the shaft and bearing stiffnesses.
5. The absolute values of bearing damping by themselves do not

indicate the type of behaviour of the rotor. The effective viscous damping values, which are functions of stiffnesses and rotational speed, actually determine the dynamic behaviour.

6. The unbalance response of a rotor bearing system is reduced with an increase in shaft stiffness compared with bearing stiffness coefficients.

7. The shaft damping has no influence on the unbalance response, when the effective viscous damping ratio of the system is high.

A fairly symmetrical rotor can be modelled as a simple rotor outlined in this section, e.g., Morton [8]. For practical configurations, however, a transfer matrix analysis as outlined in the next section will be necessary. This transfer matrix type approach was used by Kramer [9] and Rao and Sarma [10] to study the unbalance response of a rotor.

8.2 Transfer Matrix Analysis of Rotors in Fluid Film Bearings

For a general rotor we can use the transfer matrix analysis outlined in Chapter 5, to determine the out-of-balance response by considering the bearing properties at a station i. Figure 8.11 shows schematically, the stiffness and damping properties of the bearing, which leads to the

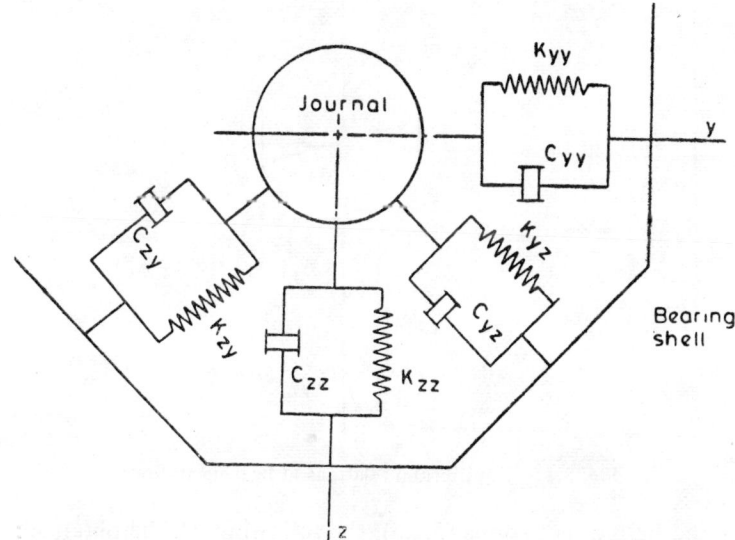

Fig. 8.11 Stiffness and damping coefficients of the fluid film in bearing

equilibrium relations given in Fig. 8.12. The effect of shaft tilt is neglected, for reasons discussed in Chapter 7. From Fig. 8.12, we get

$$V_{zi}^R = V_{zi}^L + K_{zz}w_i + K_{zy}v_i + C_{zz}\dot{w}_i + C_{zy}\dot{v}_i - m_iw_i\omega^2$$
$$V_{yi}^R = V_{yi}^L + K_{yy}v_i + K_{yz}w_i + C_{yy}\dot{v}_i + C_{yz}\dot{w}_i - m_iv_i\omega^2 \qquad (8.2.1)$$

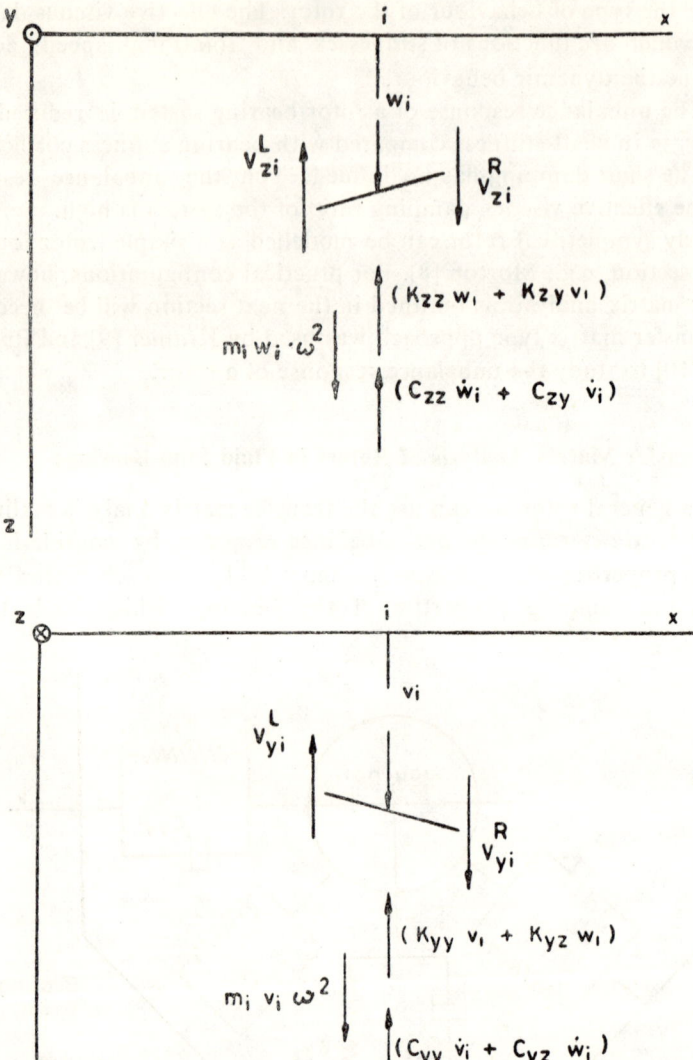

Fig. 8.12 Equilibrium relations at bearing station

With the help of equations (5.2.3), the following can be obtained:

$$V^R_{zci} = V^L_{zci} + K_{zz}w_{ci} + K_{zy}v_{ci} + C_{zz}\omega w_{si} + C_{zy}\omega v_{si} - m_i w_{ci}\omega^2$$
$$V^R_{zsi} = V^L_{zsi} + K_{zz}w_{si} + K_{zy}v_{si} - C_{zz}\omega w_{ci} - C_{zy}\omega v_{ci} - m_i w_{si}\omega^2$$
$$-V^R_{yci} = V^L_{yci} - K_{yy}v_{ci} - K_{yz}w_{ci} - C_{yy}\omega v_{si} - C_{yz}\omega w_{si} + m_i v_{ci}\omega^2$$
$$-V^R_{ysi} = -V^L_{ysi} - K_{yy}v_{si} - K_{yz}w_{si} + C_{yy}\omega v_{ci} + C_{yz}\omega w_{ci} + m_i v_{si}\omega^2$$

(8.2.2)

The deflection, slope and moment quantities are continuous at the bearing and the bearing transfer matrix therefore can be written as

follows:
$$\{\bar{S}\}_i^R = [B]_i \{\bar{S}\}_i^L \qquad (8.2.3)$$
where $[B]$ is the bearing transfer matrix, whose non zero elements are

$$B_{r,s} = 1, \ r = s = 1, 2, \ldots, 17 \qquad (8.2.4)$$

$$B_{4,1} = B_{8,5} = -K_{zz} + m\omega^2$$
$$B_{12,9} = B_{16,13} = -K_{yy} + m\omega^2 \qquad (8.2.5)$$
$$B_{4,5} = -B_{8,1} = -C_{zz}\omega$$
$$B_{4,9} = B_{8,13} = K_{zy}$$
$$B_{4,13} = -B_{8,9} = C_{zy}\omega \qquad (8.2.6)$$
$$B_{12,1} = B_{16,5} = K_{yz}$$
$$B_{12,5} = -B_{16,1} = C_{yz}\omega$$
$$B_{12,13} = -B_{16,9} = -C_{yy}\omega \qquad (8.2.7)$$

The procedure for an overhung rotor gets simplified with fluid film bearings, unlike in the case of rigid bearings as outlined in section 5.4, because of the transfer matrix definition at the flexible support.

Let the overall transfer matrix of a rotor mounted on two fluid film bearings be represented by

$$\{S\}_n^R = [U]\{S\}_1^L \qquad (8.2.8)$$

where $[U]$ is overall transfer matrix of size 17×17. Since the shear forces and bending moments are zero at both the ends, we obtain the following equation:

$$\begin{bmatrix} u_{3,1} & u_{3,2} & u_{3,5} & u_{3,6} & u_{3,9} & u_{3,10} & u_{3,13} & u_{3,14} \\ u_{4,1} & u_{4,2} & u_{4,5} & u_{4,6} & u_{4,9} & u_{4,10} & u_{4,13} & u_{4,14} \\ u_{7,1} & & & & & & & \\ u_{8,1} & & & \text{etc,} & & & & \\ u_{11,1} & & & & & & & \\ u_{12,1} & & & & & & & \\ u_{15,1} & & & & & & & \\ u_{16,1} & u_{16,2} & u_{16,5} & u_{16,6} & u_{16,9} & u_{16,10} & u_{16,13} & u_{16,14} \end{bmatrix}$$

$$\times \begin{Bmatrix} -w_c \\ \theta_c \\ -w_s \\ \theta_s \\ v_c \\ \phi_c \\ v_s \\ \phi_s \end{Bmatrix}^L_1 = -\begin{Bmatrix} u_{3,17} \\ u_{4,17} \\ u_{7,17} \\ u_{8,17} \\ u_{11,17} \\ u_{12,17} \\ u_{15,17} \\ u_{16,17} \end{Bmatrix} \qquad (8.2.9)$$

We can solve the above system of linear equations, to determine w_c, θ_c etc., at the starting station, which defines $\{S\}_1^L$. Then the state vectors at every station can be obtained to get the unbalance response of a rotor.

A computer program [10] to determine the unbalance response of a rotor can be developed using the analysis outlined above.

Example 8.6 Consider Rieger's rotor given in Fig. 8.13

With the help of the computer program, the results obtained for the response of the rotor with an unbalance of 0.072 kg cm in the plane 7, are given in Figs. 8.14 to 8.17.

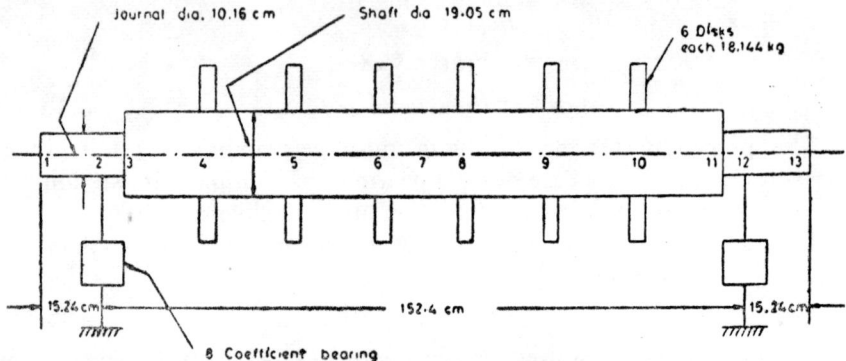

Fig. 8.13 Compressor rotor [6]

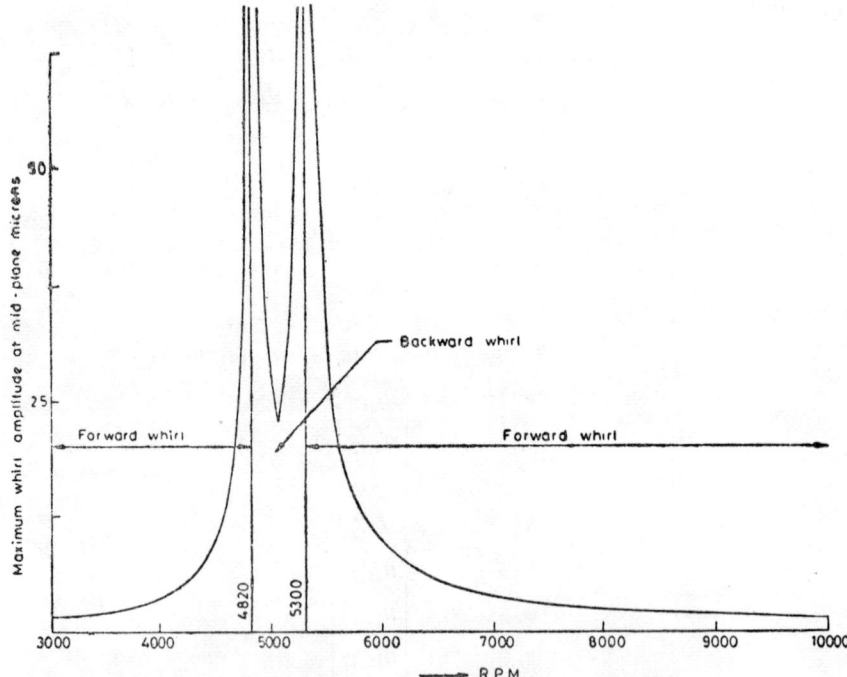

Fig. 8.14 Unbalance response with direct stiffnesses only

Rotors Mounted on Fluid Film Bearings   129

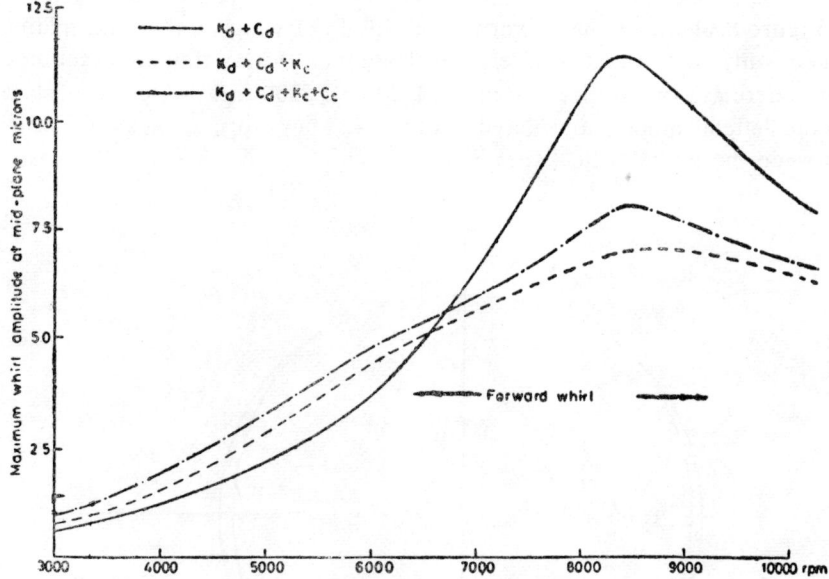

Fig. 8.15 Effect of damping and cross-coupled stiffness on the unbalance response

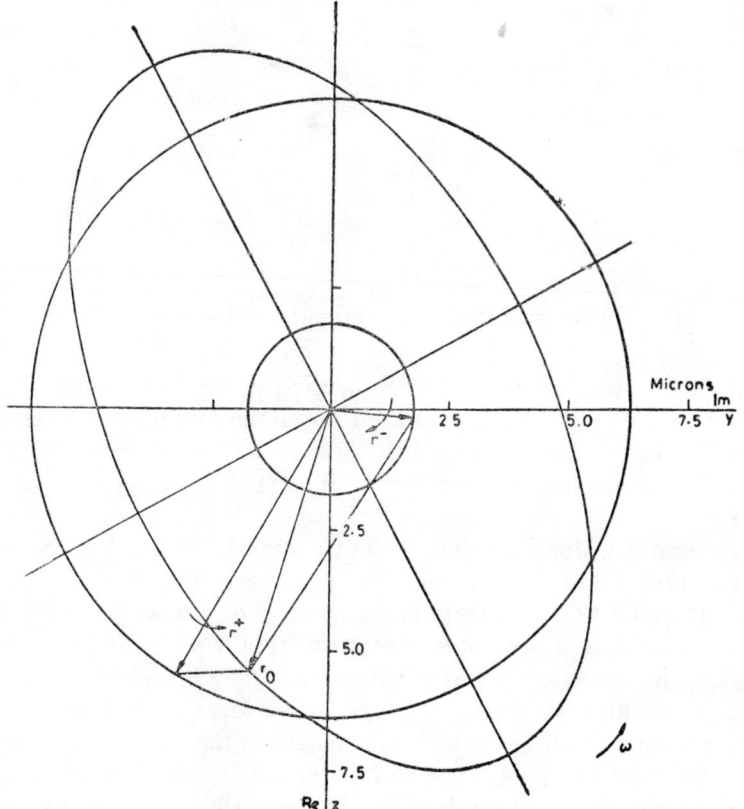

Fig. 8.16 Unbalance response at mid-plane of the rotor at 8500 rpm

Figure 8.14 shows the maximum amplitude of whirl at the mid plane, when only the direct stiffnesss coefficients of the bearing are accounted for. There are two critical speeds at 4820 and 5300 rpm, which are close to the Jeffcott model, considered in Fig. 8.4. The rotor has backward whirl between these two critical speeds.

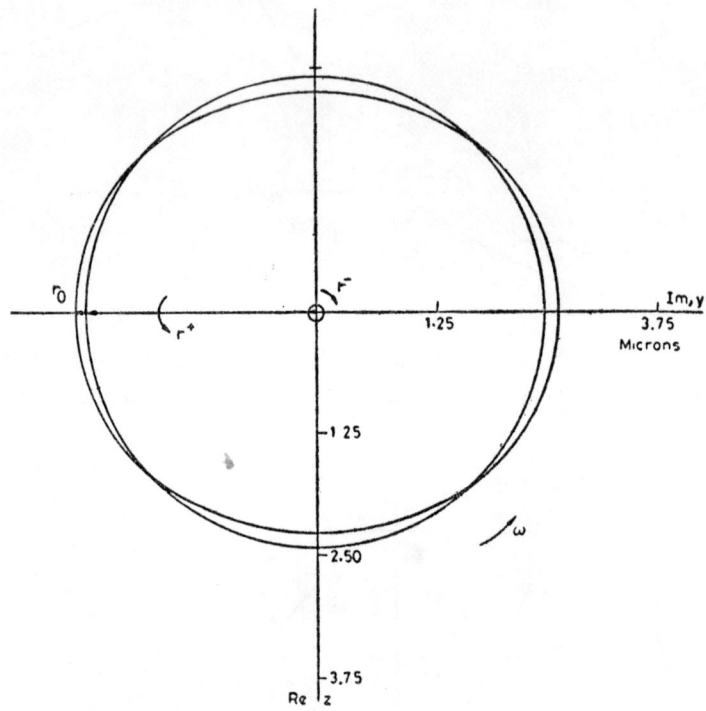

Fig. 8.17 Unbalance response at the bearing of the rotor at 8500 rpm

With different combinations of the bearing stiffness and damping coefficients, the unbalance response of the rotor at the mid plane, is given in Fig. 8.15. The rotor has forward whirl in all the region of operational speed. The results in Fig. 8.7, for the simple model, are similar to those in Fig. 8.15.

Following the procedure outlined in Chapter 5, Fig. 5.6, the response at the mid-plane and left bearing, for the rotor at 8500 rpm is given in Figs. 8.16 and 8.17. From these figures, it may be noted that the inclination of major axis of the elliptical orbit changes from point to point along the length of the rotor. This is due to the damping in the bearings. The reference position of the centre line of the rotor at time $t = 0$, r_0, also changes from point to point along the length of the rotor, as shown in Figs. 8.16 and 8.17. Figure 8.18 shows the whirl orbits at salient points along the length of the rotor and the mode shape of the rotor at 8500 rpm obtained by joining the reference positions r_0.

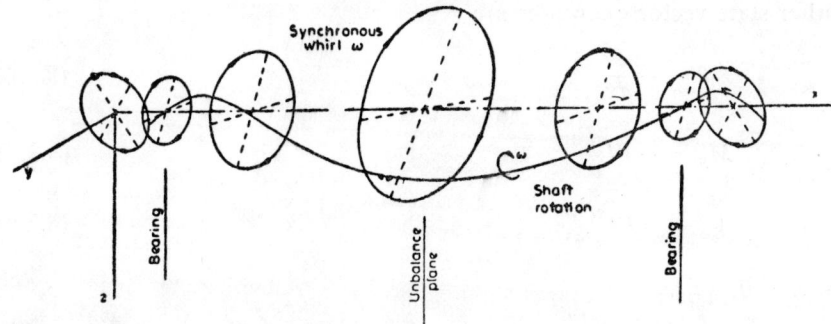

Fig. 8.18 Unbalance response at 8500 rpm

8.3 Transfer Matrix Analysis of Turbine Rotors by Distributed Elements

In the models used in section 4.3, the shaft elements are considered as massless springs and the masses of these elements are lumped at different stations. In practical rotor systems, the shaft diameters can be fairly large and it becomes useful to treat the shaft element with distributed mass and stiffness. Such an element is developed here to make the unbalance response calculations of a turbine rotor.

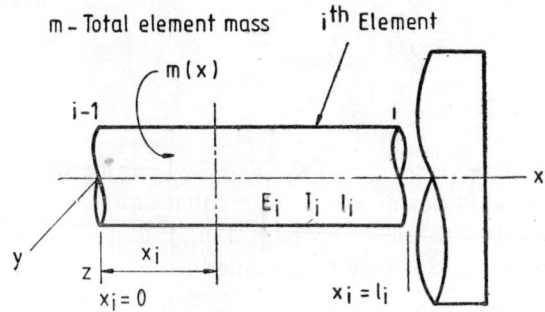

Fig. 8.19 Continuous element

In Fig. 8.19, we have the ith element with distributed mass per unit length $m(x)$ at section x. For free vibrations in x-z plane, the governing differential equation for the element is given by

$$\frac{d^4w}{dx^4} - \frac{p^2 m(x)}{EI} w = 0 \qquad (8.3.1)$$

The solution to the above equation is known to be

$$w(x) = c_1 \cosh \beta x + c_2 \sinh \beta x + c_3 \cos \beta x + c_4 \sin \beta x \qquad (8.3.2)$$

where

$$\beta^2 = \sqrt{\frac{p^2 m}{EII}} \qquad (8.3.3)$$

$m =$ mass of total element

Other state vector quantities are

$$\theta = -\frac{dw}{dx} \tag{8.3.4}$$

$$M_y = -EI\frac{d^2w}{dx^2} \tag{8.3.5}$$

$$V_z = -EI\frac{d^3w}{dx^3} \tag{8.3.6}$$

At $x = 0$, we have

$$w_0 = c_1 + c_3$$
$$\theta_0 = -(c_2 + c_4)\beta$$
$$M_{y0} = (-c_1 + c_3)EI\beta^2$$
$$V_{z0} = (-c_2 + c_4)EI\beta^3$$

Hence

$$c_1 = \tfrac{1}{2}w_0 - \frac{1}{2EI\beta^2} M_{y0}$$

$$c_2 = -\frac{1}{2\beta} \theta_0 - \frac{1}{2EI\beta^3} V_{z0}$$

$$c_3 = \tfrac{1}{2}w_0 + \frac{1}{2EI\beta^2} M_{y0}$$

$$c_4 = -\frac{1}{2\beta} \theta_0 + \frac{1}{2EI\beta^3} V_{z0} \tag{8.3.7}$$

Therefore, the state vector quantities at $x = 1$ can be written in terms of the state vector quantities at $x = 0$ from equations (8.3.2) and (8.3.4) to (8.3.6) to give the ith field matrix, taking into account the distributed mass. Equation (4.3.8) can now be modified as

$$\begin{Bmatrix} -w \\ \theta \\ M_y \\ V_z \end{Bmatrix}_i^L = \begin{bmatrix} \dfrac{ch+c}{2} & \dfrac{sh+s}{2\beta} & \dfrac{ch-c}{2EI\beta^2} & \dfrac{sh-s}{2EI\beta^3} \\ \dfrac{\beta(sh-s)}{2} & \dfrac{ch+c}{2} & \dfrac{sh+s}{2EI\beta} & \dfrac{ch-c}{2EI\beta^2} \\ \dfrac{EI\beta^2(ch-c)}{2} & \dfrac{EI\beta(sh-s)}{2} & \dfrac{ch+c}{2} & \dfrac{sh+s}{2\beta} \\ \dfrac{EI\beta^3(sh+s)}{2} & \dfrac{EI\beta^2(ch-c)}{2} & \dfrac{\beta(sh-s)}{2} & \dfrac{ch+c}{2} \end{bmatrix} \begin{Bmatrix} -w \\ \theta \\ M_y \\ V_z \end{Bmatrix}_{i-1}^R \tag{8.3.8}$$

where

$$ch = \cosh \beta l$$
$$sh = \sinh \beta l$$
$$c = \cos \beta l$$
$$s = \sin \beta l \tag{8.3.9}$$

We can include the gyroscopic effects of any disk at station *i* by using the point matrix in equation (6.1.15) and obtain the following.

$$\{S\}_i^R = [P][F]\{S\}_{i-1}^R$$
$$= [T]\{S\}_{i-1}^R \quad (8.3.10)$$

The transfer matrix $[T]$ in the above is given by

$$[T] = \tfrac{1}{2}\begin{bmatrix} (ch+c) & (sh+s)/\beta & (ch-c)/EI\beta^2 & (sh-s)/EI\beta^3 \\ \beta(sh-s) & (ch+c) & (sh+s)/EI\beta & (ch-c)/EI\beta^2 \\ [EI\beta^2(ch-c)+ & [EI\beta(sh-s)+ & [(ch+c)+ & [(sh+s)/\beta+ \\ I'\beta(sh-s)] & I'(ch+c)] & I'(sh+s)/EI\beta] & I'(ch-c)/EI\beta^2] \\ [EI\beta^3(sh+s)+ & [EI\beta^2(ch-c)+ & [\beta(sh-s)+ & [(ch+c)+ \\ m'(ch+c)] & m'(sh+s)/\beta] & m'(ch-c)/EI\beta^2] & m'(sh-s)/EI\beta^3] \end{bmatrix}$$

(8.3.11)

where

$$I' = (I_p - I_T)p^2$$
$$m' = mp^2 \quad (8.3.12)$$

The bearing matrix in equation (8.2.3) can be used along with the above and the procedure of previous section followed to obtain the unbalance response. Reference may be made to the work of Rao and Bhaskara Sarma [22] and Bhaskara Sarma [23] in this regard.

8.3.1 Modelling Aspects

A typical rotor of a turbine or compressor contains several steps, disks, tapered shafts, hollow shafts and blade rows. To model such a shaft, we define separate mass and rigidity diameters to calculate the effective mass and stiffness distribution of the elements. Typical cases are discussed below.

1. INTERMEDIATE STEP ELEMENT

When the diameter of a shaft cross-section changes abruptly, the rigidity decreases near the region of the step. To account for this decreased shaft stiffness, we define for the step shown in Fig. 8.20, with diameters D_1

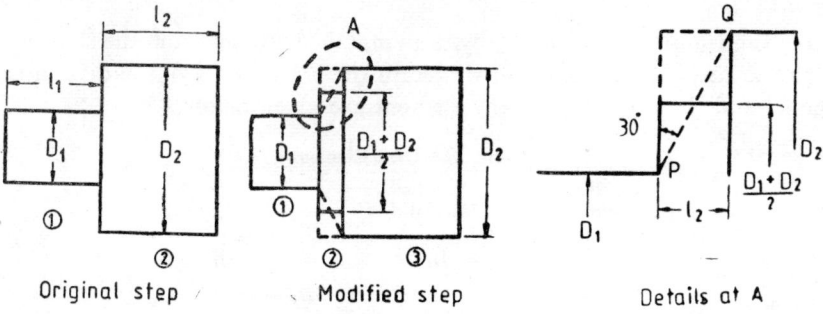

Fig. 8.20 Intermediate step element

and D_2, an additional element of width l_2, in the modified step. A line making 30 deg with the face of the step from point P is drawn to meet the surface of the shaft section 2 at Q as shown in the details at A. This defines the width of the intermediate step element 2. The entire mass of this element is retained by keeping the mass diameter as D_2, but an average value of the diameter is taken for the purpose of calculating the shaft stiffness. The mass and rigidity diameters of the three elements in the modified step are given in Table 8.1 below.

Table 8.1 Intermediate Step Element

Element	Width	Mass dia	Rigidity dia
1	l_1	D_1	D_1
2	$\dfrac{D_2 - D_1}{\sqrt{12}}$	D_2	$\dfrac{D_1 + D_2}{2}$
3	$l_2 - \dfrac{D_2 - D_1}{\sqrt{12}}$	D_2	D_2

2. **DISK ELEMENT**

In the presence of a disk on the shaft, shown in Fig. 8.21, we deter-

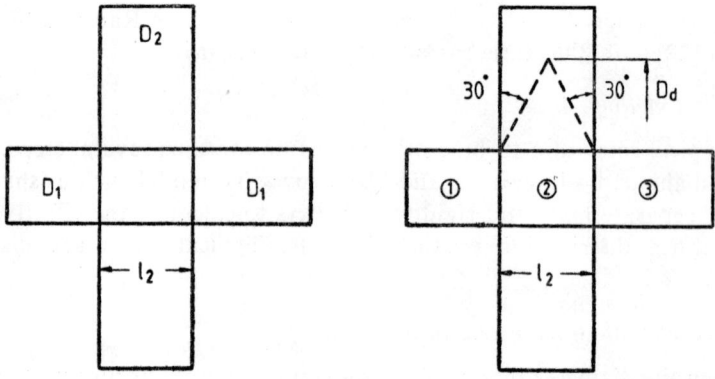

Original disk   Modified disk

Fig. 8.21 Disk element

mine the rigidity diameter D_d by drawing 30 deg lines from the left and right faces of the disk. Again we retain the entire mass of the disk in the model and the properties of the element are given below.

Table 8.2 Disk Element

Element	Mass dia	Rigidity dia
1	D_1	D_1
2	D_2	$D_d = D_1 + \sqrt{3}\, l_2$
3	D_1	D_1

3. TAPERED SHAFT

A tapered shaft is divided into a convenient number of elements, e.g., three in Fig. 8.22. In this case, the rigidity diameter is taken to be the

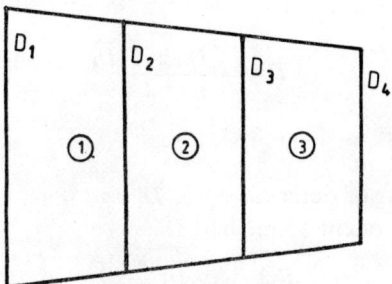

Fig. 8.22 Tapered shaft

minimum diameter of the element and the entire mass is retained. The corresponding properties of the element are given in Table 8.3.

Table 8.3 Tapered Shaft

Element	Mass dia	Rigidity dia
1	$(D_1 + D_2)\frac{1}{2}$	D_2
2	$(D_2 + D_3)\frac{1}{2}$	D_3
3	$(D_3 + D_4)\frac{1}{2}$	D_4

4. BLADE RING

A blade ring of width l on a shaft is shown in Fig. 8.23. Let the total

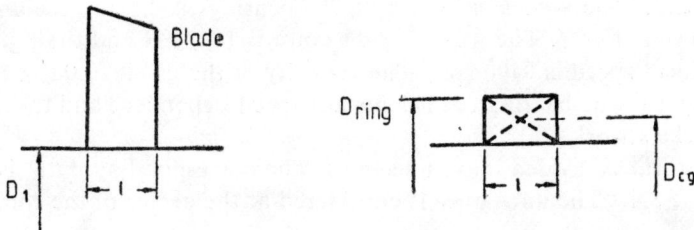

Fig. 8.23 Blade ring section

mass of the blades be m_b and shaft mass of width l be m_s. The blades are replaced by an equivalent shaft ring of dimater D_{ring}. The entire mass is retained as before, giving the mass dia as

$$D_{\text{ring}} = \sqrt{\frac{4(m_b + m_s)}{\rho l \pi}}$$

where ρ is the mass density of the shaft material. The rigidity diameter is taken as

$$D_{cg} = \frac{D_{\text{ring}} + D_1}{2}$$

5. HOLLOW SHAFT

For a hollow shaft of outer diameter D_1 and inner diameter D_2, the mass diameter of an equivalent solid shaft is

$$D_{hm} = \sqrt{D_1^2 - D_2^2}$$

The stiffness of an equivalent solid shaft is calculated by the rigidity diameter, which is

$$D_{hr} = (D_1^4 - D_2^4)^{1/4}$$

Example 8.7

Consider a 6 MW turbine rotor shown in Fig. 8.24. The rotor is mounted on four lobe bearings of 80 mm dia at the left end and 100 mm dia at the

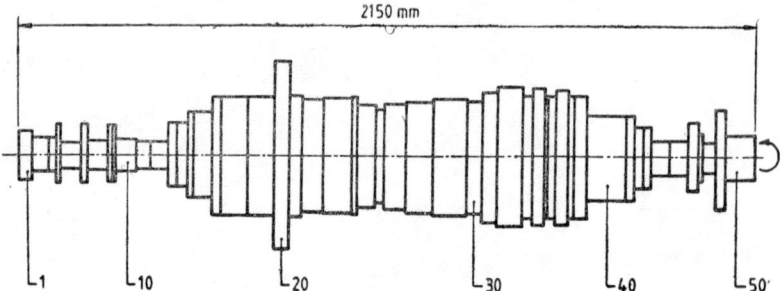

Fig. 8.24 6 MW Turbine rotor model

right end. The aspect ratio of both the bearings is 0.5 and the diametral clearance is 1.74%. The weight of the rotor is 10234 N and the rigid bearing critical speed is 8400 rpm. The viscosity of the oil is 0.02 centipoise. The eight linear bearing coefficients are speed dependent and taken from Glienicke's work [24].

The rotor is divided into 51 sections. The corresponding data is given in Table 8.4. The unbalance is considered at the center of the rotor only. Fig. 8.25 shows the unbalance response obtained for the rotor. When the bearing damping is included, the response is well damped out and the critical speed is at around 7500 rpm, which is lower than the rigid bearing critical speed 8400 rpm. With the stiffness coefficients only, the rotor exhibits the two critical speeds which are 2700 rpm and 6700 rpm.

Table 8.4 Sectional Data of 6 MW Rotor

Section No.	Length (mm)	Rigidity diameter (mm)	Mass diameter (mm)	Temperature °C
1	35.0	65.0	146.0	60
2	51.0	65.0	100.0	64
3	17.0	65.0	100.0	68
4	15.0	65.0	173.0	72
5	60.0	70.0	70.0	75
6	23.0	80.0	80.0	79
7	55.0	85.0	85.0	85
8	16.0	83.0	170.0	90
9	7.0	100.0	170.0	90
10	62.0	90.0	90.0	180
11	41.0	80.0	80.0	265
12	55.0	80.0	80.0	350
13	29.0	130.0	180.0	400
14	29.0	180.0	180.0	440
15	19.5	214.0	247.0	440
16	51.0	247.0	247.0	440
17	27.5	297.0	344.0	440
18	74.0	342.0	344.0	440
19	79.5	342.0	344.0	440
20	50.0	380.0	350.0	440
21	36.0	340.0	331.0	440
22	63.0	331.0	332.0	440
23	86.5	331.0	333.0	430
24	12.5	314.0	295.0	420
25	56.0	292.0	279.0	416
26	16.5	271.0	307.0	415
27	72.0	258.0	320.0	415
28	79.5	267.0	336.0	360
29	100.0	278.0	336.0	360
30	46.0	293.0	331.0	348
31	47.0	360.0	382.0	336.0
32	73.0	303.0	403.0	324.0
33	29.0	304.0	360.0	312.0
34	32.5	275.0	398.0	300.0
35	11.5	294.0	360.0	288.0
36	20.0	279.0	360.0	276.0
37	42.5	244.0	391.0	265.0
38	15.5	271.0	360.0	249.0
39	33.5	302.0	360.0	210.0
40	125.0	246.0	248.0	170.0

(contd.)

Table 8.4 (Contd.)

Section No.	Length (mm)	Rigidity diameter (mm)	Mass diameter (mm)	Temperature °C
41	18.5	211.0	248.0	130.0
42	28.0	180.0	180.0	90.0
43	26.0	140.0	180.0	86.0
44	55.0	100.0	100.0	82.0
45	54.0	100.0	100.0	78.0
46	30.0	95.0	209.0	74.0
47	12.0	95.0	140.0	70.0
48	40.0	90.0	90.0	65.0
49	32.0	90.0	29.0	60.0
50	82.0	88.0	135.0	60.0

Bearing details

	Length (mm)	Diameter (mm)	Location section	Type
Front bearing	40.0	80.0	11	Four lobe
Rear bearing	50.0	100.0	44	Four lobe

K_d - Direct stiffness
K_c - Coupled stiffness
C_d - Direct damping
C_c - Coupled damping

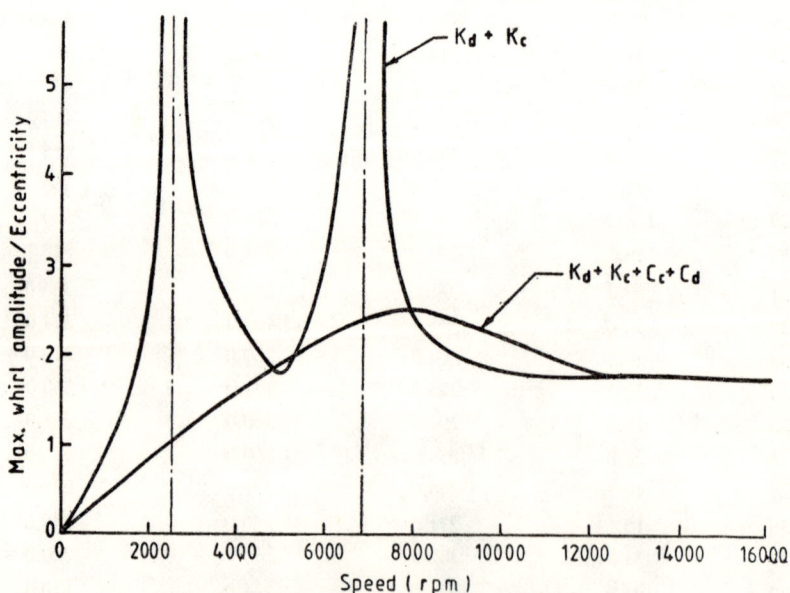

Fig. 8.25 Unbalance response of 6 MW rotor

Rotors Mounted on Fluid Film Bearings 139

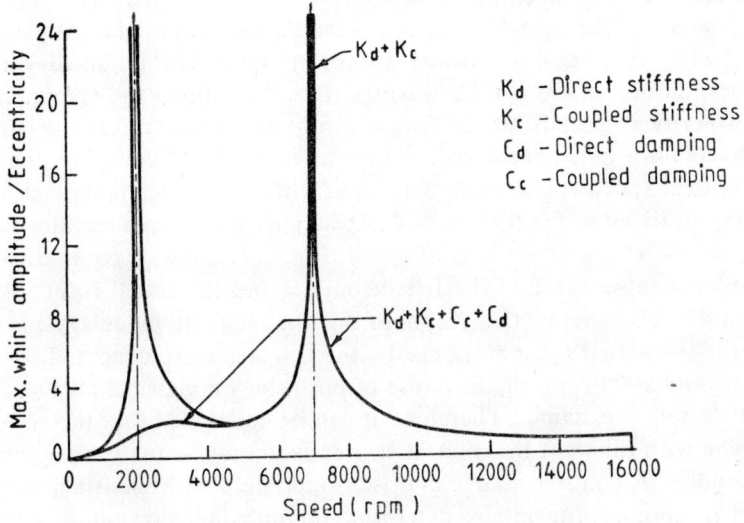

Fig. 8.26 Unbalance response of 6 MW rotor (Jeffcott Model)

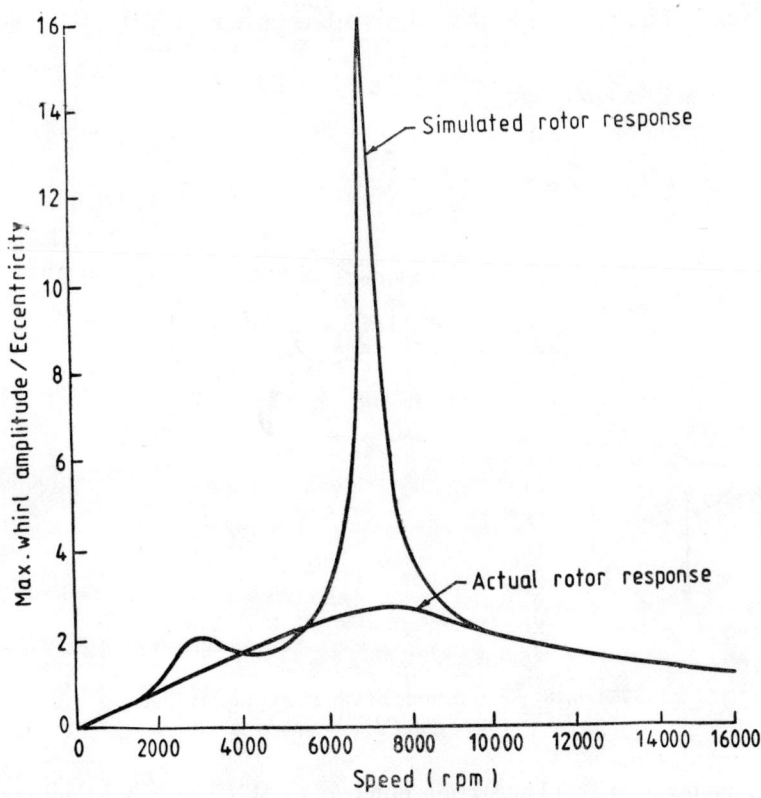

Fig. 8.27 Comparison of unbalance response of simulated rotor and actual rotor

To consider how accurate a simulated Jeffcott model will represent the response of the actual rotor, Bhaskara Sarma [23] used a single mass model with rigid bearing critical speed of 8400 rpm mounted on two identical 80 mm dia four lobe bearings. Fig. 8.26 shows the response with the two critical speeds at 2000 rpm and 7000 rpm, when the stiffness coefficients are only considered. With the inclusion of damping, the 2000 rpm critical speed is shifted to 3000 rpm with a significant reduction in its amplitude. However, the second critical speed 7000 rpm remained same exibiting itself as a major critical speed. Fig. 8.27 shows a comparison of the unbalance response of the Jeffcott model and the actual rotor. At the second critical speed, the response of the Jeffcott model, r_u/a is 16.0, whereas the actual rotor response by the transfer matrix method is only 2.5. Beyond 9000 rpm, the response of both the simulated model and the actual rotor are same. Therefore it can be concluded that the rotors of this type with different bearings at two ends cannot be analyzed by simulation studies. A transfer matrix multistation system with continuous mass model is more appropriate to determine the unbalance response of rotors with large diameters and different bearings at both the ends.

The rotor is now excited by two unbalances of 0.0461 N mm, to simulate the first mode of vibration. The response obtained at 4000 rpm is given in Fig. 8.28 with the whirl orbits given at only few selected

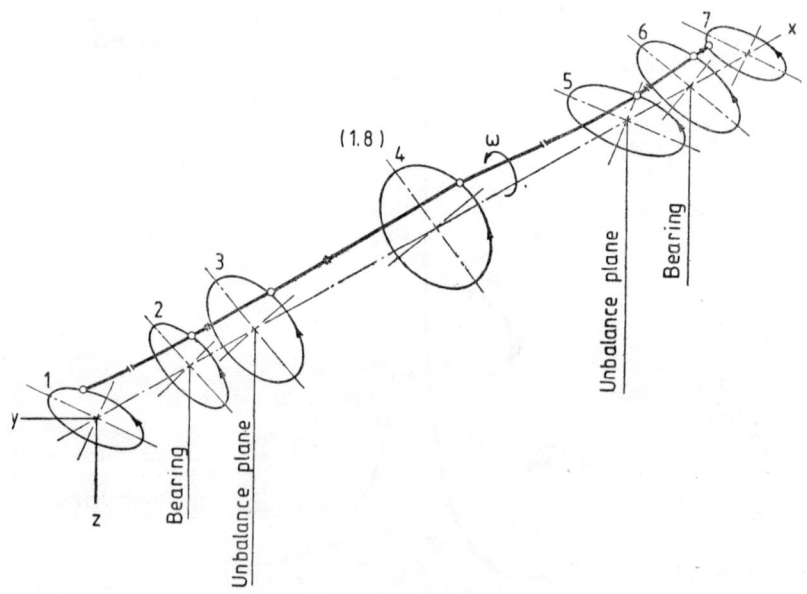

Fig. 8.28 Spatial mode of vibration of 6 MW rotor at 4000 rpm (Not to scale)

stations for clarity. (The station numbers on this diagram do not correspond to the rotor model with 51 stations.) Since the rotor is running below the first critical speed, the deflected form also represents the translatory whirl.

Fig. 8.29 shows the response in the vicinity of the critical speed 8350 rpm. This figure indicates large amplitudes of whirl, r_u/a at the mid plane = 2.5, as compared to 1.8 at 4000 rpm and also considerable phase difference between the left and right overhangs compared with the main shaft between the bearings. The conical whirl is very apparent at 10500 rpm in Fig. 8.30, which is above the first critical speed.

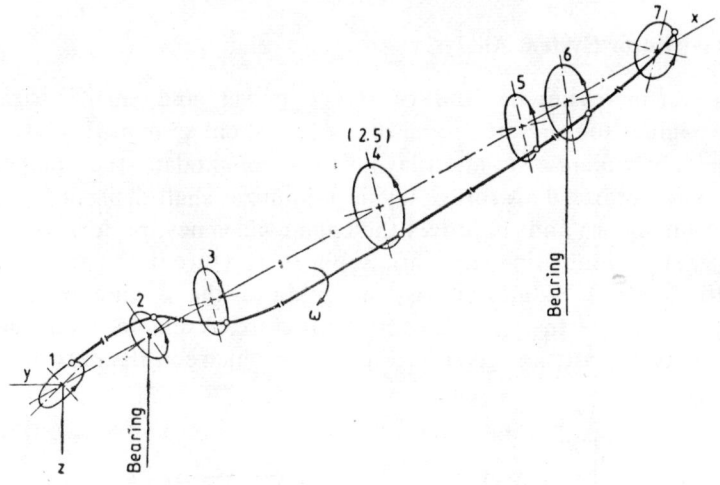

Fig. 8.29 Spatial mode of vibration of 6 MW rotor at 8350 rpm

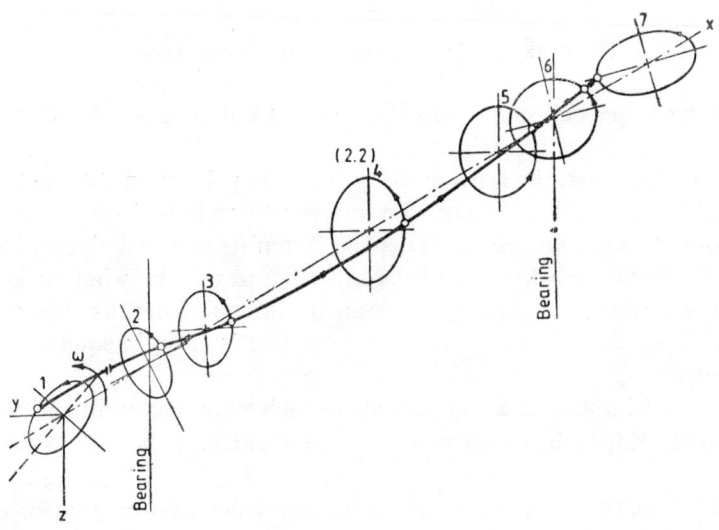

Fig. 8.30 Spatial mode of vibration of 6 MW rotor at 10500 rpm

For symmetrical rotors, we can use a Jeffcott model to predict the general performance of the rotor; however, a transfer matrix analysis will be necessary to obtain the whirling mode shape of the rotor. For rotors which do not have symmetry, particularly overhung rotors, Jeffcott model may not give accurate results and transfer matrix method may be preferred.

Additional references on this subject dealing with matrix methods [11-14], modal analysis [15, 16] and finite element analysis [17-20], are given at the end of the chapter.

8.4 Dual Rotor System Analysis

With ever increasing demand of larger power and smaller size gas turbine engines for aircraft propulsion, a two spool system with intershaft bearings is becoming a standard layout to accommodate the compressor and turbine rotors. This sort of layout minimizes shaft deflections caused by rotor unbalance and improves the engine efficiency, performance and reliability by eliminating the static support structure in the aerodynamic flow path. It also facilitates an easy mounting of the engine casing and gives rise to compactness in the overall structure. A simple mathematical model for such a system is given in Fig. 8.31a, which consists of two coaxial

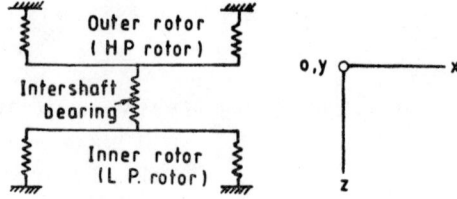

Fig. 8.31a Theoretical model of dual-rotor system

rotors, the outer and inner rotors running at different speeds, interlinked through an intershaft bearing.

In such a system, the cross-exciting vibration between the inner and outer shafts is effected through the intershaft bearing. Thus, the dynamic response of each rotor, not only depends upon its own unbalance, but also upon the dynamic behaviour of the other. And this is what makes the dynamic analysis of dual rotor system different from that of a straight rotor. Towards this end, we modify the state vector in equation (5.1.6) by considering:

1. The sine and cosine components of each state vector quantity, viz., deflection, slope, bending moment and shear force in both x-z and x-y planes.

2. Each quantity of the state vector to have two components, one corresponding to the inner rotor speed and the other to the outer rotor speed.

8.4.1 State Vector

Let the suffixes m and n represent the effects of inner and outer rotors respectively, i.e., ω_m-inner rotor speed, ω_n-outer rotor speed etc. Then, for x-z plane we have

$$w = w_{cm} \cos \omega_m t + w_{sm} \sin \omega_m t + w_{cn} \cos \omega_n t + w_{sn} \sin \omega_n t$$
$$\theta = \theta_{cm} \cos \omega_m t + \theta_{sm} \sin \omega_m t + \theta_{cn} \cos \omega_n t + \theta_{sn} \sin \omega_n t$$
$$M_y = M_{ycm} \cos \omega_m t + M_{ysm} \sin \omega_m t + M_{ycn} \cos \omega_n t + M_{ysn} \sin \omega_n t$$
$$V_z = V_{zcm} \cos \omega_m t + V_{zsm} \sin \omega_m t + V_{zcn} \cos \omega_n t + V_{zsn} \sin \omega_n t$$
(8.4.1)

Similarly for x-y plane we have

$$v = v_{cm} \cos \omega_m t + v_{sm} \sin \omega_m t + v_{cn} \cos \omega_n t + v_{sn} \sin \omega_n t$$
$$\phi = \phi_{cm} \cos \omega_m t + \phi_{sm} \sin \omega_m t + \phi_{cn} \cos \omega_n t + \phi_{sn} \sin \omega_n t$$
$$M_z = M_{zcm} \cos \omega_m t + M_{zsm} \sin \omega_m t + M_{zcn} \cos \omega_n t + M_{zsn} \sin \omega_n t$$
$$V_y = V_{ycm} \cos \omega_m t + V_{ysm} \sin \omega_m t + V_{ycn} \cos \omega_n t + V_{ysn} \sin \omega_n t$$
(8.4.2)

The state vector is modified to contain 33 quantities given in the above two equations as given below.

$$\{S\} = \begin{Bmatrix} \{S_{zm}\} \\ \{S_{zn}\} \\ \{S_{ym}\} \\ \{S_{yn}\} \\ 1 \end{Bmatrix} \quad (8.4.3)$$

where

$$\{S_{zm}\} = \begin{Bmatrix} -w_{cm} \\ \theta_{cm} \\ M_{ycm} \\ V_{zcm} \\ -w_{sm} \\ \theta_{sm} \\ M_{ysm} \\ V_{zsm} \end{Bmatrix} \quad \{S_{zn}\} = \begin{Bmatrix} -w_{cn} \\ \theta_{cn} \\ M_{ycn} \\ V_{zcn} \\ -w_{sn} \\ \theta_{sn} \\ M_{ysn} \\ V_{zsn} \end{Bmatrix}$$

$$\{S_{ym}\} = \begin{Bmatrix} v_{cm} \\ \phi_{cm} \\ M_{zcm} \\ -V_{ycm} \\ v_{sm} \\ \phi_{sm} \\ M_{zsm} \\ -V_{ysm} \end{Bmatrix} \quad \{S_{yn}\} = \begin{Bmatrix} v_{cn} \\ \phi_{cn} \\ M_{zcn} \\ -V_{ycn} \\ v_{sn} \\ \phi_{sn} \\ M_{zsn} \\ -V_{ysn} \end{Bmatrix} \quad (8.4.4)$$

8.4.2 Field Matrix

It is easy to show that the corresponding field matrix is given by

$$[\mathbf{F}]_{33\times 33} = \begin{bmatrix} [F] & 0 & 0 & 0 & 0 & 0 & 0 & 0 & 0 \\ 0 & [F] & 0 & 0 & 0 & 0 & 0 & 0 & 0 \\ 0 & 0 & [F] & 0 & 0 & 0 & 0 & 0 & 0 \\ 0 & 0 & 0 & [F] & 0 & 0 & 0 & 0 & 0 \\ 0 & 0 & 0 & 0 & [F] & 0 & 0 & 0 & 0 \\ 0 & 0 & 0 & 0 & 0 & [F] & 0 & 0 & 0 \\ 0 & 0 & 0 & 0 & 0 & 0 & [F] & 0 & 0 \\ 0 & 0 & 0 & 0 & 0 & 0 & 0 & [F] & 0 \\ 0 & 0 & 0 & 0 & 0 & 0 & 0 & 0 & 1 \end{bmatrix} \qquad (8.4.5)$$

where $[F]$ can be taken either from equation (4.3.8) or (8.3.8), depending on the need whether the shaft can be considered massless or whether one should account for the distributed mass in the elemental length.

8.4.3 Point matrix

It can be easily shown that the modified point matrix for the masses in the dual rotor system is given by

$$[\mathbf{P}]_{33\times 33} = \begin{bmatrix} [P_m] & 0 & 0 & 0 & 0 \\ 0 & [P_n] & 0 & 0 & 0 \\ 0 & 0 & [P_m] & 0 & 0 \\ 0 & 0 & 0 & [P_n] & 0 \\ 0 & 0 & 0 & 0 & 1 \end{bmatrix} \qquad (8.4.6)$$

where

$$[P_m]_{8\times 8} = \begin{bmatrix} [P] & 0 \\ 0 & [P] \end{bmatrix}_{p=\omega_m} \qquad (8.4.7)$$

$$[P_n]_{8\times 8} = \begin{bmatrix} [P] & 0 \\ 0 & [P] \end{bmatrix}_{p=\omega_n} \qquad (8.4.8)$$

and $[P]$ is given by equation (4.3.11). If the gyroscopic couple is to be taken into account, $[P]$ in the above equation can be taken from equation (6.1.15).

8.4.4 Bearing matrix

In a similar manner, it can be shown that

$$[\mathbf{B}]_{33\times 33} = \begin{bmatrix} [B_{mz1}] & 0 & [B_{mz2}] & 0 & 0 \\ 0 & [B_{nz1}] & 0 & [B_{nz2}] & 0 \\ [B_{my2}] & 0 & [B_{my1}] & 0 & 0 \\ 0 & [B_{ny2}] & 0 & [B_{ny1}] & 0 \\ 0 & 0 & 0 & 0 & 1 \end{bmatrix} \qquad (8.4.9)$$

Rotors Mounted on Fluid Film Bearings   145

In the above

$[B_{mz1}](i,i) = 1, i = 1, 2, \ldots 8.$

$[B_{mz1}](4,1)$ and $(8,5) = -K_{zz} + m\omega_m^2$

$[B_{mz1}](4,5) = -C_{zz}\omega_m$ (8.4.10)

$[B_{mz1}](8,1) = C_{zz}\omega_m$

$[B_{mz1}]$ Rest $= 0$

$[B_{mz2}](4,1)$ and $(8,5) = K_{zy}$

$[B_{mz2}](4,5) = C_{zy}\omega_m$

$[B_{mz2}](8,1) = -C_{zy}\omega_m$ (8.4.11)

$[B_{mz2}]$ Rest $= 0$

Matrices $[B_{nz1}]$ and $[B_{nz2}]$ are obtained from the above by replacing ω_m with ω_n. Matrices $[B_{my1}]$ and $[B_{my2}]$ are obtained from the above by replacing subscripts z with y and y with z. Similarly by replacing ω_m with ω_n, we can obtain $[B_{ny1}]$ and $[B_{ny2}]$.

8.4.5 Point matrix for Unbalance Mass

Figures 5.4a and 5.4b should now be read with $\omega = \omega_m$ or $\omega = \omega_n$ depending on the ith station belonging to the inner or outer rotors, Hence equations (5.2.1) and (5.2.2) also get modified with $\omega = \omega_m$ or ω_n as the case may be. With the help of equations (8.4.1) and (8.4.2) we can write for the inner rotor, the following equations.

$$\begin{aligned}
V_{zcm}^R &= V_{zcm}^L - m\omega_m^2 w_{cm} - \omega_m^2 u_z \\
V_{zsm}^R &= V_{zsm}^L - m\omega_m^2 w_{sm} + \omega_m^2 u_y \\
V_{zcn}^R &= V_{zcn}^L - m\omega_n^2 w_{cn} \\
V_{zsn}^R &= V_{zsn}^L - m\omega_n^2 w_{sn} \\
V_{ycm}^R &= V_{ycm}^L - m\omega_m^2 v_{cm} - \omega_m^2 u_y \\
V_{ysm}^R &= V_{ysm}^L - m\omega_m^2 v_{sm} - \omega_m^2 u_z \\
V_{ycn}^R &= V_{ycn}^L - m\omega_n^2 v_{cn} \\
V_{ysn}^R &= V_{ysn}^L - m\omega_n^2 v_{sn}
\end{aligned} \quad (8.4.12)$$

Hence the point matrix for the ith unbalance mass on inner rotor can be written as

$$\begin{Bmatrix} \{S_{zm}\} \\ \{S_{zn}\} \\ \{S_{ym}\} \\ \{S_{yn}\} \\ 1 \end{Bmatrix}_i^R = \begin{bmatrix} [P_m] & 0 & 0 & 0 & \{m_{zm}\} \\ 0 & [P_n] & 0 & 0 & 0 \\ 0 & 0 & [P_m] & 0 & \{m_{ym}\} \\ 0 & 0 & 0 & [P_n] & 0 \\ 0 & 0 & 0 & 0 & 1 \end{bmatrix} \begin{Bmatrix} \{S_{zm}\} \\ \{S_{zn}\} \\ \{S_{ym}\} \\ \{S_{yn}\} \\ 1 \end{Bmatrix}_i^L \quad (8.4.13)$$

146 Rotor Dynamics

where

$$\{m_{zm}\} = \begin{Bmatrix} 0 \\ 0 \\ 0 \\ -u_z\omega_m^2 \\ 0 \\ 0 \\ 0 \\ u_y\omega_m^2 \end{Bmatrix} \quad \{m_{ym}\} = \begin{Bmatrix} 0 \\ 0 \\ 0 \\ u_y\omega_m^2 \\ 0 \\ 0 \\ 0 \\ u_z\omega_m^2 \end{Bmatrix} \quad (8.4.14)$$

Notice that we added an additional 33rd column and rotor to include the unbalance terms.

The point matrix for the ith unbalance mass on the outer rotor can be obtained in a similar way by writing equations (8.4.12) for the outer rotor which gives

$$\{S\}_i^R = \begin{bmatrix} [P_m] & 0 & 0 & 0 & 0 \\ 0 & [P_n] & 0 & 0 & \{m_{zn}\} \\ 0 & 0 & [P_m] & 0 & 0 \\ 0 & 0 & 0 & [P_n] & \{m_{yn}\} \\ 0 & 0 & 0 & 0 & 1 \end{bmatrix} \{S\}_i^L \quad (8.4.15)$$

where

$$\{m_{zn}\} = \begin{Bmatrix} 0 \\ 0 \\ 0 \\ -u_z\omega_n^2 \\ 0 \\ 0 \\ 0 \\ u_y\omega_n^2 \end{Bmatrix} \quad \{m_{yn}\} = \begin{Bmatrix} 0 \\ 0 \\ 0 \\ u_y\omega_n^2 \\ 0 \\ 0 \\ 0 \\ u_z\omega_n^2 \end{Bmatrix} \quad (8.4.16)$$

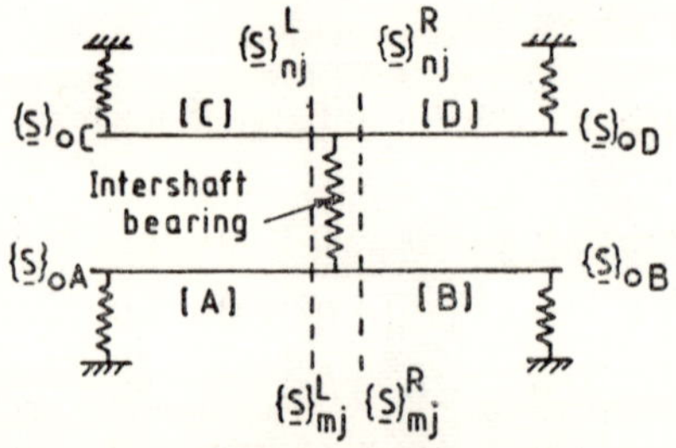

Fig. 8.31b Mathematical model of dual-rotor system

8.4.6 Junction conditions and overall system equations

In Fig. 8.31b, the junction conditions at the intershaft bearing are considered. Let $[A]$, $[B]$, $[C]$ and $[D]$ represent the overall transfer matrices considering all the stations between the starting station, zero, for the A, B, C and D shaft sections and the junction point where the intershaft bearing is located. Then

$$\begin{aligned} \{S\}_{mj}^L &= [A]\{S\}_{OA} \\ \{S\}_{OB} &= [B]\{S\}_{mj}^R \\ \{S\}_{nj}^L &= [C]\{S\}_{OC} \\ \{S\}_{OD} &= [D]\{S\}_{nj}^R \end{aligned} \quad (8.4.17)$$

At the junction of the inner rotor, we have

$$\begin{aligned} \{S\}_{mj}^R &= [I]\{S\}_{mj}^L + [\mathbf{B}]\{S\}_{mj}^L - [\mathbf{B}]\{S\}_{nj}^L \\ &= [[I] + [\mathbf{B}]]\{S\}_{mj}^L - [\mathbf{B}]\{S\}_{nj}^L \end{aligned} \quad (8.4.18)$$

Similarly at the junction of the outer rotor, we have

$$\{S\}_{nj}^R = [[I] + [\mathbf{B}]]\{S\}_{nj}^L - [\mathbf{B}]\{S\}_{mj}^L \quad (8.4.19)$$

Substituting $\{S\}_{mj}^R$ and $\{S\}_{nj}^R$ from the above in the second and fourth equations of (8.4.17), we get

$$\begin{aligned} \{S\}_{OB} &= [B][[I] + [\mathbf{B}]]\{S\}_{mj}^L - [B][\mathbf{B}]\{S\}_{nj}^L \\ \{S\}_{OD} &= [D][[I] + [\mathbf{B}]]\{S\}_{nj}^L - [D][\mathbf{B}]\{S\}_{mj}^L \end{aligned} \quad (8.4.20)$$

With the help of first and third equations of (8.4.17), the above can be written as

$$\begin{aligned} \{S\}_{OB} &= [B][[I] + [\mathbf{B}]][A]\{S\}_{OA} - [B][\mathbf{B}][C]\{S\}_{OC} \\ \{S\}_{OD} &= [D][[I] + [\mathbf{B}]][C]\{S\}_{OC} - [D][\mathbf{B}][A]\{S\}_{OA} \end{aligned} \quad (8.4.21)$$

Premultiplying the first equation by $[B]^{-1}$ and the second by $[D]^{-1}$, the above equations can be written in one matrix equation as follows.

$$\begin{bmatrix} [A_m] & [B_m] & [C_m] & [0] \\ [A_n] & [0] & [C_n] & [D_n] \end{bmatrix}_{6 \times 132} \begin{Bmatrix} \{S\}_{OA} \\ \{S\}_{OB} \\ \{S\}_{OC} \\ \{S\}_{OD} \end{Bmatrix}_{132 \times 1} = 0 \quad (8.4.22)$$

where

$$\begin{aligned} {[A_m]} &= [[I] + [\mathbf{B}]][A] \\ {[C_n]} &= [[I] + [\mathbf{B}]][C] \end{aligned} \quad (8.4.23)$$

$$\begin{aligned} {[B_m]} &= -[B]^{-1} \\ {[D_n]} &= -[D]^{-1} \end{aligned} \quad (8.4.24)$$

$$[C_m] = -[\mathbf{B}][C]$$
$$[A_n] = -[\mathbf{B}][A] \tag{8.4.25}$$

If the intershaft bearing matrix is zero, it can be seen that equation (8.4.22) reduces to two uncoupled equations for the outer and inner rotors.

Consider free ends of all shaft sections A, B, C and D for the initial state quantities, then all M_y, V_z, M_z, V_y in $\{S\}_0$ vectors are zero. The unknown quantities are w, θ, v and ϕ at the four starting stations. For each starting station, we have 16 unknowns, $\{w_{cm}, w_{cn}, w_{sm}, w_{sn}, \theta_{cm}, \theta_{cn}, \theta_{sm}, \theta_{sn}$ etc., $\phi_{sn}\}_{0A}$. Let

$$\{S_{OA}\} = \begin{Bmatrix} -w_{cm} \\ \theta_{cm} \\ -w_{sm} \\ \theta_{sm} \\ -w_{cn} \\ \theta_{cn} \\ -w_{sn} \\ \theta_{sn} \\ v_{cm} \\ \phi_{cm} \\ v_{sm} \\ \phi_{sm} \\ v_{cn} \\ \phi_{cn} \\ v_{sn} \\ \phi_{sn} \end{Bmatrix}_{OA} \tag{8.4.26}$$

and similarly $\{S_{OB}\}$, $\{S_{OC}\}$, $\{S_{OD}\}$ be the unknown quantities at the starting stations of all the four sections. Since M_y, V_z etc., terms in the vector of equation (8.4.22) are zero, we can drop 3, 4, 7, 8, 11, 12 etc., 31, 32; 36, 37 etc., 64, 65; 69, 70 etc., 97, 98; 102, 103 etc., 130, 131 columns in the 66×132 matrix of equation (8.4.22). We also note that the 33rd, 66th, 99th and 132nd elements of the vector are unity and hence when multiplied with the corresponding column quantity in the matrix, give rise to a constant quantity independent of any state vector dynamic quantity. These four constant terms can be added and taken to the right hand side of the corresponding equation. We can also delete the 33rd and 66th rows of the equation (8.4.22) as they simply give identity equations. Then the equations (8.4.22) reduce to

$$[\bar{A}]_{64 \times 64} \{S_o\}_{64 \times 1} = \{\bar{B}\}_{64 \times 1} \tag{8.4.27}$$

where

$$\{S_O\} = \begin{Bmatrix} \{S_{OA}\} \\ \{S_{OB}\} \\ \{S_{OC}\} \\ \{S_{OD}\} \end{Bmatrix} \qquad (8.4.28)$$

$$\bar{B}_i = A_{mi,33} + B_{mi,33} + C_{mi,33} + D_{mi,33} \qquad (8.4.29)$$

and $\bar{A}$ are the elements picked up from the matrix (8.4.22) as described in the above manner, by dropping columns 3, 4, 7, 8 etc., and rows 33 and 66.

The linear system of equations (8.4.27) can be solved to obtain $\{S_O\}$ in equation (8.4.32). From there, the state vector quantities at any station can be determined using the appropriate transfer matrices. A computer program based on the above has been developed in reference [25].

Example 8.8

Consider the simple dual rotor system mounted on isotropic bearings as shown in Fig. 8.32. The shaft diameter is 1.5 cm and the Young's modulus is taken as 210×10^9 N/sq m. The unbalance on each mass is in phase with the other and is equal to 0.001 kg m.

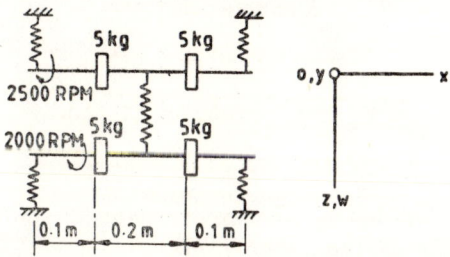

Fig. 8.32 Test model of dual-rotor system

Case 1:

Let all the four supports be rigid and that the intershaft bearing be of zero stiffness. In other words, we have two separate rotors and the computer program should give corresponding results. To simulate the rigid bearings, the stiffness is considered large, i.e., K support/K shaft = 10^6, so that the program boundary conditions need not be changed.

For the two degenerated rotors, closed from solutions can be obtained by influence coefficients, see reference [26], as

$$w_1 = \frac{[\alpha_{11} + m_2\omega^2(\alpha_{12}^2 - \alpha_{11}\alpha_{22})]P_1 + \alpha_{12}P_2}{(1 - m_1\alpha_{11}\omega^2)(1 - m_2\alpha_{22}\omega^2) - m_1m_2\alpha_{12}^2\omega^4} \qquad (8.4.30)$$

Gupta et al [25] found the values from the computer program to match with

150 *Rotor Dynamics*

the above simple solution at two speeds viz., 2000 rpm and 2500 rpm. The responses at these two speeds respectively are 0.254 mm and 0.135 mm.

Case 2:

All supports including the intershaft bearing are considered rigid, which should simulate a rigid connection between the two shafts. The inner and outer rotor speeds are 2000 rpm and 2500 rpm respectively. The results at the junction are

$$w_{cmj} = w_{cnj} = 0$$

$$w_{smj} = w_{snj} = w_{sm} \sin \omega_m t + w_{sn} \sin \omega_n t$$

$$= -0.1883 \times 10^{-3} \sin 209.44t - 0.1557 \times 10^{-2} \sin 261.88t$$

It can be seen that the results are identical for both the inner and outer rotors at the junction.

Case 3:

A general case of support and intershaft bearing stiffnesses is considered with the following data.

$$K \text{ support} = 1 \text{ MN/m}$$

$$K \text{ intershaft} = 1 \text{ MN/m}$$

$$\text{Inner rotor speed} = 1200 \text{ rpm}$$

$$\text{Outer rotor speed} = 1000 \text{ rpm}$$

In this case the response of either rotor will be affected by the unbalance of its own as well as the other rotor. Therefore, we expect the total response at any station to be the sum of two harmonic components with 1200 rpm and 1000 rpm frequencies. This situation is analogous to the case of superposition of two different harmonic signals, giving rise to a periodic beating phenomenon. In the case of a straight rotor response, as we discussed before in section 5.3, the superposition of two perpendicular harmonic motions with same angular frequency gives rise to an elliptical motion. However, for a dual rotor, the unbalance response will have two different response components belonging to the inner and outer rotors running at different speeds. The motion will be periodic and if the speeds of both the rotors are close to each other, beating phenomenon should be observed. The period of such a beating phenomenon is

$$T = \frac{2\pi}{\dfrac{\omega_n - \omega_m}{p - q}} \tag{8.4.31}$$

where p and q are smallest integers such that

$$\frac{p}{q} = \frac{\omega_n}{\omega_m} \tag{8.4.32}$$

The superposition of the response components w and v will amount to the combination of two distinct orbital motions of different angular frequencies. Hence the resulting motion will not describe any single closed orbit; but instead will exhibit periodicity whose period is given by equation (8.4.31). The number of loops in such a motion for one period is given by

$$n = \{\text{Max } [\omega_n, \omega_m]\} \frac{p-q}{\omega_n - \omega_m} \tag{8.4.33}$$

For the present case

$$\frac{p}{q} = \frac{1000}{1200} = \frac{5}{6}$$

Hence

$$T = \frac{2\pi \times 60}{2\pi \times 200} = 0.3 \text{ sec}$$

$$n = \frac{1200}{200} = 6$$

Figs. 8.33a and 8.33b show the response for the inner and outer rotors for one full period of 0.3 sec. The number of loops are found to be 6. For a speed ratio of 1300 to 1000 rpm, the response at the inner rotors for a full period of 0.6 sec is given in Fig. 8.33c. The orbit has 13 loops.

Example 8.9

Consider the dual rotor system of Fig. 8.34 supported on identical

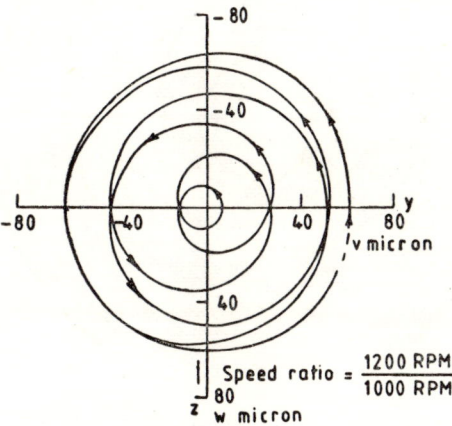

Fig. 8.33a Whirl orbit at inner rotor in one full period

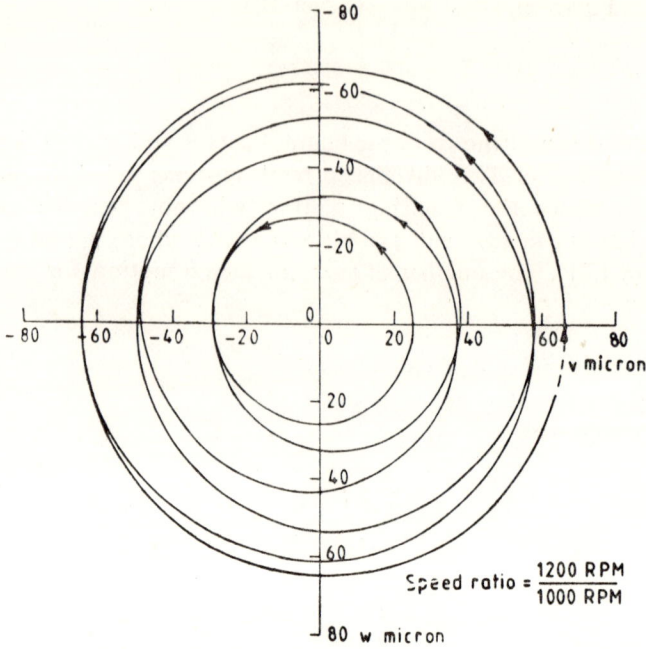

Fig. 8.33b Whirl orbit at outer rotor in one full period

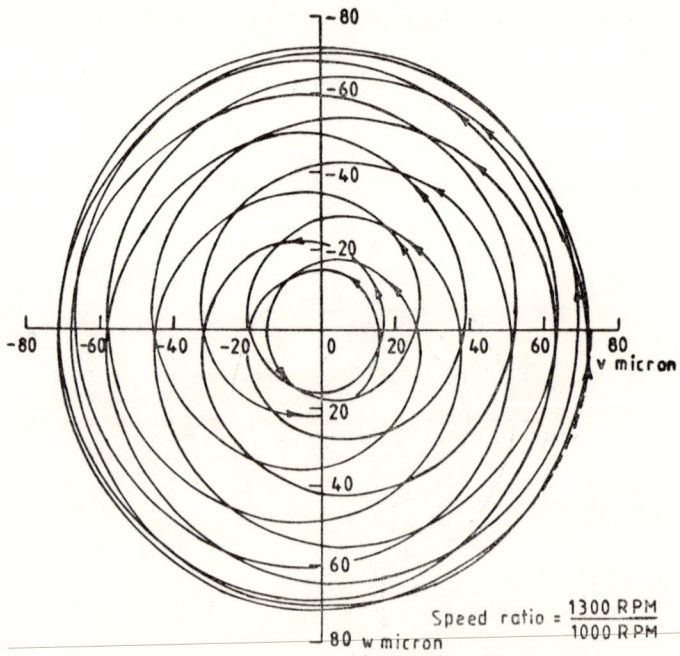

Fig. 8.33c Whirl orbit at inner rotor in full period

bearings, including the intershaft bearing such that

$$K_{yy} = K_{zz} = 17.5 \text{ MN/m}$$

$$K_{yz} = K_{zy} = 10.0 \text{ MN/m}$$

$$C_{yy} = C_{zy} = 700 \text{ KN-s/m}$$

$$C_{yz} = C_{zz} = 400 \text{ KN-s/m}$$

The speed ratio is assumed to be 4000/3000 rpm.

The response obtained is plotted in Figs. 8.35a and 8.35b. The number of loops is 4 and the time period for one full cycle is 0.06 sec.

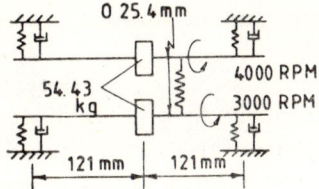

Fig. 8.34 Simple dual-rotor on hydrodynamic bearings

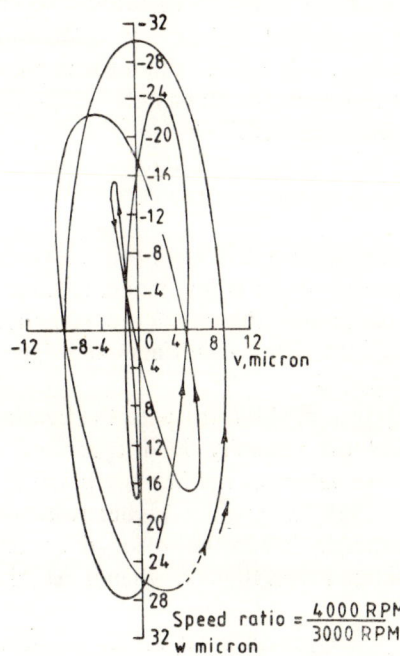

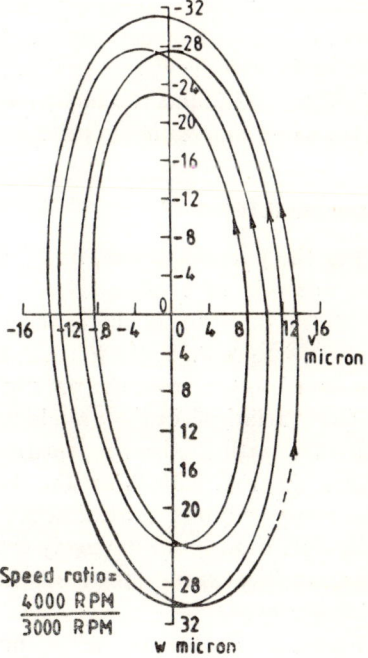

Fig. 8.35a Whirl orbit on inner rotor in one full period (with hydrodynamic bearings)

Fig. 8.35b Whirl orbit on outer rotor in one full period (with hydrodynamic bearings)

154   *Rotor Dynamics*

8.5   Optimum Design of Bearings for Minimum Unbalance Response

In this chapter, so far, we discussed methods of determining the unbalance response of rotors mounted on hydrodynamic bearings. We have seen that the bearing stiffness and damping coefficients have significant influence on the performance of the rotor, viz., the critical speeds and the unbalance response. The unbalance in the rotor can be kept to a minimum by adopting good mechanical design principles and proper balancing techniques. However, this cannot be completely eliminated. We have another option in the design of a bearing, to choose its parameters like the diameter, clearance and the oil viscosity for a given operating load and speed, that will minimize the unbalance response of a rotor and thus achieve smooth running conditions. Bhat, Rao and Sankar [27] conducted such a study using an optimization process based on the method of feasible directions [28].

The objective function in this case is, therefore, the maximum unbalance response, that can be calculated by a suitable procedure. For a Jeffcott model, equation (8.1.32) gives this quantity in a closed form as a function of the rotational speed. This expression can be used in a given operational speed range as the objective function. We can also set upper and lower limits on the design variables, e.g., bearing diameter, radial clearance and the oil viscosity and apply a constraint on the Sommerfeld number. Initially, a set of starting values are chosen for the design parameters arbitrarily, however, within the set limits with which the unbalance response of the rotor and the constraint values are calculated. The optimization procedure then suggests a new set of design parameters, which will minimize the objective function, viz., the unbalance response, at the same time satisfying the constraints imposed.

Example 8.10

For the purpose of this study, we use the rotor 2 of example 8.2. Using the analysis of section 8.1.2, a parametric study is first made, to understand the influence of the bearing parameters on the unbalance response. Figures 8.36 to 8.38 show these results. The bearing stiffness and damping data is taken from the previous chapter 7.

The variation of the unbalance response of the rotor with the bearing diameter, with the bearing clearance and oil viscosity held constant, is shown in Fig. 8.36. For the two lower values of the bearing diameter, the rotor exhibits critical speeds; however, for the two higher values of diameter, it does not have any critical speeds. When the rotor does not possess critical speeds, the response is smaller than the case when it exhibits critical speeds.

Figure 8.37 shows the unbalance response for different values of bearing clearance with the bearing diameter and the oil viscosity kept constant. For the lowest value of clearance shown, the rotor does not possess any critical speeds and as the clearance is increased, the rotor exhibits critical

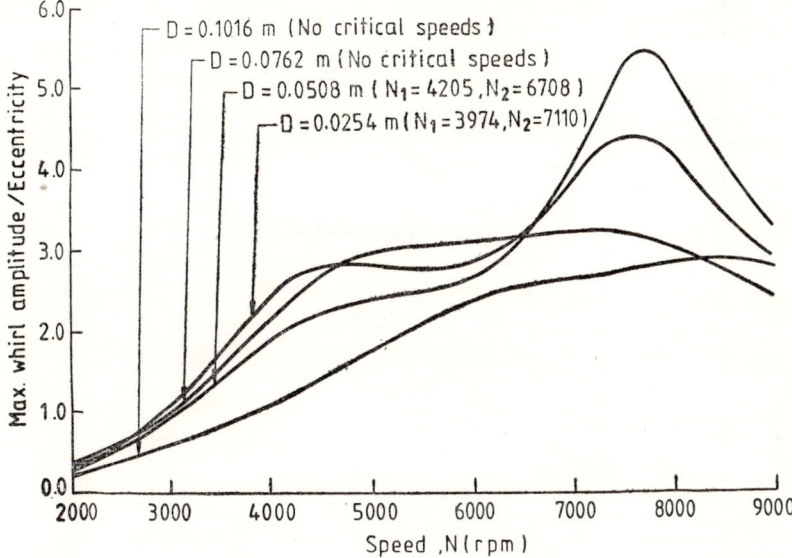

Fig. 8.36 Variation of unbalance response of rotor with bearing diameter ($c = 5.08 \times 10^{-5}$ m, $\mu = 5.68 \times 10^{-3}$ N s/m², $L/D = 0.5$)

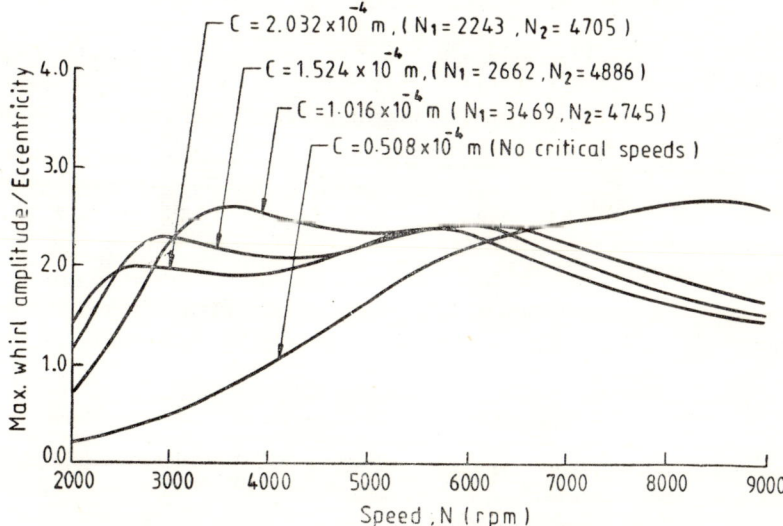

Fig. 8.37 Variation of unbalance response of rotor with bearing clearance ($D = 0.1016$ m, $\mu = 5.68 \times 10^{-3}$ Ns/m², $L/D = 0.5$)

speeds. When there are no critical speeds, the maximum unbalance response occurs around 8400 rpm, whereas when the rotor has critical speeds, the maximum response occurs at one of the critical speeds which are much lower than 8400 rpm.

The unbalance response of the rotor is shown in Fig. 8.38 for different values of oil viscosity with the bearing diameter and the clearance

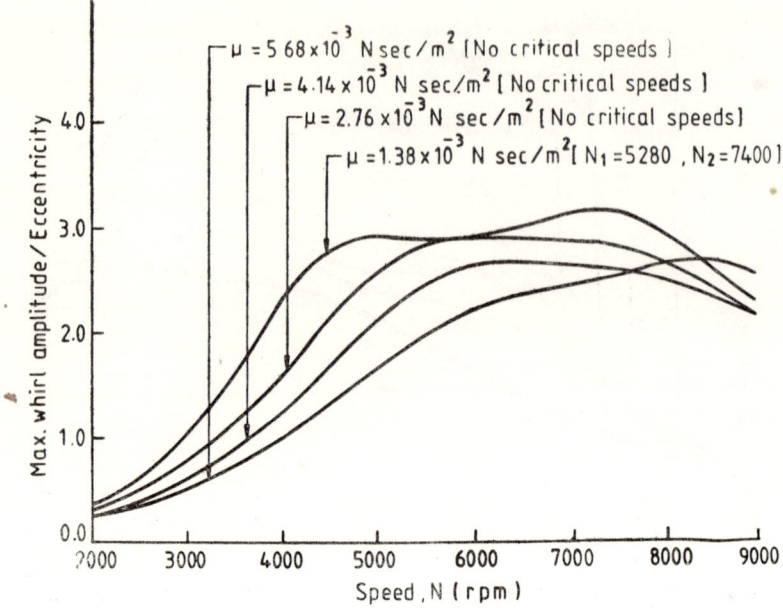

Fig. 8.38 Variation of unbalance response of rotor with viscosity ($D = 0.1016$ m, $c = 5.08 \times 10^{-5}$ m, $L/D = 0.5$)

remaining constant. For the lowest value of the viscosity shown, the rotor has two critical speeds, whereas when the viscosity is increased, it does not exhibit any critical speeds. The maximum unbalance response occurs at different speeds for different viscosity values.

The above results show that the bearing parameters significantly influence the unbalance response of the rotor. Hence the logical way to design hydrodynamic bearings for a given rotor is to choose optimum bearing parameters which minimize the maximum unbalance response in the operational speed range. The results of the optimization study are summarized in Table 8.5, which gives the starting values of the design

Table 8.5 Starting and optimum values of bearing parameters

L/D	A/B	Dia m	Clearance 10^{-4} m	Viscosity 10^{-3} Ns/m²
Plain cylindrical bearings				
0.5	A	0.0762	1.016	2.758
	B	0.1016	0.908	6.894
1.0	A	0.0762	1.016	2.758
	B	0.1016	1.327	6.894
Four axial groove bearings				
0.5	A	0.0508	2.540	1.379
	B	0.1016	1.153	6.894
1.0	A	0.0508	2.540	1.379
	B	0.1004	1.238	1.683

A—denotes starting values.
B—denotes optimized values.

parameters and the optimum values for the plain cylindrical and four axial groove bearings. The operational speed range considered is from 2000 to 9000 rpm. The unbalance response of the rotor corresponding to the starting values and the optimum values of the bearing parameters is shown in Figs. 8.39 to 8.42.

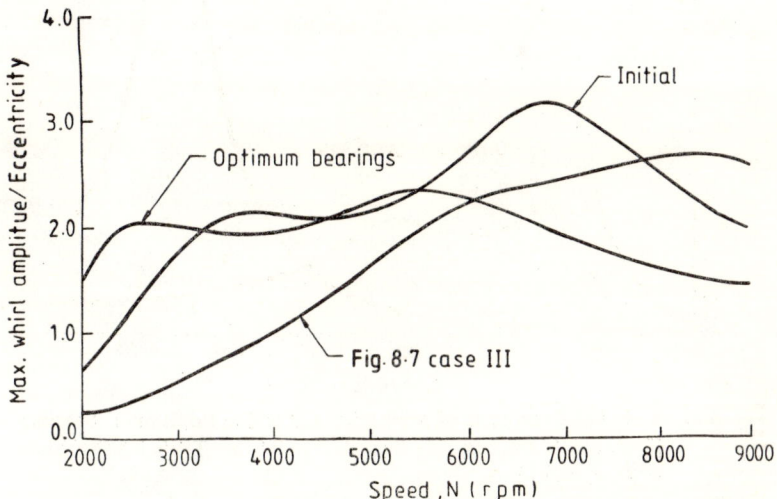

Fig. 8.39 Unbalance response of rotor with and without optimum bearings [plain cylinder bearings $L/D = 0.5$, see Table 8.5 for values of D, c, and μ]

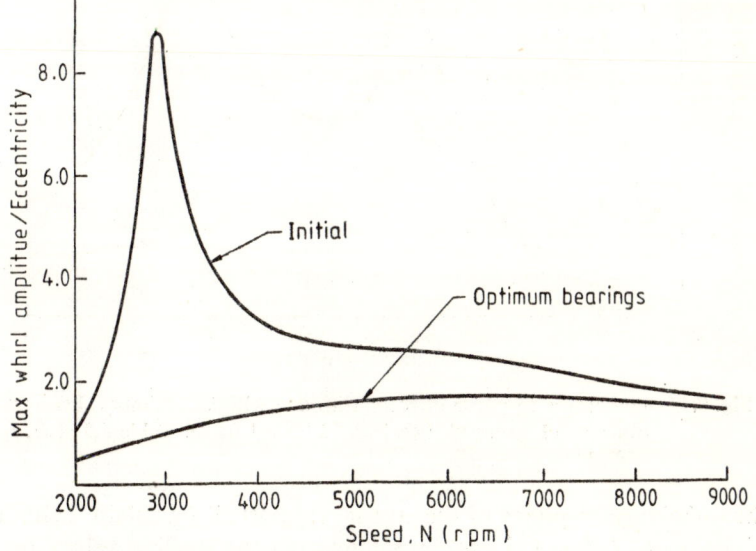

Fig. 8.40 Unbalance response of rotor with and without optimum bearings [plain cylindrical bearings $L/D = 1.0$, see Table 8.5 for values of D, c and μ]

158 *Rotor Dynamics*

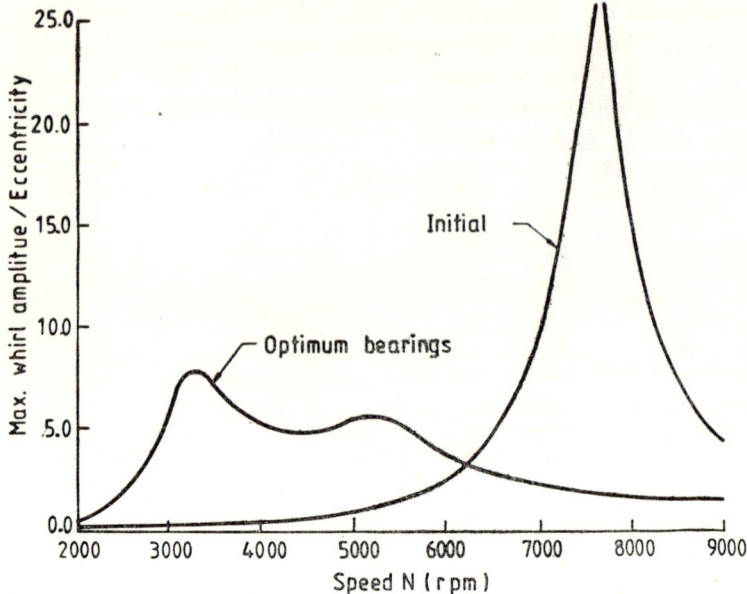

Fig. 8.41 Unbalance response of rotor with and without optimum bearings [four axial grooved bearings, $L/D = 0.5$, see Table 8.5 for values of D, c, and μ]

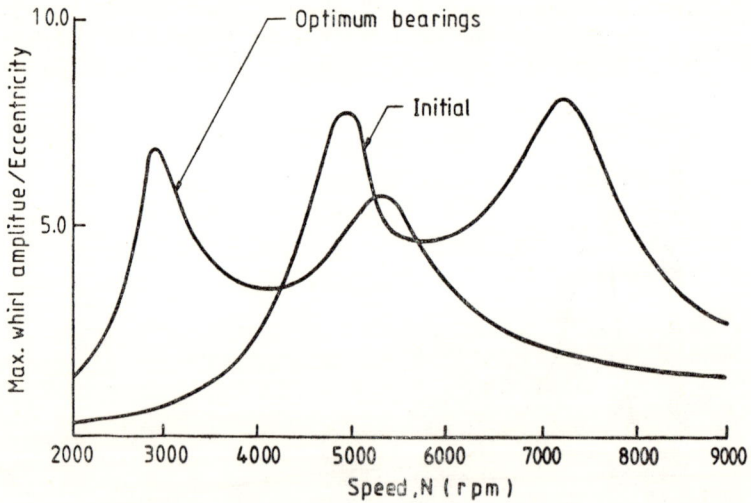

Fig. 8.42 Unbalance response of rotor with and without optimum bearings [four axial grooved bearings, $L/D = 1.0$, see Table 8.5 of D, c, and μ]

The unbalance response of the rotor supported on plain cylindrical bearings with $L/D = 0.5$, corresponding to the starting values and the optimum values of the bearing parameters, is shown in Fig. 8.39. The response from Fig. 8.7, for the case III, is also included in Fig. 8.39, for reference. For the case of starting values and optimum values of the bearing parameters, two peaks at different frequencies are observed, whereas the response in example 8.4, has only one peak. The response corresponding to the optimum bearings is less than the other two curves in Fig. 8.39, showing a 13% reduction over the original bearings. When $L/D = 1$, the results obtained are shown in Fig. 8.40. In this case, the optimum bearings have a much lower unbalance response. The maximum unbalance response in this case is reduced by 39% from the corresponding values of the original bearings.

The cases of four axial groove bearings with $L/D = 0.5$ and 1.0 are shown in Figs. 8.41 and 8.42 respectively. The optimum bearings have a much lower response compared to the bearings with which the optimization was started when $L/D = 0.5$. However, the optimum groove bearings have a higher unbalance response compared to the optimum plain cylindrical bearings. In reference [27], it was shown that a four shoe tilting pad bearing with $L/D = 0.5$, reduced the maximum unbalance response of the rotor by 55% from that corresponding to the original set of plain cylindrical bearings.

References

1. Tondl, A. Some problems of rotor dynamics, Publishing House of Czechoslovakian Academy of Science, Prague, 1965.
2. Dimintberg, F.M. Flexural vibrations of rotating shafts, Butterworth's Publishing House, London, 1961.
3. Gasch, R., and Pfutzner, H. Rotor Dynamik, Springer Verlag, Berlin, 1975.
4. Hull, E.H. Shaft whirling as influenced by stiffness asymmetry, J. of Basic Engng., ASME, Vol. 83, 1961, p. 219.
5. Rao, J.S. Synchronous whirl of a flexible rotor in hydrodynamic bearings, Mechanism and Machine Theory, Vol. 17, No. 2, 1982, p. 143.
6. Rieger, N.F. and Cundiff, R.A. Discussion on the paper by Morton, P.G., Influence of coupled asymmetric bearings on the motion of a massive flexible rotor, Proc. Instn. Mech. Engrs. Vol. 182, Pt. 1, No. 13, 1967-68, p. 271.
7. Rao, J.S., Bhat, R.B. and Sankar, T.S. Effect of damping on the synchronous whirl of a rotor on hydrodynamic bearings, Trans. Canadian Society of Mechanical Engineers, Vol. 6, No. 3, 1981, p. 155.
8. Morton, P.G. Influence of coupled asymmetric bearings on the motion of a massive flexible rotor. Proc. Instn. of Mech. Engrs., Vol. 182, Pt. 1, No. 13, 1967-68, p. 255.
9. Kramer, E. Computation of unbalance vibrations of turbo rotors, ASME 77-DET-13.
10. Rao, J.S. and Sarma, K.V.B. Out-of-balance response of turbo alternator rotors, Computer program, Bharat Heavy Electricals Ltd., Hyderabad, India, 1982.
11. Ono, K. and Tamuri, A. On the vibrations of horizontal shaft supported in oil lubricated bearings, Bull. JSME., Vol. 11, 1968, p. 813.
12. Rieger, N.F., Thomas, C.B. and Walter, W.W. Dynamic stiffness matrix approach for rotor bearing system analysis, Proc. Instn. Mech. Engrs., 1976, p. 187 (C187-76).
13. Lund, J.W. and Orcutt, F.K. Calculations and experiments on the unbalance response of a flexible rotor, J. of Engng. for Indus., ASME, Vol. 87, 1967, p. 785.
14. Lund, J.W. Stability and damped critical speeds of a flexible rotor in fluid film bearings, J. of Engng. for Indus., ASME, Vol. 96, 1974, p. 509.

15. Lund, J.W. Modal response of a flexible rotor in fluid film bearings, J. of Engng. for Indus., Vol. 96, 1974, p. 525.
16. Gunter, E.J., Choy, K.C. and Allaire, P.E. Modal analysis of turborotors using planar modes theory, Journal of Franklin Institute, Vol. 305, 1978, p. 221.
17. Ruhl, R.L. and Booker, J.F. A finite element model for distributed parameter turborotor systems, J. of Engng. for Indus., ASME, Vol. 94, 1972, p. 126.
18. Nelson, H.D. and McVaugh, J.M. The dynamics of rotor bearing systems using finite elements, J. of Engng. for Indus., ASME, Vol. 98, 1976, p. 593.
19. Lalanne, M. and Queau, J.P. Calcul par elements finis du comportment dynamique des chaines cinematiques de reducteur, Societe Nationale des Industries Aerospatiales, May 1979.
20. Rouch, K.E. and Kao, J.S. Dynamic reduction rotor dynamics by the finite element method, J. of Mech. Des, ASME, Vol. 102, 1980, p. 360.
21. Subbiah, R., Bhat, R.B., Sankar, T.S. and Rao, J.S. Backward whirl in a simple rotor supported on hydrodynamic bearings, Instability in Rotating Machinery, NASA Conf. Publication 2409, 1985, p. 145.
22. Rao, J.S. and Bhaskara Sarma, K.V. Simulation of Multishaft rotors mounted on fluid film bearings to determine unbalance response, Proc. International Conf. Power Plant Simulation, Mexico, 1984, p. 85.
23. Bhaskara Sarma, K.V., Unbalance response and stability of symmetric and asymmetric rotors, Ph.D. Thesis, Indian Institute of Technology, New Delhi, 1987.
24. Glienicke, J. Feder und dampfungs konstanten von gleitlagern fur turbomachine und deren einfluss auf das schwingungs verhalten eines einfachen rotors, Ph.D. Thesis, Karlsruhe University, 1966.
25. Gupta, K.D., Gupta, K., Rao, J.S. and Bhaskara Sarma, K.V. Dynamic response of a dual rotor system by extended transfer matrix method, Proc. of Intl. Conf. on Vibrations in Rotating Machinery, Instn. of Mech. Engrs., London, 1988, p. 599.
26. Rao, J.S. and Gupta, K., Theory and practice of mechanical vibration, John Wiley and Sons, 1985.
27. Bhat, R.B., Rao, J.S. and Sankar, T.S. Optimum journal bearing parameters for minimum rotor unbalance response in synchronous whirl, J. Mech. Des, ASME, v. 104, 1982, p. 339.
28. Vanderplaats, G.N. Structural optimization by methods of feasible directions, Computers and Structures, v. 3, 1973, p. 739.

An alternator rotor (Courtesy: BHEL, Hyderabad)

Chapter 9

Shafts with Dissimilar Moments of Area

In the previous chapters, we considered the shaft to be circular in cross section, which has same moments of area about any diameter. For rotors on horizontal shafts, besides the main critical speed, disturbing whirl amplitudes have been observed at half the critical speed. On vertical shafts such is not the case indicating that the gravity is one of the causes for this. This effect is observed more clearly with shafts of rectangular cross section, which is not equally stiff in all directions as a circular shaft is. In practice, rotors of turbogenerators, because of their construction, exhibit a marked difference between the stiffnesses in two principal directions. The disturbance that occurs at speeds equal to half the critical speed of the rotor, can take place for even a perfectly balanced disk and therefore this disturbance is due to the instability condition, rather than due to unbalance. A similar behaviour is also observed with cracked shafts. In this chapter, we will first consider stability criteria, using the variable stiffness of the shaft and then consider the whirling of an unbalanced rotor, under steady and transient conditions.

9.1 Stability of a Shaft with Dissimilar Stiffness

The shaft cross section has two principal axes about which the second moment of area is maximum and minimum. In one revolution of the shaft, the stiffness is thus maximum twice and minimum twice, i.e., the stiffness variation passes through two complete cycles. The spring constant of the shaft is minimum say $K - \Delta K$ and maximum $K + \Delta K$, with an average value of K and so we can express the stiffness variation for a shaft rotating at angular velocity ω, as

$$F_k = K + \Delta K \sin 2\omega t \qquad (9.1.1)$$

An important equation in this connection is Mathieu equation [1], which is

$$\frac{d^2 z}{d\tau^2} + (a + 16q \cos 2\tau)z = 0 \qquad (9.1.2)$$

where a and q are real numbers, which will be identified later for the shaft with variable stiffness.

The solution of the above equation, according to Floquet's theory is

$$z = C_1 \exp(\mu\tau)\phi(\tau) + C_2 \exp(-\mu\tau)\phi(-\tau) \qquad (9.1.3)$$

The solutions of equation (9.1.2) are stable, if z remains bounded as $\tau \to \infty$ and unstable, if it grows indefinitely as $\tau \to \infty$. z are neutral, if the solution is periodic. Such neutral solutions are called Mathieu functions and a and q should have a definite relationship between them. For example, if $q = 0$, the solutions are

$$\cos i\tau, \sin i\tau; i = 1, 2, 3, \ldots$$

with $a = i^2$.

For other values of q, we have the Mathieu functions denoted by

$$Ce_n(\tau, q)$$

$$Se_n(\tau, q); n = 1, 2, \ldots.$$

For these periodic solutions, other sets of boundaries can be shown to be [1]

$$a_{c1} = 1 - 8q - 8q^2 + 8q^3 \ldots$$

$$a_{s1} = 1 + 8q - 8q^2 - 8q^3 \ldots$$

$$a_{c2} = 4 + \frac{80}{3} q^2 - \frac{6104}{27} q^4 \ldots$$

$$a_{s2} = 4 - \frac{16}{3} q^2 + \frac{40}{27} q^4 \ldots$$

$$a_{c3} = 9 + 4q^2 - 8q^3 + \frac{13}{5} q^4 \ldots$$

$$a_{s3} = 9 + 4q^2 + 8q^3 + \frac{13}{5} q^4 \ldots \qquad (9.1.4)$$

These equations are shown plotted in Fig. 9.1, with the unstable regimes marked. A diagram such as Fig. 9.1, which describes the stable and unstable regimes of a system is also called as Strutt diagram.

Consider a rectangular shaft, vertically mounted and carrying a disk of mass M shown in Fig. 9.2. The shaft is permitted to have lateral motion in one plane only, by means of guides shown at the section A. Let the maximum stiffness of the shaft be $K + \Delta K$ and the minimum stiffness $K - \Delta K$. As discussed earlier, in one revolution, the stiffness alternates twice, as shown in Fig. 9.3, and the equation of motion for the disk is

$$M\ddot{z} + (K + \Delta K \sin \omega_k t)z = 0 \qquad (9.1.5)$$

where

$$\omega_k = 2\omega$$

For a disk horizontally mounted, instead as shown in Fig. 9.2, the origin gets shifted and we can write

$$\bar{z} = z + c \qquad (9.1.6)$$

$$M\ddot{\bar{z}} + (K + \Delta K \sin \omega_k t)\bar{z} = Mg + Kc + c\Delta K \sin \omega_k t \qquad (9.1.7)$$

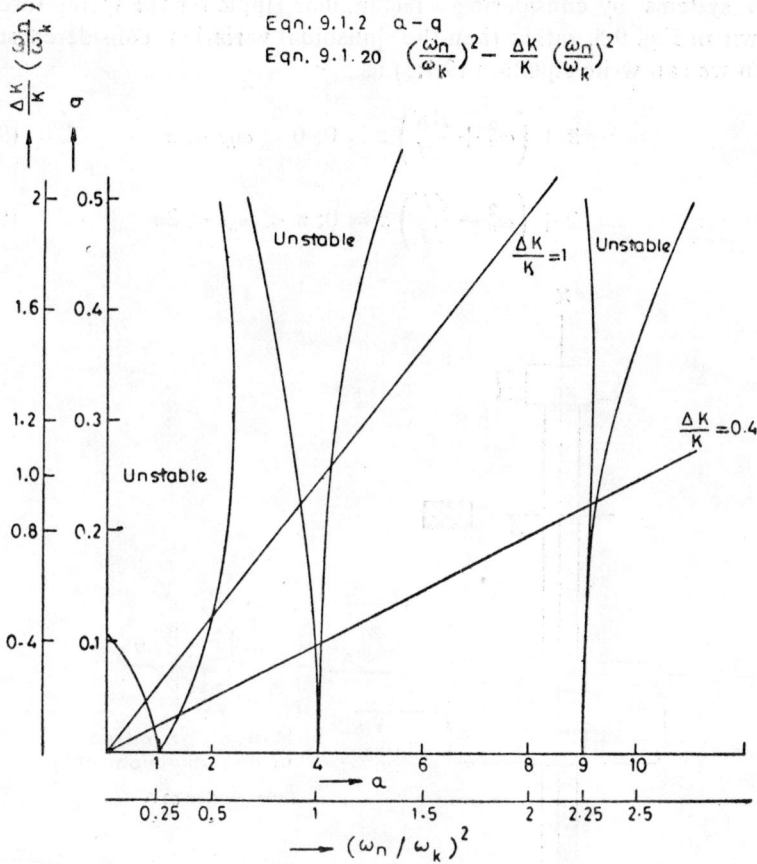

Fig. 9.1 Mathieu equation—Strutt diagram

As a first approximation, for the horizontal shaft, $c = - Mg/K$, when $\Delta K \ll K$, and equation (9.1.7) becomes (9.1.5).

Let us consider the shaft rotating at the critical speed ω_n. Referring to Fig. 9.3, we notice that for the regions 1 — 2, 3 — 4, when the amplitude is increasing, the spring force is small and when the amplitude is decreasing in regions 2 — 3, 4 — 5, the spring force is larger and hence the system is unstable. If the shaft is horizontal, however, it can be seen that the total work per cycle is zero and hence there is no instability, as in the case of vertical shaft.

A different conclusion will be reached, if we have the horizontal shaft rotating at ½ critical speed (Fig. 9.4). For lateral motion at the natural frequency, the spring force is small when the disk is going away from equilibrium, and is large when the disk is coming towards the equilibrium position. Thus we have instability at rotational speeds equal to half the lateral natural frequency. This is a typical characteristic which is not exhibited by a circular shaft.

A simple linear theory can be used to study the stability behaviour of such systems by considering a rectangular ripple for the spring force, as shown in Fig. 9.5, rather than the sinusoidal variation considered so far. Then we can write equation (9.1.5) as

$$\ddot{z} + \left(\omega_n^2 + \frac{\Delta K}{M}\right) z = 0; \quad 0 < \omega_k t < \pi \qquad (9.1.8)$$

$$\ddot{z} + \left(\omega_n^2 - \frac{\Delta K}{M}\right) z = 0; \quad \pi < \omega_k t < 2\pi \qquad (9.1.9)$$

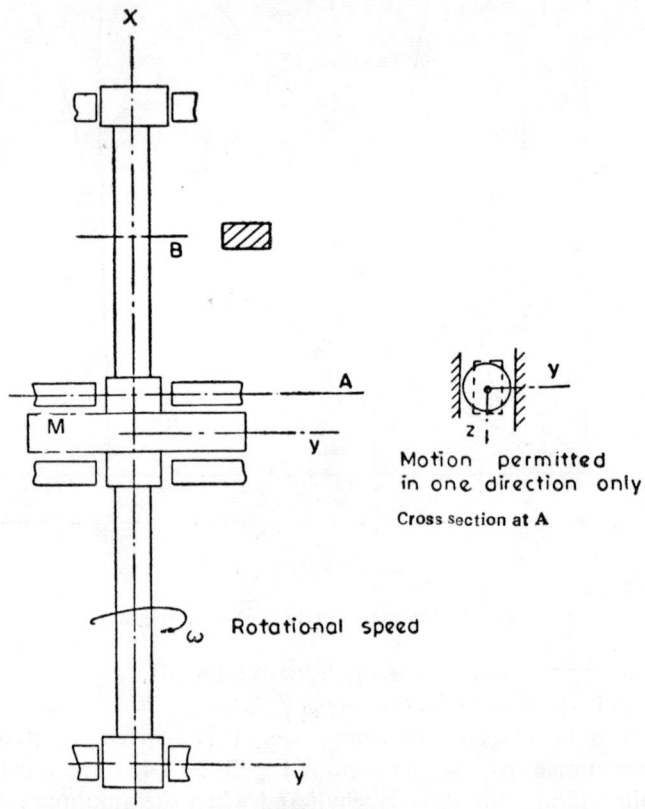

Fig. 9.2 Rectangular shaft carrying a disk M

Let

$$p_1^2 = \omega_n^2 + \frac{\Delta K}{M}$$
$$p_2^2 = \omega_n^2 - \frac{\Delta K}{M} \qquad (9.1.10)$$

Then the solution of equations (9.1.8) and (9.1.9) becomes

$$z_1 = C_1 \sin p_1 t + C_2 \cos p_1 t$$
$$z_2 = C_3 \sin p_2 t + C_4 \cos p_2 t \qquad (9.1.11)$$

Shafts with Dissimilar Moments of Area **167**

we have the following relations for a type of solution in Fig. 9.6:

$$z_1\}_{\omega_k t=0} = z_2\}_{\omega_k t=0}; \quad z_2\}_{\omega_k t=\pi} = Sz_1\}_{\omega_k t=-\pi}$$
$$\dot{z}_1\}_{\omega_k t=0} = \dot{z}_2\}_{\omega_k t=0}; \quad \dot{z}_2\}_{\omega_k t=\pi} = S\dot{z}_1\}_{\omega_k t=-\pi}$$

(9.1.12)

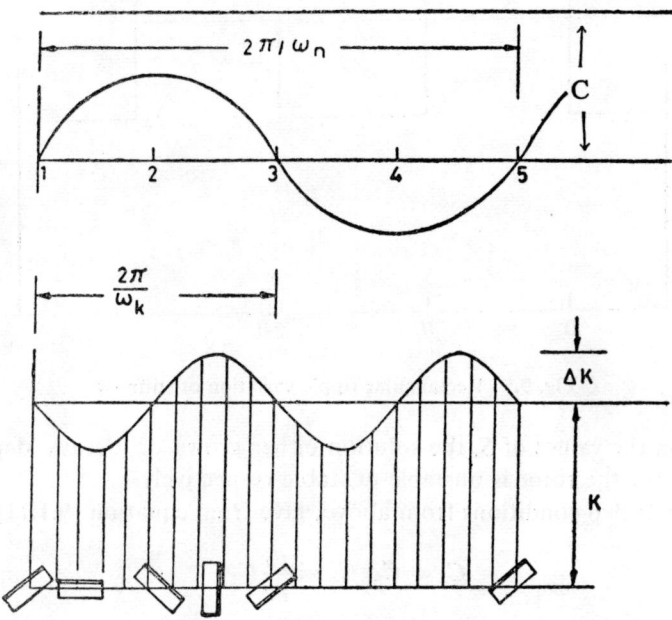

Fig. 9.3 Asymmetric shaft running at critical speed ω_n ($\omega_K = 2\omega_n$)

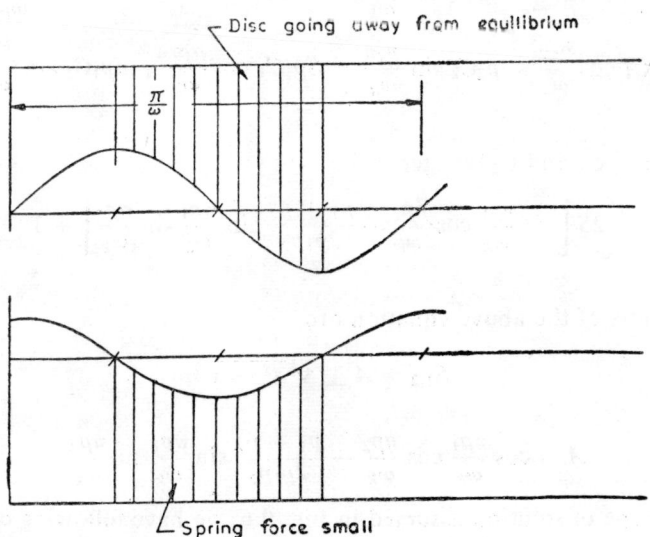

Fig. 9.4 Horizontal shaft running at ½ critical speed, $\omega_n = 2\omega = \omega_K$, $\pi/\omega = 2\pi/\omega_K = 2\pi/\omega_n$

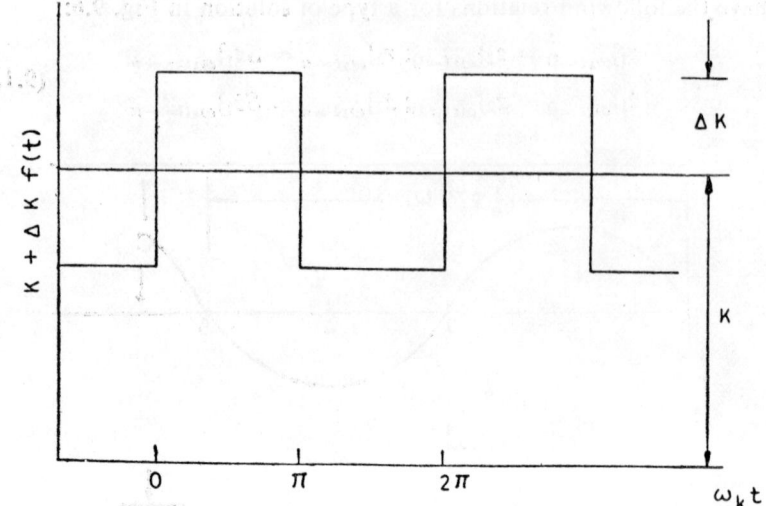

Fig. 9.5 Rectangular ripple variation of stiffness

Based on the values of S, the solution either grows or decays depending on whether the rotor is unstable or stable respectively.

The first two conditions from above, give from equation (9.1.11)

$$C_2 = C_4; \quad C_3 = \frac{p_1}{p_2} C_1 \tag{9.1.13}$$

The next two conditions in equation (9.1.12) lead to

$$\frac{p_1}{p_2} C_1 \sin \frac{p_2 \pi}{\omega_k} + C_2 \cos \frac{p_2 \pi}{\omega_k} = -SC_1 \sin \frac{p_1 \pi}{\omega_k} + SC_2 \cos \frac{p_1 \pi}{\omega_k}$$

$$p_1 C_1 \cos \frac{p_2 \pi}{\omega_k} - p_2 C_2 \sin \frac{p_2 \pi}{\omega_k} = Sp_1 C_1 \cos \frac{p_1 \pi}{\omega_k} + Sp_1 C_2 \sin \frac{p_1 \pi}{\omega_k} \tag{9.1.14}$$

Eliminating C_1 and C_2, we get

$$S^2 - 2S \left[\cos \frac{\pi p_1}{\omega_k} \cos \frac{\pi p_2}{\omega_k} - \frac{p_1^2 + p_2^2}{2 p_1 p_2} \sin \frac{\pi p_1}{\omega_k} \sin \frac{\pi p_2}{\omega_k} \right] + 1 = 0 \tag{9.1.15}$$

The roots of the above equation are

$$S_{1,2} = A \pm \sqrt{A^2 - 1} \tag{9.1.16}$$

where

$$A = \cos \frac{\pi p_1}{\omega_k} \cos \frac{\pi p_2}{\omega_k} - \frac{p_1^2 + p_2^2}{2 p_1 p_2} \sin \frac{\pi p_1}{\omega_k} \sin \frac{\pi p_2}{\omega_k} \tag{9.1.17}$$

For the type of solution assumed in Fig. 9.6, we have following cases:

Case 1. $A > 1$, clearly the solution is unstable.

Case 2. $-1 < A < 1$, roots are complex and the real component is < 1. Hence the solution is stable.

Shafts with Dissimilar Moments of Area 169

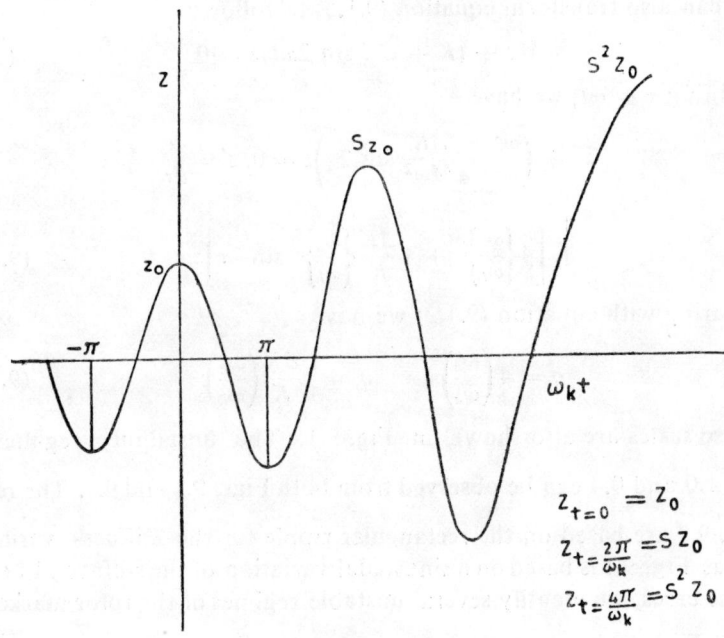

Fig. 9.6 Unstable solution

Case 3. $A < -1$, then real component < -1 and hence the solution is unstable.

Therefore the stability boundaries are

$$A = \pm 1 \qquad (9.1.18)$$

Since

$$\frac{p_1}{\omega_k} = \frac{\omega_n}{\omega_k}\left(1 + \frac{\Delta K}{K}\right)^{1/2}$$

$$\frac{p_2}{\omega_k} = \frac{\omega_n}{\omega_k}\left(1 - \frac{\Delta K}{K}\right)^{1/2}$$

and

$$\frac{p_1^2 + p_2^2}{2p_1 p_2} = \frac{1}{\left(1 + \frac{\Delta K}{K}\right)^{1/2}\left(1 - \frac{\Delta K}{K}\right)^{1/2}} \qquad (9.1.19)$$

We can plot the relation between $\left\{\frac{\omega_n}{\omega_k}\right\}^2$ and $\frac{\Delta K}{K}\left\{\frac{\omega_n}{\omega_k}\right\}^2$, for which the equation (9.1.18) is satisfied. Such a diagram given by Timoshenko [2] and Den Hartog [3] is shown in Fig. 9.7. The advantages of choosing the coordinates $\left\{\frac{\omega_n}{\omega_k}\right\}^2$ and $\frac{\Delta K}{K}\left\{\frac{\omega_n}{\omega_k}\right\}^2$ is that we can have a line representing $\frac{\Delta K}{K}$, which is the percentage of variable stiffness component, on the diagram itself and study the unstable regimes for different values of $\omega_k = 2\omega$.

We can also transform equation (9.1.5) as follows:
$$M\ddot{z} + (K + \Delta K \sin 2\omega t)z = 0 \qquad (9.1.5)$$
Defining $\tau = \omega t$, we have
$$z'' + \left(\frac{\omega_n^2}{\omega^2} + \frac{\Delta K}{M\omega^2} \sin 2\tau\right)z = 0, \; z' = \frac{d}{d\tau}$$
i.e.
$$z'' + \left[4\left\{\frac{\omega_n}{\omega_k}\right\}^2 + 4\frac{\Delta K}{K}\left\{\frac{\omega_n}{\omega_k}\right\}^2 \sin 2\tau\right]z = 0 \qquad (9.1.20)$$
Comparing with equation (9.1.2) we have
$$a = 4\left(\frac{\omega_n}{\omega_k}\right)^2; \qquad q = \frac{1}{4}\frac{\Delta K}{K}\left(\frac{\omega_n}{\omega_k}\right)^2. \qquad (9.1.21)$$

These scales are also shown in Fig. 9.1. The unstability regimes for $\frac{\Delta K}{K} = 1.0$ and 0.4 can be observed from both Figs. 9.1 and 9.7. The results of Fig. 9.7 are based on the rectangular ripple for the stiffness variation, whereas Fig. 9.1 is based on a sinusoidal variation of the stiffness. In either of these cases, we identify several unstable regimes of the rotor marked by

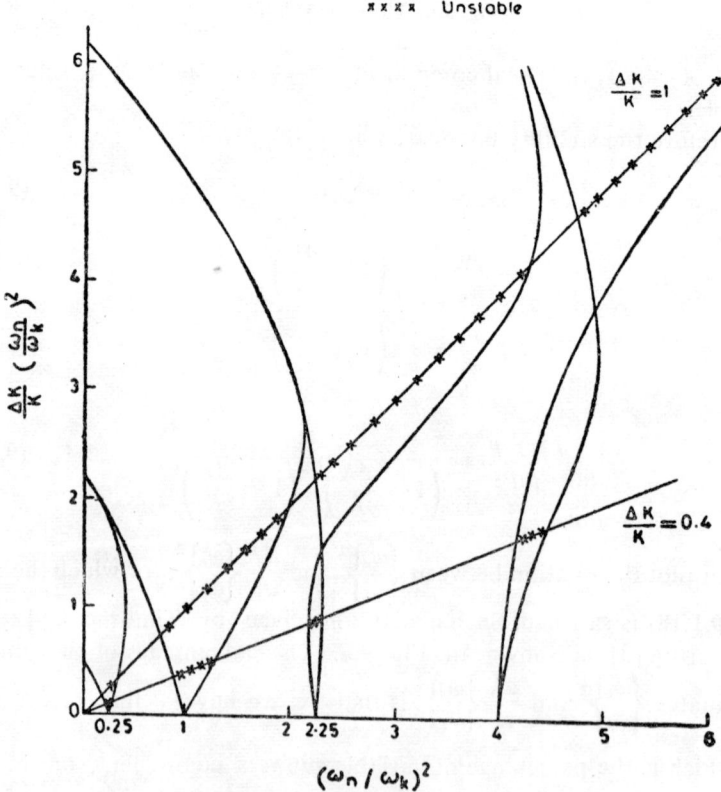

Fig. 9.7 Stability diagram—Variable elasticity shaft

a zig-zag line. The points farther from the origin on the x-axis represent low speeds of rotation and near the origin very high speeds of rotation. Thus a rotor when its speed is increased, passes through several zones of narrow regimes of instability, all of which are not particularly harmful. When the shaft speed is half the critical speed, we have $\omega_n = \omega_k$, and there is a substantial zone of instability. When $\omega_n/\omega_k = \frac{1}{2}$, i.e., the shaft is running in the critical speed zone, we have not only the unstable condition due to variable stiffness, but large vibrations due to rotor unbalance.

9.2 Whirling of a Shaft with Dissimilar Stiffnesses—Tondl [4]

The disk of Fig. 9.2 is shown in Fig. 9.8, whirling with an angular velocity ω. We use $\xi\eta$ rotating coordinates to analyse the system. The shaft centre E is defined in the stationary coordinates

$$r = z + iy \tag{9.2.1}$$

In rotating coordinates

$$\zeta = \xi + i\eta \tag{9.2.2}$$

The velocity and acceleration components of the centroid G in the rotating coordinates are

$$\begin{aligned} v_\xi &= \dot{\xi} - \omega(\eta + a_2) \\ v_\eta &= \dot{\eta} + \omega(\xi + a_1) \end{aligned} \tag{9.2.3}$$

$$\begin{aligned} a_\xi &= \ddot{\xi} - 2\omega\dot{\eta} - \omega^2(\xi + a_1) \\ a_\eta &= \ddot{\eta} + 2\omega\dot{\xi} - \omega^2(\eta + a_2) \end{aligned} \tag{9.2.4}$$

Using equation (9.2.4) and D'Alembert's principle, the inertia forces are shown in Fig. 9.8 along with the restoring forces. K_1 and K_2 are the stiffnesses of the shaft in the principal directions. Hence the equations of motion are

$$\begin{aligned} M[\ddot{\xi} - 2\omega\dot{\eta} - \omega^2(\xi + a_1)] + K_1\xi &= Mg \cos \omega t \\ M[\ddot{\eta} + 2\omega\dot{\xi} - \omega^2(\eta + a_2)] + K_2\eta &= -Mg \sin \omega t \end{aligned} \tag{9.2.5}$$

Defining

$$\omega_1^2 = K_1/M; \quad \omega_2^2 = K_2/M \tag{9.2.6}$$

$$\begin{aligned} \ddot{\xi} - 2\omega\dot{\eta} + (\omega_1^2 - \omega^2)\xi &= a_1\omega^2 + g \cos \omega t \\ \ddot{\eta} + 2\omega\dot{\xi} + (\omega_2^2 - \omega^2)\eta &= a_2\omega^2 - g \sin \omega t \end{aligned} \tag{9.2.7}$$

The solution of equations (9.2.7) consists of three parts. The free vibration solution corresponds to a perfectly balanced disk mounted on a vertical shaft. We can account for the unbalance and gravity effects separately and add all of them together to obtain the total solution. First we examine the free vibration solution

$$\begin{aligned} \xi &= \xi_0 \exp(i\lambda_0 t) \\ \eta &= \eta_0 \exp(i\lambda_0 t) \end{aligned} \tag{9.2.8}$$

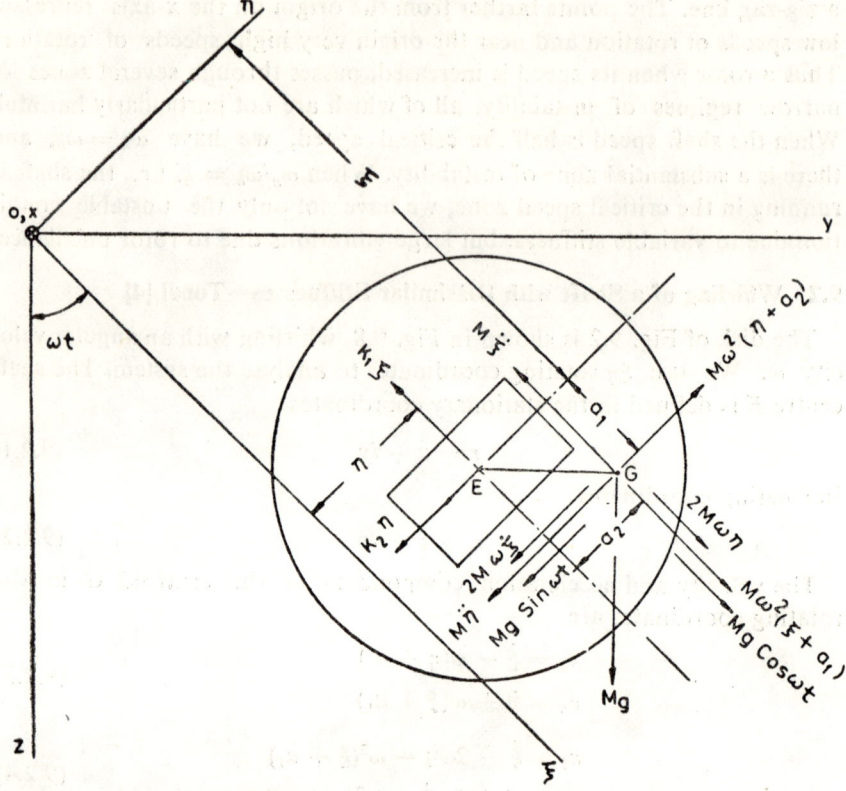

Fig. 9.8 Whirling of a rotor with a rectangular cross-section shaft

where λ_0 is the frequency of vibration of the system in rotating coordinates and the frequency of vibration in the stationary coordinates, λ, is given by

$$\lambda = \lambda_0 + \omega \qquad (9.2.9)$$

Substituting equations (9.2.8) in (9.2.7) with no forcing terms, we get

$$\begin{bmatrix} (\omega_1^2 - \omega^2 - \lambda_0^2) & -2i\omega\lambda_0 \\ 2i\omega\lambda_0 & (\omega_2^2 - \omega^2 - \lambda_0^2) \end{bmatrix} \begin{Bmatrix} \xi_0 \\ \eta_0 \end{Bmatrix} = 0 \qquad (9.2.10)$$

Hence the frequency equation can be obtained as

$$\lambda_0^4 - (\omega_1^2 + \omega_2^2 + 2\omega^2)\lambda_0^2 + (\omega_1^2 - \omega^2)(\omega_2^2 - \omega^2) = 0 \qquad (9.2.11)$$

$$\lambda_i = \lambda_{0i} + \omega \qquad (9.2.12)$$

Let

$$K_1 = K - \Delta K; \quad K_2 = K + \Delta K \qquad (9.2.13)$$

Then

$$\omega_1^2 = \omega_n^2 \left(1 - \frac{\Delta K}{K}\right); \quad \omega_2^2 = \omega_n^2 \left(1 + \frac{\Delta K}{K}\right) \qquad (9.2.14)$$

where

$$\omega_n^2 = \frac{K}{M} \qquad (9.2.15)$$

Shafts with Dissimilar Moments of Area 173

Taking the roots of equation (9.2.11) and using (9.2.12) we obtain

$$\frac{\lambda}{\omega_n} = \frac{\omega}{\omega_n} \pm \sqrt{1 + \left(\frac{\omega}{\omega_n}\right)^2 \pm \sqrt{\left[1 + \left(\frac{\omega}{\omega_n}\right)^2\right]^2 - \left[1 - \frac{\Delta K}{K} - \left(\frac{\omega}{\omega_n}\right)^2\right]\left[1 + \frac{\Delta K}{K} - \left(\frac{\omega}{\omega_n}\right)^2\right]}} \quad (9.2.16)$$

For $\Delta K/K = 0.4$, the above equation is plotted in Fig. 9.9. There are four real roots for λ/ω_n, for all ω/ω_n except when

$$1 - \frac{\Delta K}{K} < \left\{\frac{\omega}{\omega_n}\right\}^2 < 1 + \frac{\Delta K}{K} \quad (9.2.17)$$

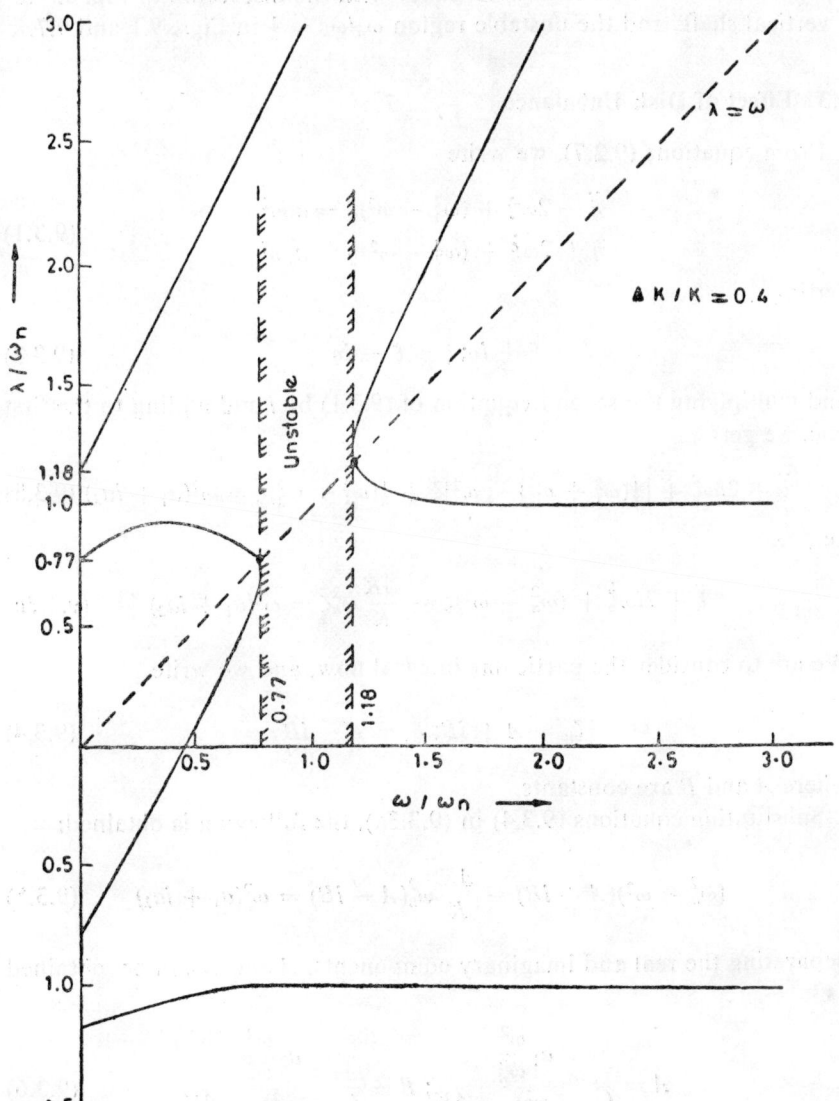

Fig. 9.9 Free whirling of a balanced vertical rotor on a rectangular shaft

at which there are only two real roots. In view of equation (9.2.14) the above inequality is

$$\omega_1 < \omega < \omega_2 \tag{9.2.18}$$

At $\omega = \omega_1$ and $\omega = \omega_2$, there are two coincident roots as shown in Fig. 9.9, which also fall on the synchronous whirl line. For ω in the region of equation (9.2.18), the two complex roots of equation (9.2.16) have positive real parts and hence the oscillatory solution grows. Thus the system is unstable, even in the absence of unbalance, in the region indicated by equation (9.2.18). This is in accordance with the discussion of Fig. 9.3 for a vertical shaft, and the unstable region $\omega_n/\omega_k = \frac{1}{2}$ in Figs. 9.1 and 9.7.

9.3 Effect of Disk Unbalance

From equations (9.2.7), we write

$$\ddot{\xi} - 2\omega\dot{\eta} + (\omega_1^2 - \omega^2)\xi = a_1\omega^2$$
$$\ddot{\eta} + 2\omega\dot{\xi} + (\omega_2^2 - \omega^2)\eta = a_2\omega^2 \tag{9.3.1}$$

Setting

$$\zeta = \xi + i\eta; \quad \bar{\zeta} = \xi - i\eta \tag{9.3.2}$$

and multiplying the second equation of (9.3.1) by i and adding to the first one, we get

$$\ddot{\zeta} + 2i\omega\dot{\zeta} + [\tfrac{1}{2}(\omega_1^2 + \omega_2^2) - \omega^2]\zeta + \tfrac{1}{2}(\omega_1^2 - \omega_2^2)\bar{\zeta} = \omega^2(a_1 + ia_2) \tag{9.3.3}$$

i.e.,

$$\ddot{\zeta} + 2i\omega\dot{\zeta} + (\omega_n^2 - \omega^2)\zeta - \frac{\Delta K}{K}\omega_n^2\bar{\zeta} = \omega^2(a_1 + ia_2) \tag{9.3.3a}$$

We are to consider the particular integral now, and we write

$$\zeta_u = A + iB; \quad \bar{\zeta}_u = A - iB \tag{9.3.4}$$

where A and B are constants.

Substituting equations (9.3.4) in (9.3.3a), the following is obtained:

$$(\omega_n^2 - \omega^2)(A + iB) - \frac{\Delta K}{K}\omega_n^2(A - iB) = \omega^2(a_1 + ia_2) \tag{9.3.5}$$

Separating the real and imaginary components, A and B can be obtained as

$$A = \frac{a_1 \dfrac{\omega^2}{\omega_n^2}}{\left(1 - \dfrac{\omega^2}{\omega_n^2}\right) - \dfrac{\Delta K}{K}}; \quad B = \frac{a_2 \dfrac{\omega^2}{\omega_n^2}}{\left(1 - \dfrac{\omega^2}{\omega_n^2}\right) + \dfrac{\Delta K}{K}} \tag{9.3.6}$$

Shafts with Dissimilar Moments of Area 175

The whirl radius $r_u = \sqrt{A^2 + B^2}$ is

$$r_u = \frac{\omega^2}{\omega_n^2} \frac{\sqrt{a_1^2\left[1 - \frac{\omega^2}{\omega_n^2} + \frac{\Delta K}{K}\right]^2 + a_2^2\left[1 - \frac{\omega^2}{\omega_n^2} - \frac{\Delta K}{K}\right]^2}}{[1 - (\omega^2/\omega_n^2)]^2 - (\Delta K/K)^2} \qquad (9.3.7)$$

We notice that the whirl radius is infinite when

$$\frac{\omega^2}{\omega_n^2} = 1 - \frac{\Delta K}{K} \text{ or } 1 + \frac{\Delta K}{K} \qquad (9.3.8)$$

which defines the two critical speeds according to equation (9.2.14). For a vertical rotor, the response is unstable between the two critical speeds. For a horizontal rotor, however, the response is not unstable between the two critical speeds, and the orbit is circular given by

$$r = |r_u| \exp(i\omega t) \qquad (9.3.9)$$

The response of the system due to unbalance of a horizontal shaft is given in Fig. 9.10, for $\Delta K/K = 0.4$.

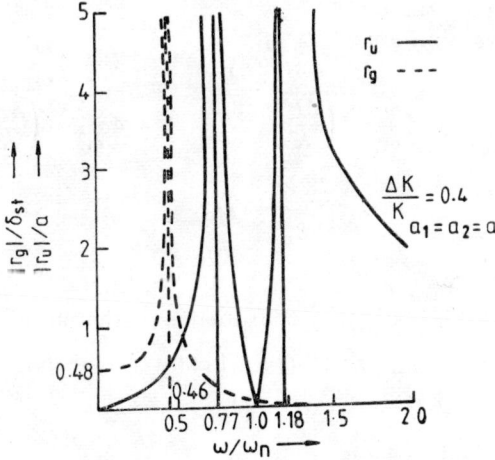

Fig. 9.10 Unbalance response of a rotor on an asymmetric shaft (horizontal)

9.4 Effect of Gravity on a Balanced Disk

The effect of gravity plays an important role in the dynamic behaviour of a horizontal shaft with dissimilar stiffness, as discussed earlier in Fig. 9.4. For a perfectly balanced disk ($a_1 = a_2 = 0$), we have from equation (9.2.7)

$$\begin{aligned} \ddot{\xi} - 2\omega\dot{\eta} + (\omega_1^2 - \omega^2)\xi &= g\cos\omega t \\ \ddot{\eta} + 2\omega\dot{\xi} + (\omega_2^2 - \omega^2)\eta &= -g\sin\omega t \end{aligned} \qquad (9.4.1)$$

With the help of equation (9.3.2), the above equations can be reduced to

$$\ddot{\zeta} + 2i\omega\dot{\zeta} + (\omega_n^2 - \omega^2)\zeta - \frac{\Delta K}{K}\omega_n^2\bar{\zeta} = g\exp(-i\omega t) \qquad (9.4.2)$$

The solution of the above equation for the particular integral is

$$\zeta_g = A \exp(i\omega t) + B \exp(-i\omega t)$$
$$\bar{\zeta}_g = \bar{A} \exp(-i\omega t) + \bar{B} \exp(i\omega t) \quad (9.4.3)$$

where A and B are now complex constants and $\bar{A}$ and $\bar{B}$ are their conjugates. Hence equation (9.4.2) gives

$$-A\omega^2 - 2A\omega^2 + (\omega_n^2 - \omega^2)A - \frac{\Delta K}{K}\omega_n^2 \bar{B} = 0$$
$$-B\omega^2 + 2B\omega^2 + (\omega_n^2 - \omega^2)B - \frac{\Delta K}{K}\omega_n^2 \bar{A} = g \quad (9.4.4)$$

We set $A = \bar{A}$ and $B = \bar{B}$ in the above equation and obtain the following:

$$\begin{bmatrix} (\omega_n^2 - 4\omega^2) & -\frac{\Delta K}{K}\omega_n^2 \\ -\frac{\Delta K}{K}\omega_n^2 & \omega_n^2 \end{bmatrix} \begin{Bmatrix} A \\ B \end{Bmatrix} = \begin{Bmatrix} 0 \\ g \end{Bmatrix} \quad (9.4.5)$$

From the above,

$$\frac{A}{\delta_{st}} = \frac{\frac{\Delta K}{K}}{1 - 4\frac{\omega^2}{\omega_n^2} - \left(\frac{\Delta K}{K}\right)^2}; \quad \frac{B}{\delta_{st}} = \frac{1 - 4\frac{\omega^2}{\omega_n^2}}{1 - 4\frac{\omega^2}{\omega_n^2} - \left(\frac{\Delta K}{K}\right)^2} \quad (9.4.6)$$

where

$$\delta_{st} = \frac{Mg}{K} \quad (9.4.7)$$

is the static deflection.

The solution in the stationary coordinates is, therefore,

$$r_g = \zeta_g \exp(i\omega t) = A \exp(2i\omega t) + B \quad (9.4.8)$$

and

$$\frac{r_g}{\delta_{st}} = \frac{1 - 4\frac{\omega^2}{\omega_n^2}}{1 - 4\frac{\omega^2}{\omega_n^2} - \left(\frac{\Delta K}{K}\right)^2} + \frac{\frac{\Delta K}{K}}{1 - 4\frac{\omega^2}{\omega_n^2} - \left(\frac{\Delta K}{K}\right)^2} \exp(2i\omega t) \quad (9.4.9)$$

The response is at twice the rotational speed and resonance takes place when

$$\omega = \tfrac{1}{2}\omega_n \sqrt{1 - \left(\frac{\Delta K}{K}\right)^2} \quad (9.4.10)$$

The second term of equation (9.4.9) is also shown plotted in Fig. 9.10, which shows resonance at $\omega = 0.46\,\omega_n$.

For considering the stability of the second harmonic, we modify the free vibration system equation, with ξ and η coordinates rotating at 2ω, by virtue of the forced vibration of the horizontal shaft that can occur at 2ω

frequency. Then

$$\ddot{\xi} - 4\omega\dot{\eta} + (\omega_1^2 - 4\omega^2)\xi = 0$$
$$\ddot{\eta} + 4\omega\dot{\xi} + (\omega_2^2 - 4\omega^2)\eta = 0 \qquad (9.4.11)$$

Using equation (9.2.8), the frequency equation becomes

$$\lambda_0^4 - \lambda_0^2(\omega_1^2 + \omega_2^2 + 8\omega^2) + (\omega_1^2 - 4\omega^2)(\omega_2^2 - 4\omega^2) = 0 \qquad (9.4.12)$$

Since $\lambda = \lambda_0 + 2\omega$, we get

$$\frac{\lambda}{\omega_n} = \frac{2\omega}{\omega_n}$$

$$\pm \sqrt{1 + 4\frac{\omega^2}{\omega_n^2} \pm \sqrt{\left(1 + 4\frac{\omega^2}{\omega_n^2}\right)^2 - \left(1 - \frac{\Delta K}{K} - 4\frac{\omega^2}{\omega_n^2}\right)\left(1 + \frac{\Delta K}{K} - 4\frac{\omega^2}{\omega_n^2}\right)}} \qquad (9.4.13)$$

In general there are four real roots for λ/ω_n, except when

$$1 - \frac{\Delta K}{K} < 4\frac{\omega^2}{\omega_n^2} < 1 + \frac{\Delta K}{K} \qquad (9.4.14)$$

Under these conditions, there are two real roots and two complex roots which give rise to oscillatory solutions. However the real parts being positive, these solutions are unstable. Equation (9.4.14) is

$$\tfrac{1}{2}\omega_1 < \omega < \tfrac{1}{2}\omega_2 \qquad (9.4.15)$$

In this region, any possible disturbance at second harmonic of rotational speed would make the rotor unstable.

For an additional study, use may be made of [5] to [16] given at the end of the chapter.

9.5 Transient Response by Time Marching Technique

So far in this chapter, the bearings are considered rigid. In practice, it is important to consider the influence of hydrodynamic bearings on the behaviour of asymmetric rotating shafts, as in the case of alternator rotors in turbogenerator sets. For the present, we restrict ourselves to Jeffcott models.

When we considered the asymmetric shafting on rigid bearings, it was easier to write the equations of motion in rotating coordinates. However, when the bearing forces are to be combined to the asymmetric shaft, it becomes useful to write the equations in the fixed coordinate system. The solution of such equations can be obtained by time marching techniques such as modified Euler's method.

9.5.1 Equations of motion

Consider the Jeffcott model of a rotor with rectangular cross-section shaft mounted on fluid film bearings as shown in Fig. 9.11. Figure 9.12

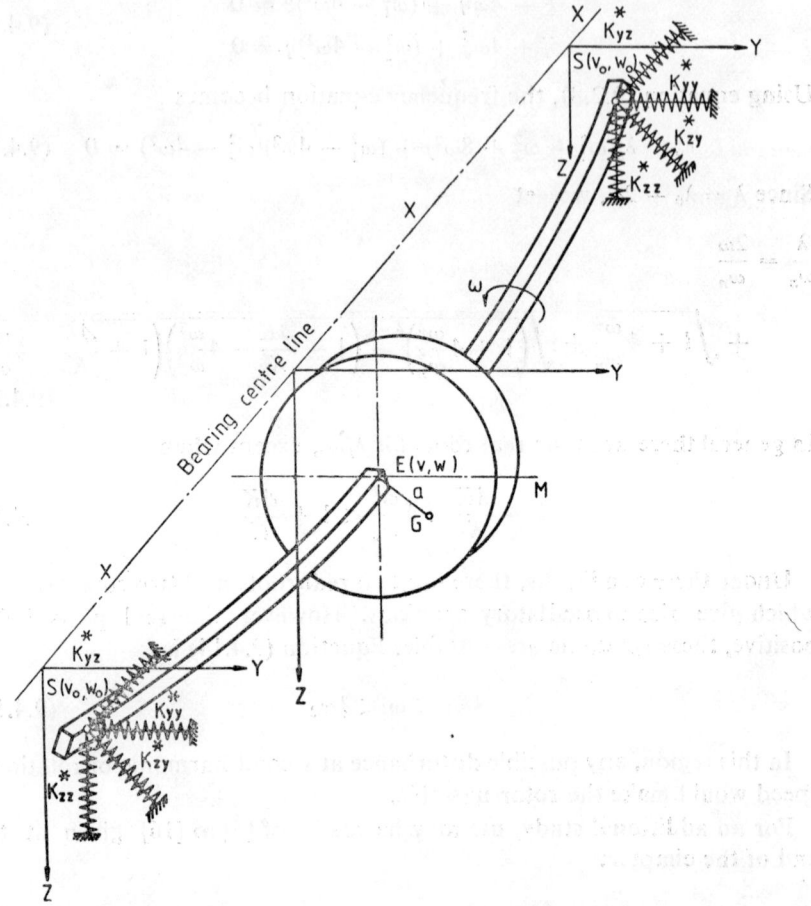

Fig. 9.11 A Jeffcott rotor with asymmetric cross-section mounted on fluid film bearings

gives the equilibrium relations taking into account the shaft deflections in the bearings. Let e_ξ, e_η, e_z and e_y represent the unit vectors along the ξ, η rotating coordinates and z, y stationary coordinates respectively. The equations of motion are

$$M(\ddot{w}e_z + \ddot{v}e_y) + K_1(\xi - \xi_0)e_\xi + K_2(\eta - \eta_0)e_\eta$$
$$= Mge_z + Ma\omega^2(e_\xi \cos\Omega + e_\eta \sin\Omega) \qquad (9.5.1)$$

$$K_1(\xi - \xi_0)e_\xi + K_2(\eta - \eta_0)e_\eta = F_z e_z + F_y e_y \qquad (9.5.2)$$

The bearing forces F_z and F_y are assumed in the above as

$$F_z = 2(K_{zz}w_0 + C_{zz}\dot{w}_0 + K_{zy}v_0 + C_{zy}\dot{v}_0)$$
$$F_y = 2(K_{yy}v_0 + C_{yy}\dot{v}_0 + K_{yz}w_0 + C_{yz}\dot{w}_0) \qquad (9.5.3)$$

Shafts with Dissimilar Moments of Area 179

The rotating coordinate quantities can be expressed in the fixed coordi-

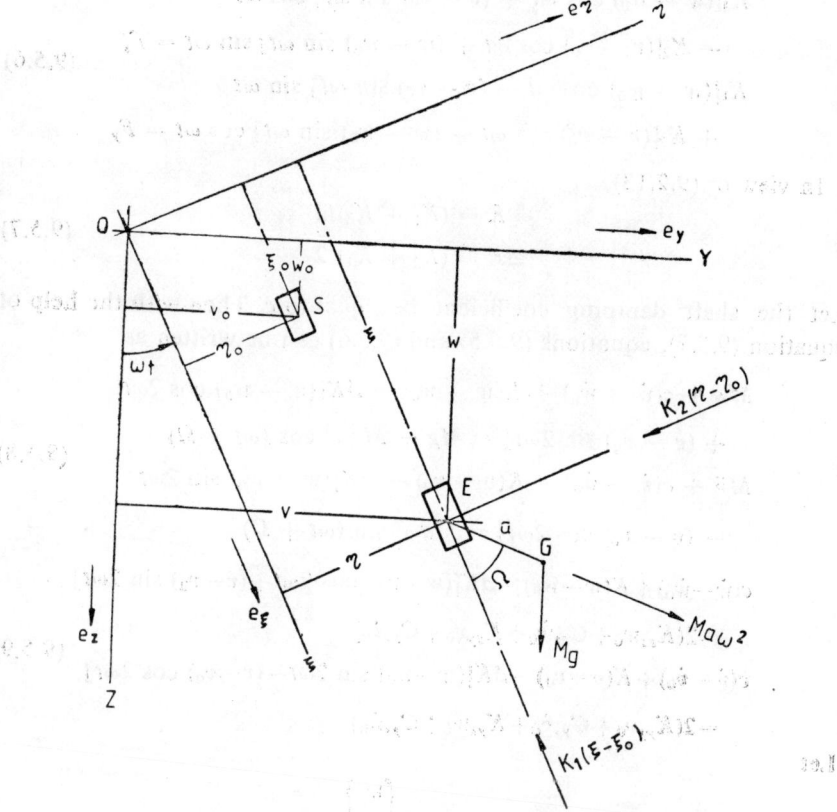

Fig. 9.12 Equilibrium relations of the shaft

nates using the following relations.

$$e_\xi = e_z \cos \omega t + e_y \sin \omega t$$
$$e_\eta = e_y \cos \omega t - e_z \sin \omega t$$
$$\xi = w \cos \omega t + v \sin \omega t$$
$$\eta = v \cos \omega t - w \sin \omega t$$
(9.5.4)

Substituting equations (9.5.4) in (9.5.1), we can write the equation entirely in the inertial coordinate system y-z. Separating e_z and e_y components, we get

$$M\ddot{w} + K_1[\tfrac{1}{2}(w - w_0)(1 + \cos 2\omega t) + \tfrac{1}{2}(v - v_0) \sin 2\omega t]$$
$$+ K_2[\tfrac{1}{2}(w - w_0)(1 - \cos 2\omega t) - \tfrac{1}{2}(v - v_0) \sin 2\omega t]$$
$$= Mg + Ma\omega^2 \cos (\omega t + \Omega)$$
$$M\ddot{v} + K_1[\tfrac{1}{2}(v - v_0)(1 - \cos 2\omega t) + \tfrac{1}{2}(w - w_0) \sin 2\omega t]$$
$$+ K_2[\tfrac{1}{2}(v - v_0)(1 + \cos 2\omega t) - \tfrac{1}{2}(w - w_0) \sin 2\omega t]$$
$$= Ma\omega^2 \sin (\omega t + \Omega)$$
(9.5.5)

180 *Rotor Dynamics*

Similarly equation (9.5.2) gives

$$K_1[(w-w_0)\cos\omega t + (v-v_0)\sin\omega t]\cos\omega t$$
$$- K_2[(v-v_0)\cos\omega t + (w-w_0)\sin\omega t]\sin\omega t = F_z \quad (9.5.6)$$
$$K_1[(w-w_0)\cos\omega t + (v-v_0)\sin\omega t]\sin\omega t$$
$$+ K_2[(v-v_0)\cos\omega t + (w-w_0)\sin\omega t]\cos\omega t = F_y$$

In view of (9.2.13),

$$K = (K_1 + K_2)/2$$
$$\Delta K = (K_2 + K_1)/2 \quad (9.5.7)$$

Let the shaft damping coefficient be equal to c. Then with the help of equation (9.5.7), equations (9.5.5) and (9.5.6) can be written as

$$M\ddot{w} + c(\dot{w} - \dot{w}_0) + K(w - w_0) - \Delta K[(w - w_0)\cos 2\omega t$$
$$+ (v - v_0)\sin 2\omega t] = Mg + Ma\omega^2 \cos(\omega t + \Omega) \quad (9.5.8)$$
$$M\ddot{v} + c(\dot{v} - \dot{v}_0) + K(v - v_0) - \Delta K[(w - w_0)\sin 2\omega t$$
$$- (v - v_0)\cos 2\omega t] = Ma\omega^2 \sin(\omega t + \Omega)$$

$$c(\dot{w}-\dot{w}_0)+K(w-w_0)-\Delta K[(w-w_0)\cos 2\omega t+(v-v_0)\sin 2\omega t]$$
$$=2(K_{zz}w_0+C_{zz}\dot{w}_0+K_{zy}v_0+C_{zy}\dot{v}_0]$$
$$c(\dot{v}-\dot{v}_0)+K(v-v_0)-\Delta K[(w-w_0)\sin 2\omega t-(v-v_0)\cos 2\omega t] \quad (9.5.9)$$
$$=2(K_{yy}v_0+C_{yy}\dot{v}_0+K_{yz}w_0+C_{yz}\dot{w}_0]$$

Let

$$\{U\} = \begin{Bmatrix} w \\ w_0 \\ v \\ v_0 \end{Bmatrix} \quad (9.5.10)$$

then, equations (9.5.8) and (9.5.9) can be written as

$$[\bar{M}]\{\ddot{U}\} + [\bar{C}]\{\dot{U}\} + [\bar{K}]\{U\} = \{\bar{F}\} \quad (9.5.11)$$

where

$$[\bar{M}] = \begin{bmatrix} M & 0 & 0 & 0 \\ 0 & 0 & 0 & 0 \\ 0 & 0 & M & 0 \\ 0 & 0 & 0 & 0 \end{bmatrix}$$

$$[\bar{C}] = \begin{bmatrix} c & -c & 0 & 0 \\ c & -c & 0 & 0 \\ 0 & 0 & c & -c \\ 0 & 0 & c & -c \end{bmatrix}$$

$$[\bar{K}] = \begin{bmatrix} K_{c-} & -K_{c-} & -\Delta K_s & \Delta K_s \\ K_{c-} & -K_{c-} & -\Delta K_s & \Delta K_s \\ -\Delta K_s & \Delta K_s & K_{c+} & -K_{c+} \\ -\Delta K_s & \Delta K_s & K_{c+} & -K_{c+} \end{bmatrix}$$

$$K_{c-} = K - \Delta K \cos 2\omega t$$
$$K_{c+} = K + \Delta K \cos 2\omega t$$
$$\Delta K_s = K \sin 2\omega t$$

$$\{\bar{F}\} = \begin{Bmatrix} Mg + Ma\omega^2 \cos(\omega t + \Omega) \\ F_z \\ Ma\omega^2 \sin(\omega t + \Omega) \\ F_y \end{Bmatrix}$$

9.5.2 Average acceleration method

The most suitable method to solve the above set of equations in (9.5.11) is the constant average acceleration Newmark method. In a given interval of time, the acceleration is taken as constant equal to an average value between the beginning and completion of time for this interval. Let $\{U_t\}$ represent the state vector quantities of (9.5.10) at any instant of time t. Then the velocity and displacement at an interval Δt later are given by

$$\{\dot{U}_{t+\Delta t}\} = \{\dot{U}_t\} + \tfrac{1}{2}(\{\ddot{U}_t\} + \{\ddot{U}_{t+\Delta t}\}) \Delta t \qquad (9.5.12)$$

$$\{U_{t+\Delta t}\} = \{U_t\} + \{\dot{U}_t\} \Delta t + \tfrac{1}{4}(\{\ddot{U}_t\} + \{\ddot{U}_{t+\Delta t}\})(\Delta t)^2 \qquad (9.5.13)$$

In writing the above equations, the acceleration in the interval Δt is taken as average value between t and $t + \Delta t$, i.e.,

$$\text{Average acceleration} = \tfrac{1}{2}(\{\ddot{U}_t\} + \{\ddot{U}_{t+\Delta t}\})$$

Hence the name average acceleration method. From equation (9.5.13), we have,

$$\{\ddot{U}_{t+\Delta t}\} = \{-\ddot{U}_t\} - \{\dot{U}_t\}\frac{4}{\Delta t} + (\{U_{t+\Delta t}\} - \{U_t\})\frac{4}{(\Delta t)^2} \qquad (9.5.14)$$

Substituting the above in equation (9.5.12), we get

$$\{\dot{U}_{t+\Delta t}\} = \{-\dot{U}_t\} + (\{U_{t+\Delta t}\} - \{U_t\})\frac{2}{\Delta t} \qquad (9.5.15)$$

Substituting above two equations in (9.5.11) for time $t + \Delta t$, we get

$$[\bar{M}]\left(\{-\ddot{U}_t\} - \{\dot{U}_t\}\frac{4}{\Delta t} + (\{U_{t+\Delta t}\} - \{U_t\})\frac{4}{(\Delta t)^2}\right)$$
$$+ [\bar{C}]\left(\{-\dot{U}_t\} + (\{U_{t+\Delta t}\} - \{U_t\})\frac{2}{\Delta t}\right) + [\bar{K}]\{U_{t+\Delta t}\} = \{\bar{F}_{t+\Delta t}\}$$
$$(9.5.16)$$

Rewriting the above

$$\left([\bar{M}]\frac{4}{(\Delta t)^2} + [\bar{C}]\frac{2}{\Delta t} + [\bar{K}]\right)\{U_{t+\Delta t}\} = \{\bar{F}_{t+\Delta t}\}$$
$$+ [\bar{M}]\left(\{\ddot{U}_t\} + \{\dot{U}_t\}\frac{4}{\Delta t} + \{U_t\}\frac{4}{(\Delta t)^2}\right) + [\bar{C}]\left(\{\dot{U}_t\} + \{U_t\}\frac{2}{\Delta t}\right) \quad (9.5.17)$$

i.e.

$$[\bar{A}]\{U_{t+\Delta t}\} = \{\bar{F}_{t+\Delta t}\} + \{G_t\} \quad (9.5.18)$$

where

$$[\bar{A}] = [\bar{M}]\frac{4}{(\Delta t)^2} + [\bar{C}]\frac{2}{\Delta t} + [\bar{K}] \quad (9.5.19)$$

$$\{G_t\} = [\bar{M}]\left(\{\ddot{U}_t\} + \{\dot{U}_t\}\frac{4}{\Delta t} + \{U_t\}\frac{4}{(\Delta t)^2}\right) + [\bar{C}]\left(\{\dot{U}_t\} + \{U_t\}\frac{2}{\Delta t}\right) \quad (9.5.20)$$

In equation (9.5.18). $[\bar{A}]$ and $\{G_t\}$ are defined completely at time t. $\{\bar{F}\}$ is not determinable at time $t + \Delta t$, since F_z and F_y in equation (9.5.3) are dependent on $w_0, v_0, \dot{w}_0, \dot{v}_0$ at the bearing at time $t + \Delta t$. We ignore this problem by taking these quantities at time t, for the first iteration and solve equation (9.5.18), to determine $\{U\}$, at $t + \Delta t$ and then $\{\dot{U}\}$, $\{\ddot{U}\}$ from (9.5.15) and (9.5.14) respectively. The values of $\{U\}$ and $\{\dot{U}\}$ thus determined at $t + \Delta t$ are then used in the second iteration to determine $\{\bar{F}\}$ more accurately this time than in the first iteration. We repeat this process of iteration until $\{U\}$ determined at $t + \Delta t$ converges to a desired accuracy. At this stage we can note that F_z and F_y in equation (9.5.3) need not be limited to a linear case.

Rao and Bhaskara Sarma [17] developed a computer program based on the above analysis, to determine the transient response of a Jeffcott type rotor, with asymmetric stiffnesses and mounted on linear 8 coefficient bearings. They found that a small amount of damping in the shaft is necessary to obtain a converged solution in a reasonable number of iterations. When no damping was used, the solution did not converge to a satisfactory value even after 110 iterations. In a typical example such a solution converged in less than 20 iterations, if a small value of damping ratio 0.0001 is used in the analysis. The solution converges much faster, say in 5 iterations, if the damping ratio value is 0.1. The shaft damping ratio is therefore a critical value to be chosen in obtaining reasonably good solution for the undamped case, consistent with the computer time that one can spend on this procedure.

Another factor involved in the numerical solution is the time step. It was found in a typical case, $\Delta t = 0.4 \times 10^{-3}$ seconds to be optimal considering the accuracy and computer time factors. It should be noted that there is unfortunately no specific rule or guidance in arriving at such factors effecting the numerical solution. One should rely upon his own experience in solving such problems.

Example 9.1

Consider a rotor of mass 54.3 kg with a rigid bearing critical speed 4820 rpm. The asymmetry of the shaft is $\Delta K/K = 0.4$. Both the support bearings are rigid. There is no unbalance and the response due to gravity alone is to be considered.

The response obtained from the computer program at 1000 rpm, corresponding to $\omega/\omega_n = 0.207$ is shown in Fig. 9.13. When disturbed from

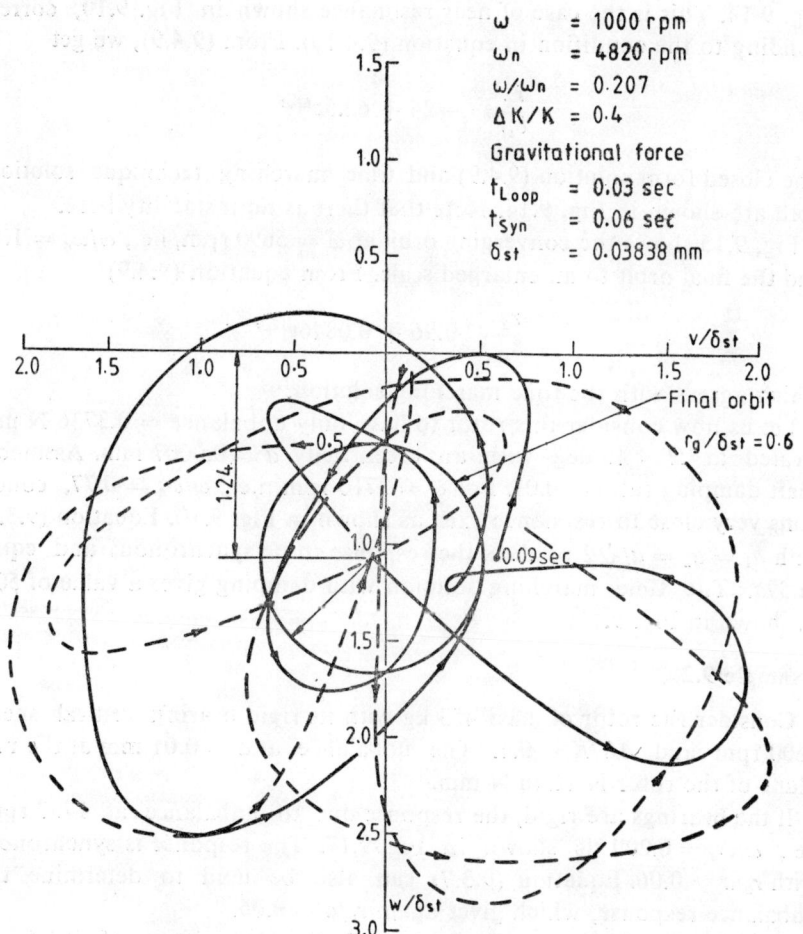

Fig. 9.13 Orbital response of an asymmetric shaft on rigid bearings due to gravity $\omega/\omega_n = 0.207$

equilibrium, the rotor initially whirls with large amplitudes before settling down to a steady state response. The final orbit as seen in Fig. 9.13 is circular in nature. It was found that time taken for completion of one loop was 0.03 sec, which is equal to half the synchronous time 0.06 secs.

From equation (9.4.9) we see that the solution has two parts, the first term gives the shaft deflection from the bearing axis and the second one, the whirl orbit radius at twice the rotational frequency, i.e.,

$$\frac{r_g}{\delta_{st}} = \frac{1 - 4(0.207)^2}{1 - 4(0.207)^2 - (0.4)^2} + \frac{0.4}{1 - 4(0.207)^2 - (0.4)^2} e^{i2\omega t}$$
$$= 1.24 + 0.6 e^{i2\omega t}$$

It can be seen from Fig. 9.13, the time marching solution gives the same result as above.

For $\omega/\omega_n = 0.46$, i.e., $\omega = 2217$ rpm, the solution obtained is shown in Fig. 9.14. This is the case of near resonance shown in Fig. 9.10, corresponding to the condition in equation (9.4.10). From (9.4.9), we get

$$\frac{r_g}{\delta_{st}} = -24 + 62.5 e^{i2\omega t}$$

The closed form solution (9.4.9) and time marching technique solution, both are shown in Fig. 9.14. Note that there is no instability here.

Fig. 9.15 shows the converging orbit at $\omega = 5690$ rpm, i.e., $\omega/\omega_n = 1.18$ and the final orbit to an enlarged scale. From equation (9.4.9)

$$\frac{r_g}{\delta_{st}} = 0.96 + 0.0846 e^{i2\omega t}$$

which agrees with the time marching solution.

Let us now consider this rotor to have only unbalance = 0.3736 N mm located at $\Omega = 45$ deg with an eccentricity $a = 0.0007$ mm. Assume a shaft damping ratio = 0.01. For $\omega = 3710$ rpm, i.e., $\omega/\omega_n = 0.77$, conditions very close to resonance exist as shown in Fig. 9.10. Equation (9.3.7) with $a_1 = a_2 = a/\sqrt{2}$ predicts the response to be synchronous and equal to $59a$. The time marching solution with damping gives a value of $50a$, as shown in Fig. 9.16.

Example 9.2

Consider the rotor of mass 453 kg with its rigid bearing critical speed 8600 rpm and $\Delta K/K = 0.4$. The unbalance at $a = 0.01$ mm at the mid plane of the rotor is 44.46 N mm.

If the bearings are rigid, the response due to unbalance at 1797 rpm, i.e., $\omega/\omega_n = 0.209$ is shown in Fig. 9.17. The response is synchronous with $r_u/a = 0.06$. Equation (9.3.7) can also be used to determine the unbalance response, which gives again $r_u/a = 0.06$.

Let this rotor be mounted on plain cylindrical bearings of 101.6 mm dia with a length of 50.88 mm. The clearance is 0.1016 mm and the oil viscosity at the operating temperature of the bearing is taken as 0.00568 Ns/m. In the analysis the bearing damping is neglected, but all the stiffness values are included.

If $\Delta K = 0$, the analysis of section 8.1.1 is applicable and the peak to peak amplitude, see Fig. 8.4, is 0.936 for a speed of 3000 rpm. The time marching method predicts a solution of similar value as shown in Fig. 9.18.

Shafts with Dissimilar Moments of Area

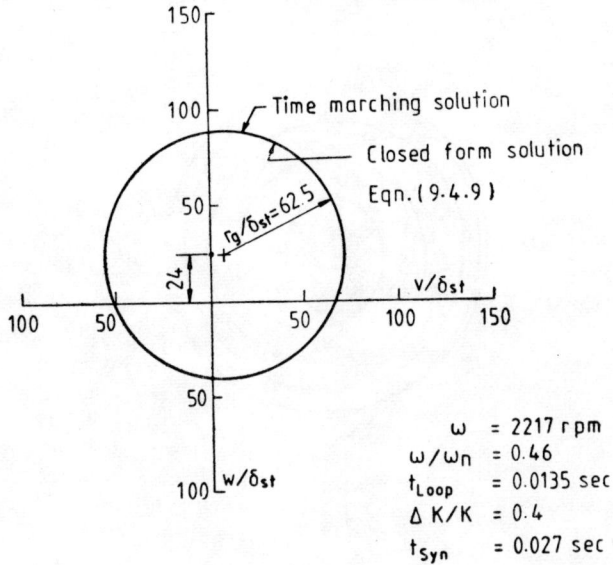

Fig. 9.14 Orbital response of an asymmetric shaft on rigid bearings due to gravity $\omega/\omega_n = 0.46$

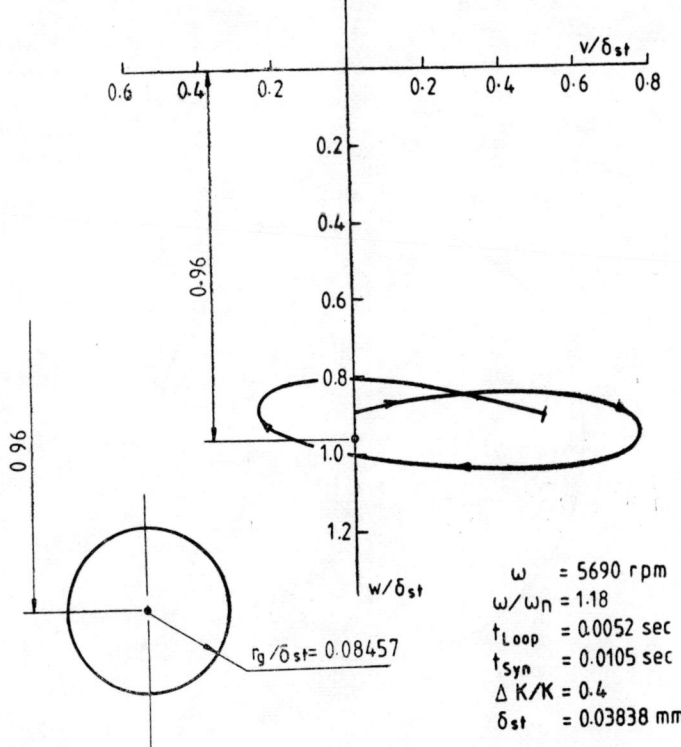

Fig. 9.15 Orbital response of an asymmetric rotor on rigid bearings due to gravity $\omega/\omega_n = 1.18$

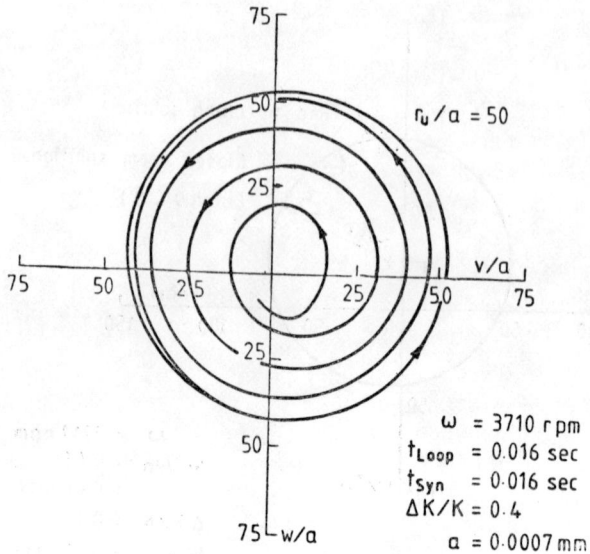

Fig. 9.16 Orbital response of an asymmetric shaft mounted on rigid bearings due to unbalance force $\omega/\omega_n = 0.77$

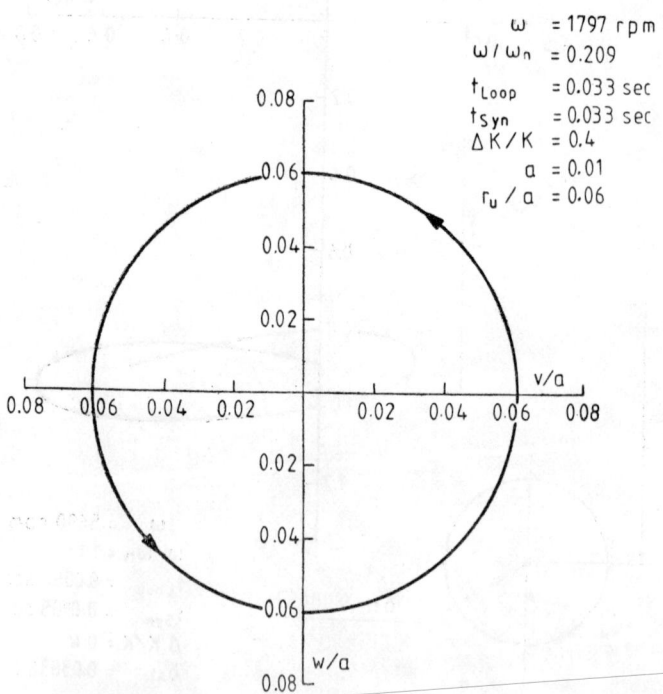

Fig. 9.17 Orbital response of an asymmetric shaft mounted on rigid bearings due to unbalance force $\omega/\omega_n = 0.209$

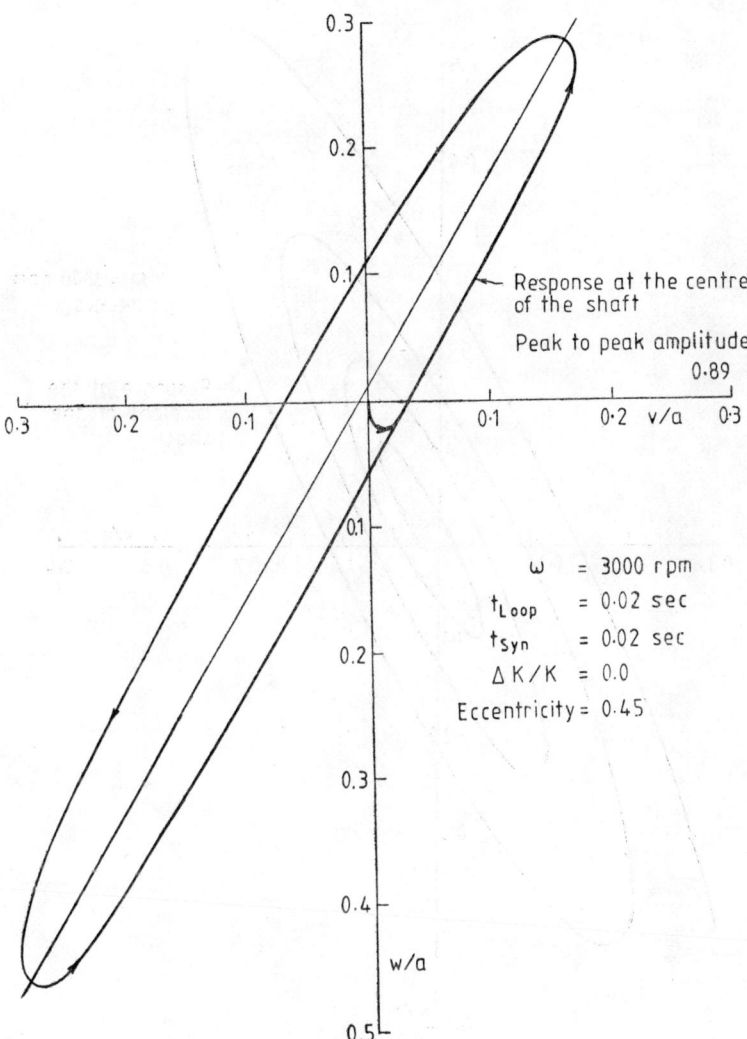

Fig. 9.18 Whirl orbit of a symmetric shaft mounted on fluid film bearings (3000 rpm)

When the asymmetry $\Delta K/K = 0.4$ is considered, the response obtained for 3870 rpm is found to be unstable as given in Fig. 9.19. The value of ω/ω_n for this case is 0.45, which falls in the instability zone corresponding to $\omega_n = \omega_k = 2\omega$ in Figs. 9.1 and 9.7. When the speed is raised to 4300 rpm, i.e., $\omega/\omega_n = 0.5$, the orbit becomes stable again as shown in Fig. 9.20.

9.6 Instability Due to a Transverse Crack

It is commonly observed that high speed and heavy duty rotor shafts develop transverse cross-sectional cracks due to fatigue at some time

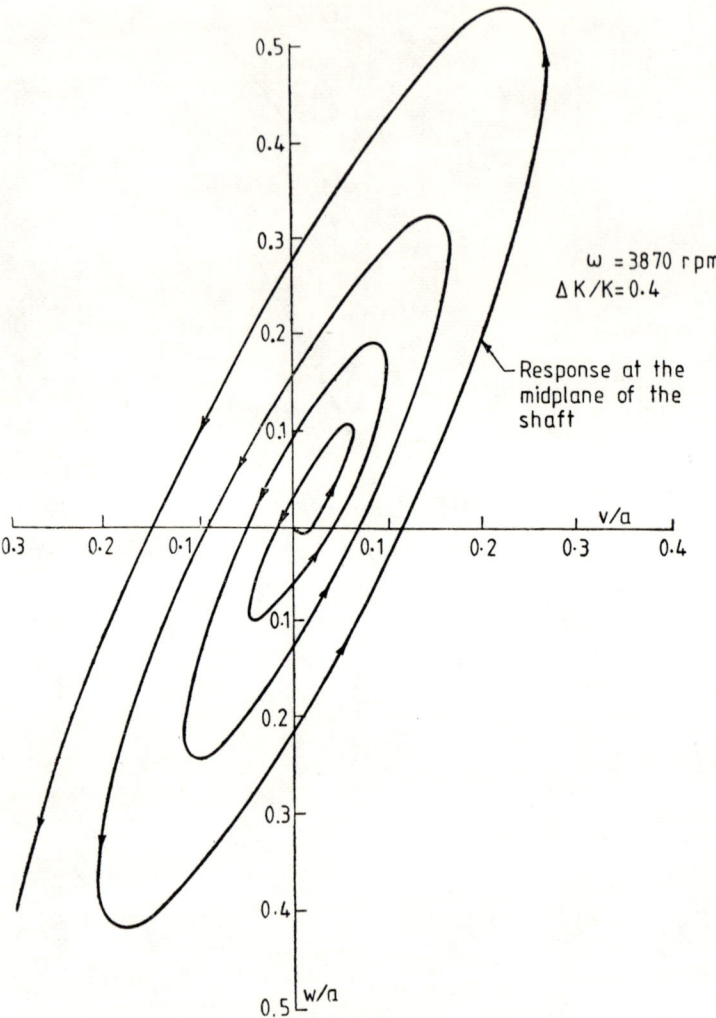

Fig. 9.19 Orbital response of an asymmetric shaft mounted on fluid film bearings $\omega/\omega_n = 0.45$

during their life period. A study of initiation of crack and its growth is a separate subject in itself. At this time, we consider the shafts that have developed cracks, specifically cross-sectional transverse cracks, due to fatigue.

Fig. 9.21 shows a simple symmetric shaft mounted on rigid bearings with a transverse cross-sectional crack. Let K be the stiffness of the uncracked shaft and the corresponding deflection due to the weight of the rotor be $\delta_{st} = Mg/K$. When the shaft is rotated, the rotor deflection will vary due to changes in the stiffness of the shaft as a function of the location of the crack. When the crack is occupying the lowermost position, say 0 deg, the shaft offers minimum stiffness. In this position, the

Shafts with Dissimilar Moments of Area 189

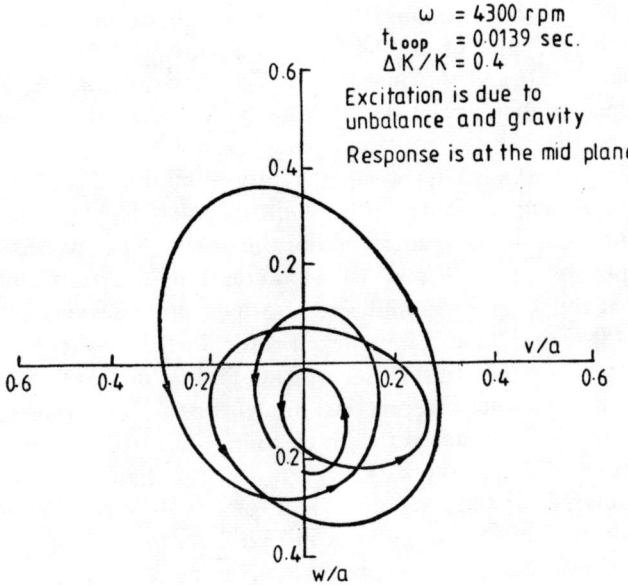

Fig. 9.20 Orbital response of an asymmetric shaft mounted on fluid film bearings $\omega/\omega_n = 0.5$

crack is located amongst the shaft fibers which are in tension, consequently it is wide open. As the shaft rotates further in a counter clockwise direction, to the 90 deg position, a portion of the crack is in compression and hence it tends to get closed as shown by the double hatched area in Fig. 9.21. Consequently, the shaft stiffness increases and the rotor deflection decreases. When the shaft is in the 180 deg position, it can be seen

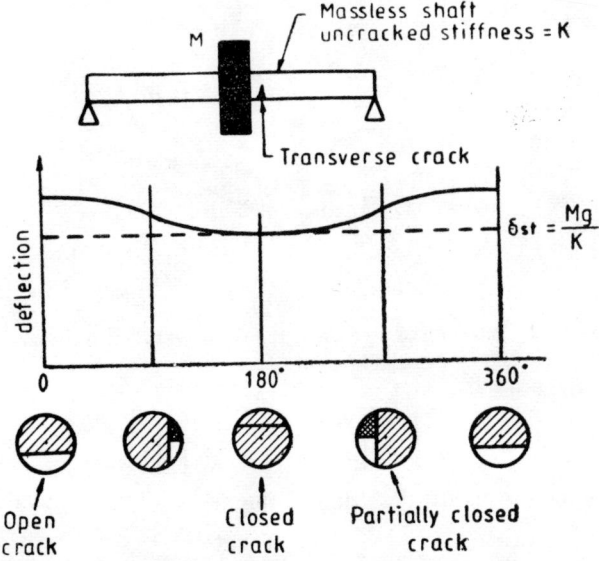

Fig. 9.21 Vertical deflection of the shaft under gravity

that the crack is located amongst the fibres which are in compression and hence it is completely closed. The stiffness in this position corresponds to the uncracked shaft and the deflection is δ_{st} as shown in Fig. 9.21. The crack begins to open as the shaft is rotated further, until it is wide open once again at the 360 deg position.

The stiffness of the shaft is significantly diminished only when the crack is of sufficient depth. For shallow crack depths, the stiffness is decreased in a limited area surrounding the crack. The periodic closing and opening of the crack is called "breathing" action and an exact modelling of this is quite complicated. Mayes and Davies [18] briefly discussed the compliance of a cracked rotor, based on the work of Paris and Sih [19], using the stress concentration factors of Harris [20]. Gasch [21] adopted a simpler but practical model for the breathing action as a step function. He considered the crack to be completely open, when it is located below the neutral axis in tension for half a revolution and completely closed for the remaining half of revolution when it is above the neutral axis of bending. Thus in Fig. 9.22, for a crack parallel to η axis in the rotating coordinates, we can write the following.

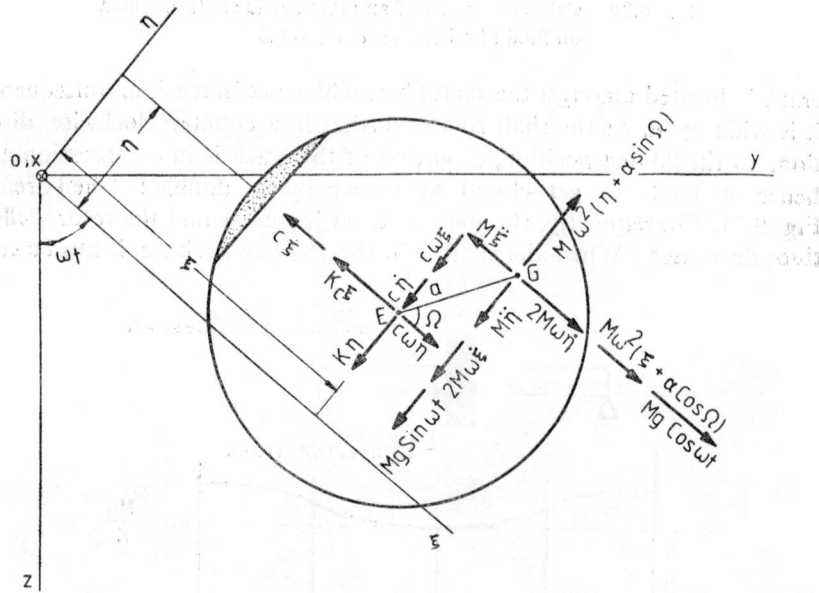

Fig. 9.22 Whirling of a rotor with a cross-sectional transverse crack parallel to η axis

$$\begin{Bmatrix} F_\xi \\ F_\eta \end{Bmatrix} = \begin{bmatrix} K - \Delta K_c & 0 \\ 0 & K \end{bmatrix} \begin{Bmatrix} \xi \\ \eta \end{Bmatrix} \quad (9.6.1)$$

where the reduction in stiffness due to the crack, ΔK_c is taken as

$$\begin{aligned} \Delta K_c \neq 0, \quad \xi > 0 \\ \Delta K_c = 0, \quad \xi < 0 \end{aligned} \quad (9.6.2)$$

Shafts with Dissimilar Moments of Area

This simpler model for breathing action above can be represented by a hinge which closes under compression and opens under tension as shown in Fig. 9.23. For such a system of forces, with c as the equivalent viscous

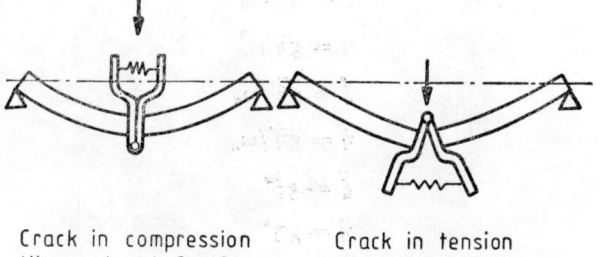

Crack in compression Crack in tension
Hinge closed $\xi < 0$ Hinge open $\xi > 0$
Stiffness = K Stiffness = $K_c = K - \Delta K_c$

Fig. 9.23 Crack model—Breathing action

damping coefficient, the equations of motion of the cracked shaft in Fig. 9.22, with the help of equations (9.2.3) and (9.2.4), can be written as

$$M[\ddot{\xi} - 2\omega\dot{\eta} - \omega^2(\xi + a \cos \Omega)] + c(\dot{\xi} - \omega\eta) + K_c\xi = Mg \cos \omega t$$
$$M[\ddot{\eta} + 2\omega\dot{\xi} - \omega^2(\eta + a \sin \Omega)] + c(\dot{\eta} + \omega\xi) + K\eta = -Mg \sin \omega t$$

(9.6.3)

where

$$K_c = K - \Delta K_c, \quad \xi > 0$$
$$= K \quad \xi < 0 \quad (9.6.4)$$

Equations (9.6.3) in matrix form are

$$\begin{bmatrix} M & 0 \\ 0 & M \end{bmatrix}\begin{Bmatrix} \ddot{\xi} \\ \ddot{\eta} \end{Bmatrix} + \begin{bmatrix} c & (-2M\omega) \\ (2M\omega) & c \end{bmatrix}\begin{Bmatrix} \dot{\xi} \\ \dot{\eta} \end{Bmatrix}$$
$$+ \begin{bmatrix} (K_c - M\omega^2) & (-c\omega) \\ c\omega & (K - M\omega^2) \end{bmatrix}\begin{Bmatrix} \xi \\ \eta \end{Bmatrix}$$
$$= \begin{Bmatrix} Mg \cos \omega t \\ -Mg \sin \omega t \end{Bmatrix} + \begin{Bmatrix} Ma\omega^2 \cos \Omega \\ Ma\omega^2 \sin \Omega \end{Bmatrix} \quad (9.6.5)$$

Let

$$\omega_n = \sqrt{K/M}$$
$$\omega_{nc} = \sqrt{K_c/M}$$
$$\zeta = 2cM\omega_n$$
$$\eta_{st} = Mg/K$$
$$= g/\omega_n^2$$
$$\bar{\xi} = \xi/\eta_{st}$$
$$\bar{\eta} = \eta/\eta_{st}$$

$$\tau = \omega_n t$$
$$\dot{} = d/d\tau \tag{9.6.6}$$

Then

$$\xi = g\bar{\xi}/\omega_n^2$$
$$\eta = g\bar{\eta}/\omega_n^2$$
$$\dot{\xi} = g\bar{\xi}'/\omega_n$$
$$\dot{\eta} = g\bar{\eta}'/\omega_n$$
$$\ddot{\xi} = g\bar{\xi}''$$
$$\ddot{\eta} = g\bar{\eta}'' \tag{9.6.7}$$

Substituting equations (9.6.7) in (9.6.5) and dividing throughout by Mg, and rearranging, we get

$$[I]\begin{Bmatrix}\bar{\xi}''\\\bar{\eta}''\end{Bmatrix} + \begin{bmatrix}2\zeta & -2\dfrac{\omega}{\omega_n}\\2\dfrac{\omega}{\omega_n} & 2\zeta\end{bmatrix}\begin{Bmatrix}\bar{\xi}'\\\bar{\eta}'\end{Bmatrix} + \begin{bmatrix}\left(\dfrac{\omega_{nc}^2}{\omega_n^2} - \dfrac{\omega^2}{\omega_n^2}\right) & \left(-2\zeta\dfrac{\omega}{\omega_n}\right)\\2\zeta\dfrac{\omega}{\omega_n} & \left(1 - \dfrac{\omega^2}{\omega_n^2}\right)\end{bmatrix}\begin{Bmatrix}\bar{\xi}\\\bar{\eta}\end{Bmatrix}$$

$$= \begin{Bmatrix}\cos\dfrac{\omega}{\omega_n}\tau\\-\sin\dfrac{\omega}{\omega_n}\tau\end{Bmatrix} + \begin{Bmatrix}\dfrac{a}{\eta_{st}}\dfrac{\omega^2}{\omega_n^2}\cos\Omega\\\dfrac{a}{\eta_{st}}\dfrac{\omega^2}{\omega_n^2}\sin\Omega\end{Bmatrix} \tag{9.6.8}$$

Equations (9.6.4) are similar to the asymmetric shaft equations (9.2.5), however they are nonlinear, because of the definition of K_c in (9.6.4). Hence closed form solutions are not possible. Gasch [21] used an analogue computer to solve equations (9.6.8). His results are briefly discussed here.

He used the following parameters.

$$\dfrac{\omega_{nc}}{\omega_n} = 1 \quad \text{(hinge closed)}$$

$$= 0.9 \quad \text{(hinge open)}$$

$$\zeta = 0.005$$

The response due to gravity alone, with $a = 0$, is shown in Fig. 9.24. The resonance peak at $\omega/\omega_n = 0.5$, is clearly visible. This is similar to the unbalance response of an asymmetric shaft given in Fig. 9.10. Also the unstable zone in the range $0.9 < \omega/\omega_n < 1.0$ can be recognized. However, in the cracked shaft response, there are other secondary resonances at $\omega/\omega_n = 1/3, 1/4$ etc. and $2/3$, which are subharmonics. They do not occur with an asymmetric shaft. Gasch found the orbits of the whirl generally elliptical, except where the subharmonics occur. Typically for ω/ω_n in $1/3$ and $2/3$ regions, the whirls obtained are shown in

Figs. 9.25a and b respectively. These are higher harmonics typical to the influence of nonlinearity of the crack.

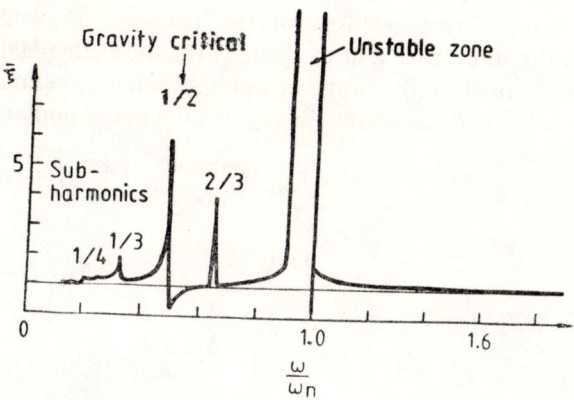

Fig. 9.24 Response of a cracked shaft under gravity

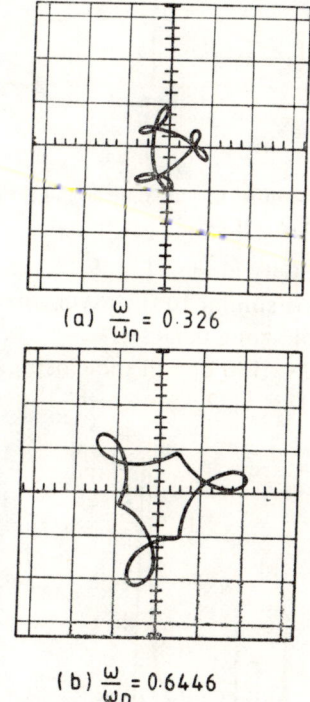

Fig. 9.25 Typical shaft whirls at subharmonics

Gasch also noted that with increased damping, the secondary resonance peaks disappear and the unstable zone near $\omega/\omega_n = 1$ becomes smaller and the gravity resonance peak at $\omega/\omega_n \approx 1/2$ less sharp. Increasing the crack depth has shown just the opposite effect giving rise to more pronounced secondary resonances, larger unstable region near $\omega/\omega_n = 1$ and sharper gravity resonance.

In the studies on the unbalance, Gasch found that the location of unbalance has interesting features of the response. Depending on the angle between the unbalance and the hinge axis, the centrifugal unbalance forces tend either to close the hinge or open it. When $\Omega = 0$, $a/\eta_{st} = 0.5$, the response obtained is shown in Fig. 9.26. This is similar to the res-

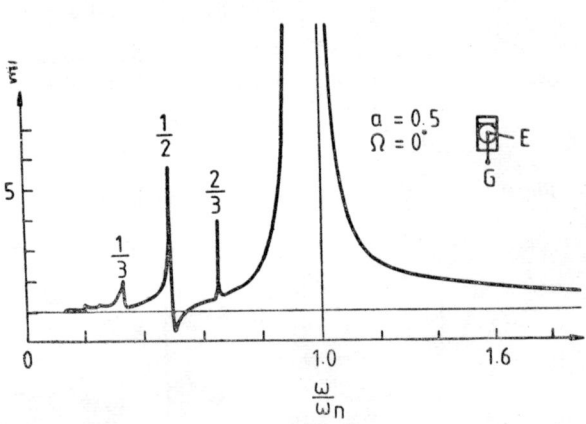

Fig. 9.26 Response due to unbalance and gravity unbalance position $\Omega = 0$

ponse obtained due to gravity alone, Fig. 9.24, excepting that the unbalance response which is similar to the asymmetric shaft in Fig. 9.10, is interrupted by the unstable zone $0.9 < \omega/\omega_n < 1.0$. When the unbalance location is changed to $\Omega = 180$ deg in the closing position, the response obtained is shown in Fig. 9.27, in which the unstable zone almost disappeared.

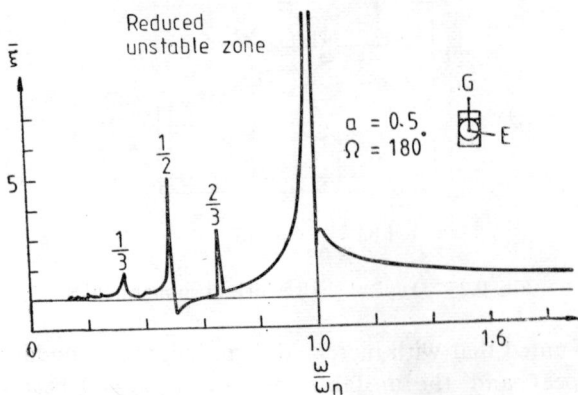

Fig. 9.27 Response due to unbalance and gravity unbalance position $\Omega = 180°$

Additional reference on this subject [22 to 36] are given at the end of the chapter for a further study.

References

1. Hayashi, T. Nonlinear oscillations in physical systems, McGraw-Hill Book Co., 1965.
2. Timoshenko, S.P. Vibration problems in engineering, D. Van Nostrand Co. Inc., 1955.
3. Den Hartog, J.P. Mechanical Vibration, McGraw-Hill Book Co., 1956.
4. Tondl, A. Some problems in rotor dynamics, Publishing House Czechoslovakian Academy of Sciences, Prague, 1965.
5. Taylor, H.D. Critical speed behaviour of unsymmetrical shafts, J. of Appld. Mech., ASME, Vol. 84, 1945, p. A77.
6. Crandall, S.H. and Brosens, P.J. On the stability of rotation of a rotor with rotationally unsymmetric inertia and stiffness properties, J. of Appld. Mech., ASME, 1961, p. 567.
7. Brosens, P.J. Vibrations of unsymmetrical rotors, J. of Appld. Mech., ASME, 1961, p. 355.
8. Yamamoto, T. and Ota, H. On the unstable vibrations of a shaft carrying an unsymmetrical rotor, JSME, Vol. 31, 1964, p. 525.
9. Yamamoto, T., et al. On the forced vibrations of the shaft carrying an unsymmetrical rotor: Forced vibrations having the circular frequencies differing from the rotating angular velocity of the shaft, Bull. JSME, Vol. 9, 1966, p. 58.
10. Yamamoto, T., et al. On the unstable vibrations of a shaft with unsymmetrical stiffness carrying an unsymmetrical rotor, J. Appld. Mech., ASME, 1968, p. 313.
11. Iwatsubo, T. Vibration of asymmetric shaft, JSME, Vol. 37, 1971, p. 1503.
12. Yamamoto, T. and Kono, K. Forced vibrations of a rotor with rotating inequality, Bull. JSME, Vol. 14, 1971, p. 1059.
13. Ota, H. and Kono, K. Unstable vibrations induced by rotationally unsymmetric inertia and stiffness properties, Bull. JSME, Vol. 14, 1971, p. 29.
14. Arnold, R.C. and Haft, E.E. Stability of unsymmetrical rotating cantilever shaft carrying an unsymmetrical rotor, J. of Engng. for Indus., ASME, 1972, p. 243.
15. Forrai, L. Vibrations of a rotating asymmetric shaft carrying two disks supported in asymmetric bearings, Vibrations in rotating machinery conf., I. Mech. E., 1976, p. 43.
16. Inagaki, T., Kanki, H. and Shiraki, K. Response analysis of a general asymmetric rotor-bearing system, ASME 79-DET-84.
17. Rao, J.S. and Bhaskara Sarma, K.V. On transient dynamics of rotors with asymmetric cross-section supported on fluid film bearings, Proc. 4th Intl. Conf. on Modal Analysis, 1986, p. 1110.
18. Mayes, I.W. and Davies, W.G.R. The vibrational behaviour of a rotating shaft system containing a transverse crack. Proc. Intl. Conf. Vibrations in Rotating Machinery, I. Mech. E., 1976, p. 53.
19. Paris, P.C. and Sih, G.C. Stress analysis of cracks, Fracture toughness testing and applications, ASTM, 1964, p. 58.
20. Harris, D.O. Stress intensity factors for hollow circumferentially notched round bars, J. Basic Engng., ASME, 89D, p. 40.
21. Gasch, R. Dynamic behaviour of a simple rotor with a cross-sectional crack, Proc. Intl. Conf. Vibrations in Rotating Machinery, I. Mech. E., 1976, p. 123.
22. Mayes, I.W. and Davies, W.G.R. A method of calculating the vibrational behaviour of coupled rotating shafts containing a transverse crack, Proc. Intl. Conf. on Vibrations in Rotating Machinery, I. Mech. E., 1980, p. 13.

23. Grabowski, B. The vibrational behaviour of a turbine rotor containing a transverse crack, J. Mech. Des., ASME, v. 102, 1980, p. 140.
24. Nilsson, L.R.K. On the vibrational behaviour of a cracked rotor bearing system, Proc. Intl. Conf. Rotor dynamics problems in power plants, IFToMM, 1982, p. 515.
25. Grabowski, B. and Mahrenholtz, O. Theoretical and experimental investigations of shaft vibrations in turbomachinery excited by cracks, Intl. Conf. Rotor dynamics problems in power plants, IFToMM, 1982, p. 507.
26. Ingaki, T., Kanki, H. and Shiraki, K. Transverse vibrations of a general cracked rotor bearing system, J. Mech. Des. ASME, v. 104, 1982, p. 343.
27. Dimarogonas, A.D. and Papadopoulos, C.A. Vibration of cracked shafts in bending, J. Sound and Vib. v. 91, 1983, p. 583.
28. Bachschmid, N. A method for calculating the dynamic behaviour of cracked shafts, Proc. VI IFToMM world congress, New Delhi, 1983, p. 1343.
29. Bachschmid, N., Diana, G. and Pizzigoni, B. The influence of unbalance on cracked rotors, Proc. Intl. Conf. Vibrations in Rotating Machinery, I. Mech. E., 1984, p. 193.
30. Schmid, J. and Kramer, E. Vibration behaviour of a rotor with a cross-sectional crack, Proc. Intl. Conf. Vibrations in Rotating Machinery, 1984, p. 183.
31. Davies, W.G.R. and Mayes, I.W. The vibration behaviour of a multishaft, multibearing system in the presence of a propagating transverse crack, J. Vib. Acous. Stress Rel. Des. ASME, v. 106, 1984, p. 146.
32. Mayes, I.W. and Davies, W.G.R. Analysis of the response of a multirotor bearing system containing a transverse crack in a rotor, J. Vib. Acous. Stress Rel. Des. ASME, v. 106, 1984, p. 139.
33. Nelson, H.D. and Nataraj, C. The dynamics of a rotor system with a cracked shaft, J. Vib. Acous Stress Rel. Des. ASME, v. 108, 1986, p. 187.
34. Gasch, R., Person, M. and Wertz, B. Dynamic behaviour of the Laval rotor with a cracked hollow shaft—a comparison of crack models, Proc. Intl. Conf. Vibrations in Rotating Machinery, I. Mech. E., 1988, p. 463.
35. Wen, B.C. and Wang, Y.B. Theoretical research, calculation and experiments of cracked shaft dynamics response, Proc. Intl. Conf. Vibrations in Rotating Machinery, I. Mech. E., 1988, p. 473.
36. Tamura, A., Iwata, Y. and Sato, H. Unstable vibrations of a rotor with a transverse crack, Proc. Intl. Conf. Vibrations in Rotating Machinery, 1988, p. 647.

Failed tilting pad bearing (Courtesy BHEL, Hyderabad)

Chapter 10

Instability Due to Fluid Film Forces and Hysteresis

The instability of a rotor with dissimilar stiffness that occurs around half the critical speed has been studied in the previous chapter. Instability of a rotor can also occur at speeds above the critical speed, due to oil film forces and internal shaft hysteresis. For high speed rotors these phenomena are very important, which are considered in this chapter.

10.1 Instability of Rotors Mounted on Fluid Film Bearings

The instability of a rotor is a self-excited vibration arising out of fluid film forces and is distinct from the large amplitudes of vibration caused by residual unbalance. This phenomenon is also known as oil whip or oil whirl and has been first observed by Newkirk and Taylor [1]. To understand this phenomenon let us consider a lightly loaded journal bearing, i.e., the pressure developed in the film is insignificant, the journal centre operates close to the bearing centre and the eccentricity is very small compared to the radial clearance. Since the pressure induced flow is assumed to be negligible, the velocity profile of the film in the clearance space is linear with a maximum value ωR at the journal surface, as shown in Fig. 10.1. The flow into the wedge of the journal bearing is

$$F_i = \tfrac{1}{2}LR\omega(C + e) \tag{10.1.1}$$

The flow out of the wedge is

$$F_0 = \tfrac{1}{2}LR\omega(C - e) \tag{10.1.2}$$

If pressure is developed in the film, when the bearing is operating under steady conditions, the flow-in is reduced and flow-out is increased by the pressure induced flow, which balances F_i and F_0 for maintaining the flow continuity. However, if the load is small, in the absence of pressure a small whirl velocity is induced to maintain the flow balance. If the instantaneous angular velocity is ν for the journal centre j, then induced velocity is $e\nu$ as shown in Fig. 10.1. By lifting off the journal from its steady state position, the film volume increases by

$$F = 2LRe\nu \tag{10.1.3}$$

where $2LR$ is the projected area of the bearing. Therefore

$$\tfrac{1}{2}LR\omega(C + e) = \tfrac{1}{2}LR\omega(C - e) + 2LRe\nu \tag{10.1.4}$$

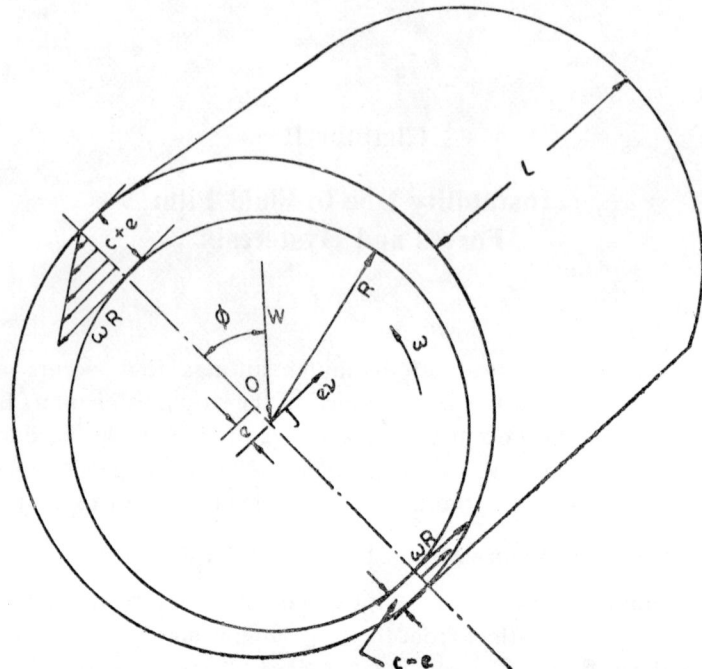

Fig. 10.1 Lightly loaded journal bearing

and

$$\nu = \tfrac{1}{2}\omega \qquad (10.1.5)$$

Hence the rotor tries to whirl at a frequency of half the speed of rotation to maintain the flow balance. We notice that if $\nu > \tfrac{1}{2}\omega$, the outward flow is more and therefore pressure is developed in the film and the bearing becomes stable. If, however $\nu < \tfrac{1}{2}\omega$, the flow-in is more, the bearing loses load carrying capacity and continues to whirl, in order to create more space for the excessive oil coming in to the wedge. The rotor thus loses load carrying capacity and becomes unstable. The frequency of whirl in the rotors under such conditions is observed to be around 0.46 to 0.48 rotational speed.

If the rotor is flexible, the instability caused by oil whirl causes large amplitudes due to resonant conditions generated by the rotor speed above the first critical. This is explained in Fig. 10.2, which is a Campbell type of diagram, showing schematically, the system resonant frequency as a function of rotor speed. The dashed lines show the synchronous whirl and the first critical speed, where the rotational speed is equal to the system natural frequency. The half frequency whirl line is also shown in Fig. 10.2. Since the journal is stable for frequency of whirl more than half the rotational speed, we can assume positive damping in the system above the half frequency line and negative damping below this. Where the system frequency and half frequency whirl lines meet, we have the

Instability due to Fluid Film Forces **201**

instability threshold, which is known as oil whip. This is obviously an important consideration in the design of the rotor and thus attracted the attention of many workers in rotor dynamics. Robertson [2] made a theoretical investigation and found the rotor inherently unstable at all speeds. Stability criteria were considered by several people for flexible rotors [3, 4]. Rieger and Thomas [5] made detailed computer studies and compared their results with experimental observations of threshold instability speeds of several others. Lund [6] gave design charts to

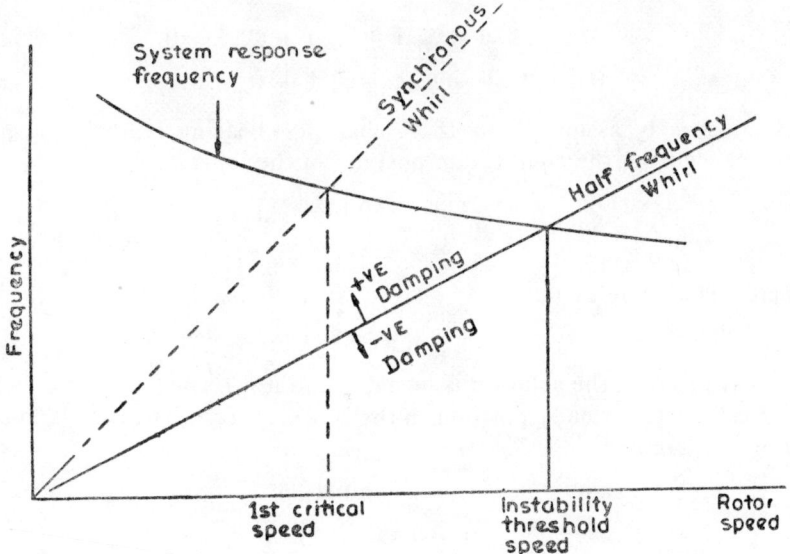

Fig. 10.2 Instability threshold speed—oil whip

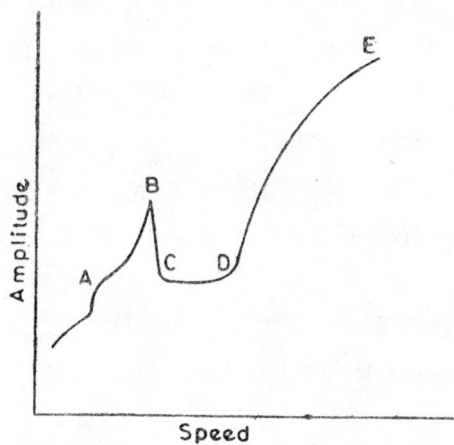

Fig. 10.3 General oil whip trend. A—half critical speed; B—resonant whip; D—threshold of instability half frequency whirl; E—oil whip.

predict the threshold instability speeds of rigid and flexible rotors on different types of bearings. The general oil whip trend observed experimentally [7] is given in Fig. 10.3, conforming to the discussions of Fig. 10.2. Analytical methods of determining the threshold speeds of instability are given in the following [8].

10.2 Rigid Rotor Instability

For a rigid rotor of mass $2M$, running in fluid film bearings, the equations of motion are

$$M\ddot{z} + K_{zz}z + K_{zy}y + C_{zz}\dot{z} + C_{zy}\dot{y} = 0$$
$$M\ddot{y} + K_{yy}y + K_{yz}z + C_{yy}\dot{y} + C_{yz}\dot{z} = 0 \quad (10.2.1)$$

Symmetry is assumed, so that mass per bearing is M. For a small disturbance from the equilibrium position of the journal.

$$z = z \exp(\lambda t)$$
$$y = y \exp(\lambda t) \quad (10.2.2)$$

where λ is a complex number

$$\lambda = \alpha + i\nu. \quad (10.2.3)$$

If α is positive, the solution is unstable and if it is negative, the whirl orbit reverts to a steady position in the bearing circle. The threshold condition is therefore

$$z = z \exp(i\nu t)$$
$$y = y \exp(i\nu t) \quad (10.2.4)$$

where ν is whirl frequency of the rotor under disturbed conditions from the equilibrium.

Substituting equations (10.2.4) in (10.2.1), we get the following equation

$$\begin{bmatrix} (K_{zz} - M\nu^2 + i\nu C_{zz}) & (K_{zy} + i\nu C_{zy}) \\ (K_{yz} + i\nu C_{yz}) & (K_{yy} - M\nu^2 + i\nu C_{yy}) \end{bmatrix} \begin{Bmatrix} z \\ y \end{Bmatrix} = 0 \quad (10.2.5)$$

For nontrivial values of z and y, we have:

$$(K_{zz} - M\nu^2)(K_{yy} - M\nu^2) - \nu^2 C_{zz}C_{yy} - K_{yz}K_{zy} + \nu^2 C_{zy}C_{yz}$$
$$+ i\nu[C_{zz}(K_{yy} - M\nu^2) + C_{yy}(K_{zz} - M\nu^2)] - i\nu(C_{yz}K_{zy} + K_{yz}C_{zy}) = 0$$
$$(10.2.6)$$

Separating real and imaginary parts, we have:

$$\frac{(K_{zz} - M\nu^2)(K_{yy} - M\nu^2) - K_{yz}K_{zy}}{C_{zz}C_{yy} - C_{zy}C_{yz}} = \nu^2 \quad (10.2.7)$$

$$\frac{(C_{zz}K_{yy} + C_{yy}K_{zz} - C_{yz}K_{zy} - C_{zy}K_{yz})}{C_{zz} + C_{yy}} = M\nu^2 \quad (10.2.8)$$

Multiplying numerator and denominator of left hand side by $(C/W)^2$

and dividing both sides of equation (10.2.7) by ω^2, we get:

$$\frac{\left(\bar{K}_{zz} - \frac{Mv^2C}{W}\right)\left(\bar{K}_{yy} - \frac{Mv^2C}{W}\right) - \bar{K}_{yz}\bar{K}_{zy}}{\bar{C}_{zz}\bar{C}_{yy} - \bar{C}_{zy}\bar{C}_{yz}} = \frac{v^2}{\omega^2} \qquad (10.2.9)$$

Similarly multiplying numerator and denominator of left hand side by $C\omega/W$ and then both sides of equation (10.2.8) by C/W, we obtain:

$$\frac{(\bar{C}_{zz}\bar{K}_{yy} + \bar{C}_{yy}\bar{K}_{zz} - \bar{C}_{yz}\bar{K}_{zy} - \bar{K}_{zy}\bar{K}_{yz})}{\bar{C}_{zz} + \bar{C}_{yy}} = \frac{Mv^2C}{W} \qquad (10.2.10)$$

In both equations (10.2.9), (10.2.10), the stiffness and damping coefficients are non-dimensionalised.

$$\bar{K}_{zz} = \frac{K_{zz}C}{W}; \quad \bar{K}_{zy} = \frac{K_{zy}C}{W}$$

$$\bar{K}_{yy} = \frac{K_{yy}C}{W}; \quad \bar{K}_{yz} = \frac{K_{yz}C}{W} \qquad (10.2.11)$$

$$\bar{C}_{zz} = \frac{C_{zz}C\omega}{W}; \quad \bar{C}_{zy} = \frac{C_{zy}C\omega}{W}$$

$$\bar{C}_{yy} = \frac{C_{yy}C\omega}{W}; \quad \bar{C}_{yz} = \frac{C_{yz}C\omega}{W} \qquad (10.2.12)$$

For a given Sommerfeld number of the bearing, equation (10.2.10) can be written as

$$K = \frac{Mv^2C}{W} \qquad (10.2.13)$$

Hence equation (10.2.9) becomes

$$\frac{(\bar{K}_{zz} - K)(\bar{K}_{yy} - K) - \bar{K}_{yz}\bar{K}_{zy}}{\bar{C}_{zz}\bar{C}_{yy} - \bar{C}_{zy}\bar{C}_{yz}} = \frac{KW}{MC\omega^2} \qquad (10.2.14)$$

where

$$K = \frac{(\bar{C}_{zz}\bar{K}_{yy} + \bar{C}_{yy}\bar{K}_{zz} - \bar{C}_{yz}\bar{C}_{zy} - \bar{C}_{zy}\bar{K}_{yz})}{\bar{C}_{zz} + \bar{C}_{yy}} \qquad (10.2.15)$$

From the definition of Sommerfeld number,

$$\omega^2 = \frac{4\pi^2 W^2 S^2}{\mu^2 D^2 L^2}\left(\frac{C}{R}\right)^4 \qquad (10.2.16)$$

writing

$$\chi = \frac{\sqrt{CMW}}{\mu DL \left(\frac{R}{C}\right)^2} \qquad (10.2.17)$$

equation (10.2.14) becomes, with the help of (10.2.16),

$$\chi = \frac{1}{2\pi S}\sqrt{\frac{K(\bar{C}_{zz}\bar{C}_{yy} - \bar{C}_{zy}\bar{C}_{yz})}{(\bar{K}_{zz} - K)(\bar{K}_{yy} - K) - \bar{K}_{yz}\bar{K}_{zy}}} \qquad (10.2.18)$$

Using the values of stiffness and damping coefficients from Figs. 7.5 to

7.8, the relation of χ and S in equation (10.2.18) is plotted in Fig. 10.4 for a plain cylindrical bearing. For a rigid rotor of mass $2M$, the parameter χ is first determined, with W as the load for each bearing. Then S can be determined from Fig. 10.4, which gives the threshold instability speed.

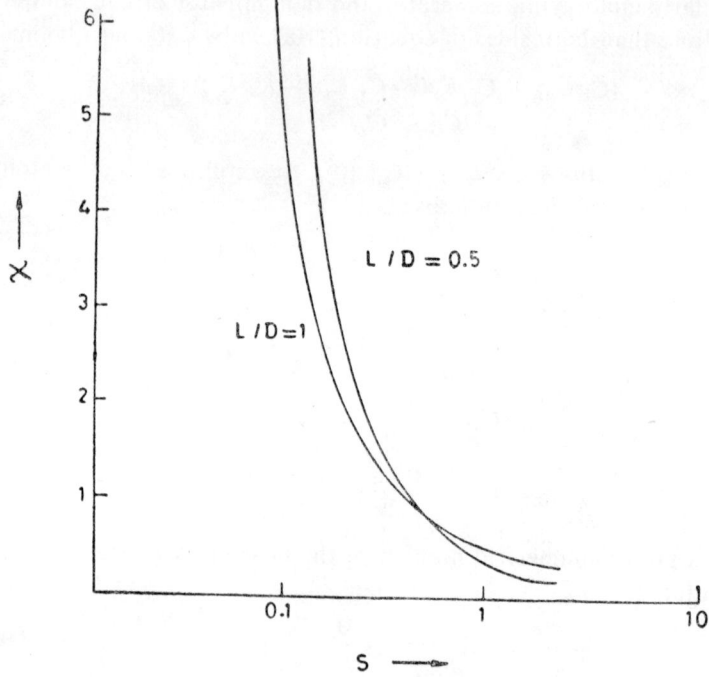

Fig. 10.4 Stability of a rigid rotor in a plain journal bearing

For conical whirl of a rigid rotor of length l, the equivalent mass M_e at the bearing is given by,

$$2\left[\tfrac{1}{2}M_e\left(\frac{l\dot\theta}{2}\right)^2\right] = \tfrac{1}{2}\left[I_T - \frac{\gamma}{\omega}I_p\right]\dot\theta^2 \qquad (10.2.19)$$

where I_T and I_p are transverse and polar mass moments of inertia of the rotor about the centre of gravity and θ is the cone angle.

If the effective mass at the bearing is more than the bearing mass, then the conical whirl critical speed is lower than translatory whirl critical speed and the instability threshold must be calculated with reference to the conical whirl. From equation (10.2.19), the effective mass is given by

$$M_e = \frac{2\left[I_T - \frac{\gamma}{\omega}I_p\right]}{l^2} \simeq \frac{2[I_T - 0.5\,I_p]}{l^2} \qquad (10.2.20)$$

The above can be written as:

$$M_e = \frac{I}{l^2} \qquad (10.2.21)$$

where
$$I = 2(I_T - 0.5I_p) \tag{10.2.22}$$

For translatory whirl instability, $M_e < M$
$$\frac{I}{Ml^2} < 1 \tag{10.2.23}$$

For a cylindrical rotor of radius R
$$I_T = \tfrac{1}{12} 2M(3R^2 + l^2)$$
$$I_p = \tfrac{1}{2} 2MR^2$$

and
$$\frac{I}{Ml^2} = \tfrac{1}{3}$$

Hence, there will be translatory whirl instability. Generally, it is the translatory whirl that is more important to calculate the threshold speed, however, a check should be made by using equation (10.2.23), to determine which mode occurs first.

10.3 Instability of a Flexible Rotor

For a flexible rotor, since the first critical speed is to be considered translatory or conical, it is convenient to consider an equivalent rotor as shown in Fig. 10.5. The shaft carries two discs M each separated by a distance ξl symmetrically with centre.

For translatory whirl:
$$\begin{aligned} Z_1 - Z &= F_{AZ}\alpha_{AA} + F_{BZ}\alpha_{AB} \\ Y_1 - Y &= F_{AY}\alpha_{AA} + F_{BY}\alpha_{AB} \end{aligned} \tag{10.3.1}$$

where α_{AA}, α_{AB}, etc... are influence coefficients. Because of symmetry $F_{AZ} = F_{BZ} = F_Z$; $F_{AY} = F_{BY} = F_Y$ and
$$\begin{aligned} Z_1 - Z &= F_Z(\alpha_{AA} + \alpha_{AB}) \\ Y_1 - Y &= F_Y(\alpha_{AA} + \alpha_{AB}) \end{aligned} \tag{10.3.2}$$

For the second mode, which is conical whirl:
$$\begin{aligned} Z_1 - \xi Z &= F_Z(\alpha_{AA} - \alpha_{AB}) \\ Y_1 - \xi Y &= F_Y(\alpha_{AA} - \alpha_{AB}) \end{aligned} \tag{10.3.3}$$

For a given flexible rotor to be equivalent to the model in Fig. 10.5a, we choose, for translatory and conical whirls respectively:
$$\xi = 1; \quad \alpha = \alpha_{AA} + \alpha_{AB} \tag{10.3.4}$$

$$\xi = \sqrt{\frac{I}{Ml^2}}; \quad \alpha = \alpha_{AA} - \alpha_{AB} \tag{10.3.5}$$

where I is given in (10.2.22).

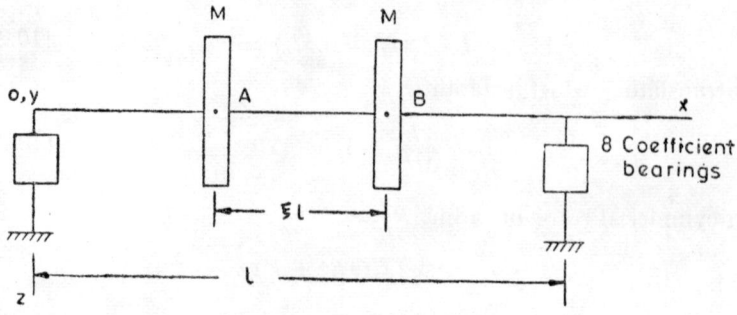

(a) Flexible rotor on oil film bearings

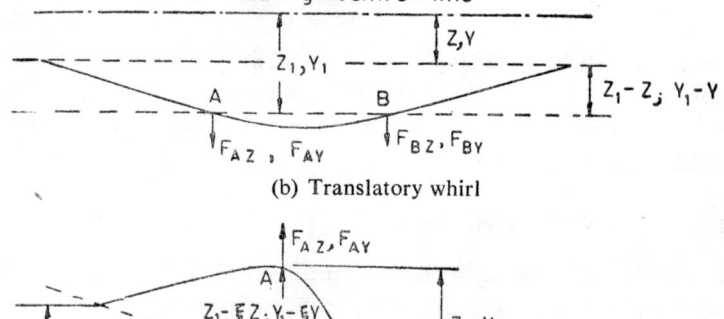

(b) Translatory whirl

(c) Conical whirl

Fig. 10.5 Relations of flexible rotor in translatory and conical whirls

Then we can combine equations (10.3.2) and (10.3.3) as:

$$Z_1 - \xi Z = F_Z \alpha$$
$$Y_1 - \xi Y = F_Y \alpha \tag{10.3.6}$$

Substituting $F_Z = -M\ddot{Z}_1$; $F_Y = -M\ddot{Y}_1$

$$M\alpha \ddot{Z}_1 + Z_1 = \xi Z$$
$$M\alpha \ddot{Y}_1 + Y_1 = \xi Y \tag{10.3.7}$$

For a whirl frequency ν

$$Z_1 = Z_1 \exp(i\nu t); \quad Z = Z \exp(i\nu t)$$
$$Y_1 = Y_1 \exp(i\nu t); \quad Y = Y \exp(i\nu t) \tag{10.3.8}$$

Solution of (10.3.7) therefore gives:

$$Z_1 = \frac{\xi Z}{1 - \alpha M \nu^2} = \frac{\xi Z}{1 - (\nu/\omega_n)^2}$$

$$Y_1 = \frac{\xi Y}{1 - (\nu/\omega_n)^2} \qquad (10.3.9)$$

where

$$\omega_n^2 = \frac{1}{\alpha M} \qquad (10.3.10)$$

α can be evaluated by first calculating the rigid bearing critical speed of the rotor.

The force balance at bearings for translatory whirl gives:

$$F_Z = \frac{Z_1 - \xi Z}{\alpha} = K_{zz}Z + K_{zy}Y + C_{zz}\dot{Z} + C_{zy}\dot{Y}$$

$$F_Y = \frac{Y_1 - \xi Y}{\alpha} = K_{yy}Y + K_{yz}Z + C_{yy}\dot{Y} + C_{yz}\dot{Z} \qquad (10.3.11)$$

The moment balance for conical whirl (2nd mode) gives:

$$F_Z \xi l = \frac{Z_1 - \xi Z}{\alpha} \xi l = (K_{zz}Z + K_{zy}Y + C_{zz}\dot{Z} + C_{zy}\dot{Y})l$$

$$F_Y \xi l = \frac{Y_1 - \xi Y}{\alpha} \xi l = (K_{yy}Y + K_{yz}Z + C_{yy}\dot{Y} + C_{yz}\dot{Z})l \qquad (10.3.12)$$

Equations (10.3.11) and (10.3.12) can be written as one set, in view of equations (10.3.4) and (10.3.5) as

$$\frac{Z_1 - \xi Z}{\alpha} \xi = K_{zz}Z + K_{zy}Y + C_{zz}\dot{Z} + C_{zy}\dot{Y}$$

$$\frac{Y_1 - \xi Y}{\alpha} \xi = K_{yy}Y + K_{yz}Z + C_{yy}\dot{Y} + C_{yz}\dot{Z} \qquad (10.3.13)$$

Using equations (10.3.8) and (10.3.9) in the above, we get

$$\begin{bmatrix} \left(K_{zz} - \frac{\xi^2}{\alpha} \frac{\nu^2/\omega_n^2}{1 - \nu^2/\omega_n^2} + i\nu C_{zz}\right) & (K_{zy} + i\nu C_{zy}) \\ (K_{yz} + i\nu C_{yz}) & \left(K_{yy} - \frac{\xi^2}{\alpha} \frac{\nu^2/\omega_n^2}{1 - \nu^2/\omega_n^2} + i\nu C_{yy}\right) \end{bmatrix} \begin{Bmatrix} Z \\ Y \end{Bmatrix} = 0 \qquad (10.3.14)$$

Forming the determinant equation and separating real and imaginary parts, we get:

$$\frac{\left[K_{zz} - \frac{\xi^2}{\alpha} \frac{\nu^2/\omega_n^2}{1 - \nu^2/\omega_n^2}\right]\left[K_{yy} - \frac{\xi^2}{\alpha} \frac{\nu^2/\omega_n^2}{1 - \nu^2/\omega_n^2}\right] - K_{yz}K_{zy}}{C_{zz}C_{yy} - C_{yz}C_{zy}} = \nu^2 \qquad (10.3.15)$$

$$\frac{(C_{zz}K_{yy} + C_{yy}K_{zz} - C_{yz}K_{zy} - C_{zy}K_{yz})}{C_{zz} + C_{yy}} = \frac{\xi^2}{\alpha} \frac{\nu^2/\omega_n^2}{1 - \nu^2/\omega_n^2} \qquad (10.3.16)$$

As in section 10.2, we define K by equation (10.2.15), then

$$K = \frac{\xi^2}{\alpha} \frac{v^2/\omega_n^2}{1 - v^2/\omega_n^2} \cdot \frac{C}{W} \qquad (10.3.17)$$

Equation (10.3.15), then becomes

$$\frac{(\bar{K}_{zz} - K)(\bar{K}_{yy} - K) - \bar{K}_{yz}\bar{K}_{zy}}{\bar{C}_{zz}\bar{C}_{yy} - \bar{C}_{yz}\bar{C}_{zy}} = \frac{K\rho}{(K + \rho)4\pi^2\xi^2 S^2 \chi^2} \qquad (10.3.18)$$

where

$$\rho = \frac{\xi^2 C}{W\alpha} \qquad (10.3.19)$$

From (10.3.18), we get:

$$\chi = \frac{1}{2\pi S\xi} \sqrt{\frac{K\rho}{(K + \rho)} \frac{\bar{C}_{zz}\bar{C}_{yy} - \bar{C}_{zy}\bar{C}_{yz}}{\{(\bar{K}_{zz} - K)(\bar{K}_{yy} - K) - \bar{K}_{yz}\bar{K}_{zy}\}}} \qquad (10.3.20)$$

We can use Fig. 10.4, to redefine the dimensionless mass parameter of the bearing as follows. If we treat the rotor as rigid, initially, the values of χ as in (10.2.18) can be determined for chosen values of S where the instability is expected. Let these values be:

$$S_i \to \bar{\chi}_i; \quad i = 1, 2, \ldots$$

then

$$\bar{\chi}_i = \frac{1}{2\pi S_i} \sqrt{\frac{K(\bar{C}_{zz}\bar{C}_{yy} - \bar{C}_{zy}\bar{C}_{yz})}{(\bar{K}_{zz} - K)(\bar{K}_{yy} - K) - \bar{K}_{yz}\bar{K}_{zy}}} \qquad (10.3.21)$$

Hence

$$\chi_i = \bar{\chi}_i \sqrt{\frac{\rho}{\xi^2(K_i + \rho)}} \qquad (10.3.22)$$

Determine K from (10.2.15) for different values of S in the region of interest and replot χ from (10.3.22). This new curve will define the stability curve for the flexible rotor. An example will illustrate the process.

Example 10.1

Consider the Jeffcott model of compressor rotor of Rieger [5] of mass 453.6 kg, mounted on two plain cylindrical hydrodynamic bearings 10.16 cm diameter and 5.08 cm long with a diametral clearance 0.01016 cm. The viscosity of the oil at the operating temperature is 5.68×10^{-3} N sec/m². The rigid bearing critical speed is 8600 rpm.

χ and K are plotted for the rotor in Fig. 10.6, as a function of S. The value of ρ for translatory whirl is

$$\rho = \frac{0.00508 \times 226.8 \times 900^2}{226.8 \times 980} = 4.2$$

The nondimensional parameter of the bearing is

$$\frac{\sqrt{CMW}}{\mu DL \left(\dfrac{R}{C}\right)^2} = 0.17$$

The rigid rotor threshold speed is given by $S = 2.9$, which corresponds to 13200 rpm. X is next plotted as shown in Fig. 10.6 again, which gives the flexible rotor threshold speed, $S = 2.0$, i.e., 9100 rpm.

10.4 Instability Threshold by Transfer Matrix Method

In general, the threshold speed of instability of a rotor can be determined by using the transfer matrix procedure. We use only 16×16 size matrices with the unbalance terms removed from the general transfer matrices considered earlier in Chapters 5 and 8. The rotor is disturbed from equilibrium by assigning a whirl frequency $v = 0.4$ to $0.5\,\omega$, say

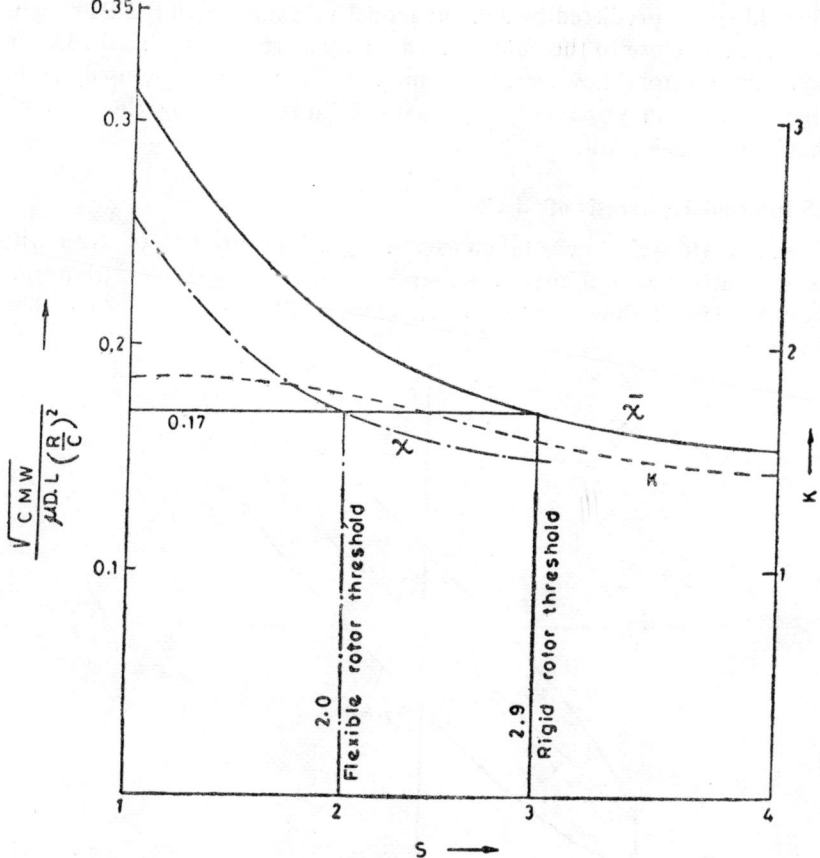

Fig. 10.6 Instability of a flexible rotor in fluid film bearings

$0.46\,\omega$. When the boundary conditions are satisfied for a rotor mounted

on fluid film bearings, we obtain equation (8.2.9) with the right hand side equal to zero. The overall transfer matrix is obtained by replacing ω in point matrix and bearing matrix by ν. For non-trivial values of the state vector $\{S\}_1^L$, in equation (8.2.9), the determinant of the 8×8 matrix in this equation should be zero. Lund [9] used this procedure to determine the instability threshold speed of a rotor.

A computer program can be developed to determine the instability threshold speed of a rotor in fluid film bearings. It may be noted that the determinant does not change sign at the threshold instability speed, because of the presence of a double root and proper precaution should be taken to extract this root in the numerical procedure used for the computer program.

Example 10.2

Consider the rotor shown in Fig. 8.13. The threshold speed for the onset of instability obtained from a computer program is 8651 rpm. The threshold speed predicted by Jeffcott model in example 10.1 is 9100 rpm, which is quite close to the value for the actual rotor in Fig. 8.13. For unsymmetric rotors, however, the computer program may be used, as the Jeffcott model may be very approximate in representing the dynamic behaviour of such a rotor.

10.5 Internal Hysteresis of Shafts

Idealised Hooke's law is taken as a straight line variation of stress with strain. In actual case of alternating stresses, there is a small deviation from this linear law as shown in Fig. 10.7. Thus a fibre of a rotating shaft

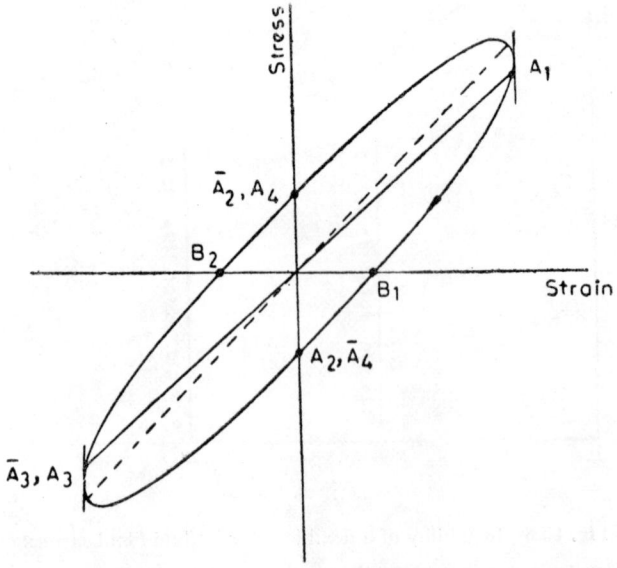

Fig. 10.7 Hysteresis loop

experiencing alternating tension and compression moves on the ellipse, from the maximum strain position A_1 to maximum negative strain position A_3 through the zero stress position B_1 and zero strain position A_2 and completes the cycle through B_2 and A_4. The deviation from Hooke's law is exaggerated in the figure and is always run through the clockwise direction as shown in Fig. 10.7.

In Fig. 10.8, we have vertical shaft with a disc at its centre (the disc not shown in the figure). B is the bearing centre and S is the shaft centre, for the full line position of the shaft, AA is neutral line of strain as all the fibres above it are in tension and below it are in compression. For synchronous whirl, the point A_1 which has maximum tension, always remains under maximum strain conditions and similarly the points A_2, A_3 and A_4 with zero strain, maximum compressive strain and zero strain respectively, for all positions of the shaft. It may be noted that the neutral stress line is not same as neutral strain line. The zero stress positions B_1 and B_2 are given in Fig. 10.7.

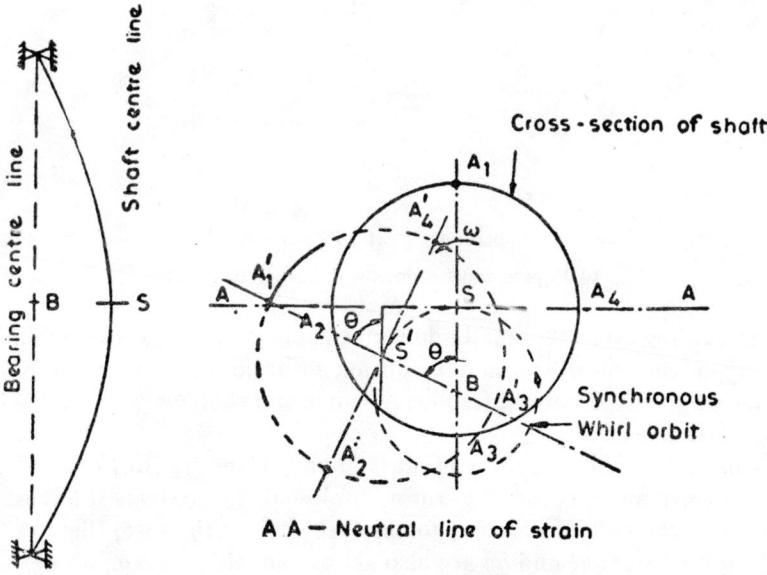

Fig. 10.8 Synchronous whirl of a vertical shaft

For non-synchronous whirl less than the rotational speed, Den Hartog [10] has shown that self-excited vibration occurs. Figure 10.9a shows this condition. The point A_1 instead of remaining in the maximum tension position A_1' as in synchronous whirl, now tries to catch up with A_2', thus having a point B_1 of zero stress as shown in Fig. 10.9a, corresponding to Fig. 10.7. The elastic force for these conditions has a horizontal component, aiding the whirl of S about B, thus generating an unstable condition. For whirl frequency more than the rotational speed, as can be observed from

Fig. 10.9b, the horizontal component opposes the whirl motion, which dampens any transient condition.

The whirling motion under the transient conditions will always be at the natural frequency, i.e., the critical speed of the shaft. Hence when the shaft is running at a speed below the critical, the whirling frequency ν will be greater than ω, thus there is no possibility of hysteresis whirl. Above the critical speed, hysteresis may generate unstable conditions, with whirl frequency equal to the critical speed. The internal hysteresis in the shaft

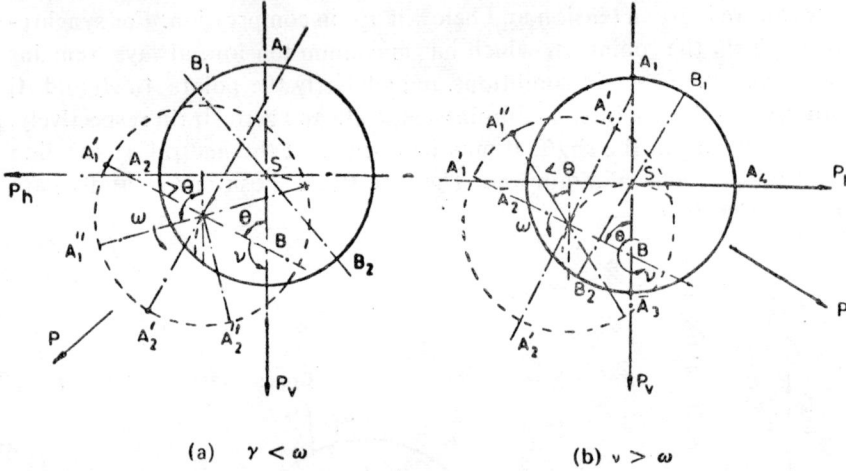

(a) $\nu < \omega$ (b) $\nu > \omega$

$A'_1 A'_2$—Synchronous whirl positions $A''_1 A''_2$—Nonsynchronous whirl position

Fig. 10.9 Non-synchronous whirl of a vertical shaft

material is however very small and is not likely to cause such unstabilities. But when the rotor has a loose shrink fit or joint, it can develop self-excited whirl at the natural frequency, when the shaft is running above the critical speed.

We consider a Jeffcott rotor of mass M shown in Fig. 10.10, to determine the unstability speed of a rotor subjected to hysteresis forces. In addition to the viscous damping forces $C(\dot{\xi} - \omega\eta)$, $C(\dot{\eta} + \omega\xi)$, the hysteresis damping forces $h\dot{\xi}$ and $h\dot{\eta}$ are also shown in this figure, along with other forces of inertia and shaft elasticity. The equations of motion in rotating coordinates are [11]:

$$M(\ddot{\xi} - 2\omega\dot{\eta} - \omega^2\xi) + C(\dot{\xi} - \omega\eta) + h\dot{\xi} + K\xi = Ma_1\omega^2 + Mg \cos \omega t$$
$$M(\ddot{\eta} + 2\omega\dot{\xi} - \omega^2\eta) + C(\dot{\eta} + \omega\xi) + h\dot{\eta} + K\eta = Ma_2\omega^2 - Mg \sin \omega t$$
(10.5.1)

Defining

$$\zeta = \xi + i\eta \qquad (10.5.2)$$

equations (10.5.1) become

$$M(\ddot{\zeta} + 2i\omega\dot{\zeta} - \omega^2\zeta) + C(\dot{\zeta} + i\omega\zeta) + h\dot{\zeta} + K\zeta = Ma\omega^2 + Mg \exp(-i\omega t)$$
(10.5.3)

Setting

$$\frac{C}{M} = 2\delta_v; \quad \frac{h}{M} = 2\delta_h \text{ and } \frac{K}{M} = \omega_n^2 \qquad (10.5.4)$$

we get

$$\ddot{\zeta} + 2i\omega\dot{\zeta} - \omega^2\zeta + 2\delta_v(\dot{\zeta} + i\omega\zeta) + 2\delta_h\dot{\zeta} + \omega_n^2\zeta = a\omega^2 + g \exp(-i\omega t) \qquad (10.5.5)$$

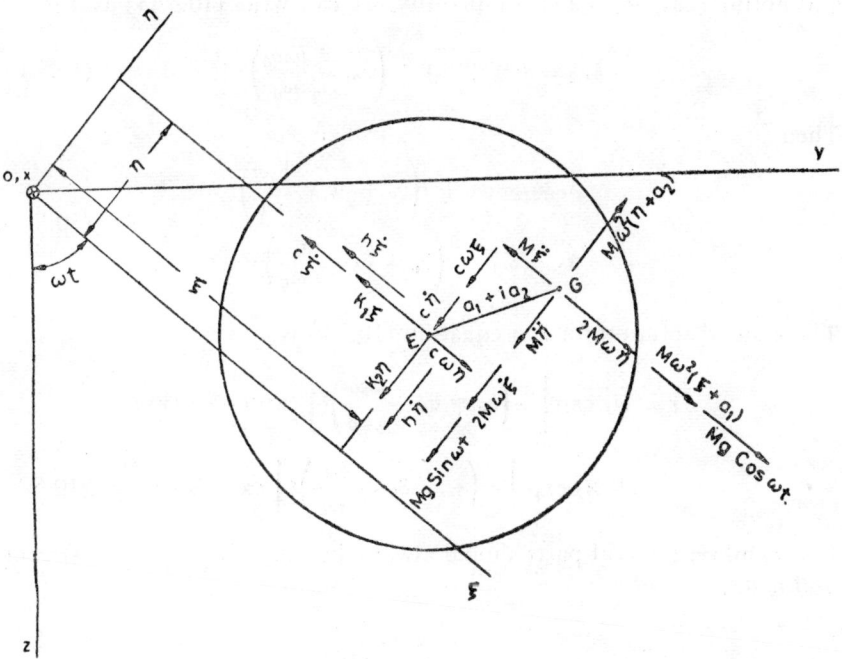

Fig. 10.10 Shaft cross-section in rotating coordinates with hysteresis damping

In stationary coordinates

$$r = \zeta \exp(i\omega t) \qquad (10.5.6)$$

Therefore

$$\zeta = r \exp(-i\omega t)$$
$$\dot{\zeta} = \dot{r} \exp(-i\omega t) - i\omega r \exp(-i\omega t)$$
$$\ddot{\zeta} = \ddot{r} \exp(-i\omega t) - 2i\omega \dot{r} \exp(-i\omega t) - \omega^2 r \exp(-i\omega t) \qquad (10.5.7)$$

and equation (10.5.5) becomes

$$\ddot{r} + 2\delta_v \dot{r} + 2\delta_h(\dot{r} - i\omega r) + \omega_n^2 r = a\omega^2 \exp(i\omega t) + g \qquad (10.5.8)$$

The transient solution of the above equation is

$$r = A \exp(i\lambda t) \qquad (10.5.9)$$

Hence

$$\lambda^2 - 2i\lambda(\delta_v + \delta_h) - \omega_n^2 + 2i\omega\delta_h = 0 \tag{10.5.10}$$

$$\lambda_{1,2} = i(\delta_v + \delta_h) \pm [\omega_n^2 - (\delta_v + \delta_h)^2 - 2i\omega\delta_h]^{1/2} \tag{10.5.11}$$

Writing

$$\omega_n^2 - (\delta_v + \delta_h)^2 = \omega_0^2 \tag{10.5.12}$$

and noting that $\omega\delta_h$ is a small quantity, we can write (10.5.11) as

$$\lambda_{1,2} = i(\delta_v + \delta_h) \pm \left(\omega_0 - \frac{i\omega\delta_h}{\omega_0}\right) \tag{10.5.13}$$

Then

$$\begin{aligned}\lambda_1 &= -\omega_0 + i\left(\delta_v + \delta_h + \frac{\delta_h\omega}{\omega_0}\right) \\ \lambda_2 &= \omega_0 + i\left(\delta_v + \delta_h - \frac{\delta_h\omega}{\omega_0}\right)\end{aligned} \tag{10.5.14}$$

The general solution for r in equation (10.5.9) is

$$\begin{aligned}r = &A_1 \exp\left[-\left(\delta_v + \delta_h + \frac{\delta_h\omega}{\omega_0}\right)t\right] \exp(-i\omega_0 t) \\ &+ A_2 \exp\left[-\left(\delta_v + \delta_h - \frac{\delta_h\omega}{\omega_0}\right)t\right] \exp(i\omega_0 t)\end{aligned} \tag{10.5.15}$$

For stability, the real parts should always be negative in the exponent and hence

$$\delta_v + \delta_h > \frac{\delta_h\omega}{\omega_0} \tag{10.5.16}$$

Hence the rotor is always stable, if

$$\omega < \omega_0\left(1 + \frac{\delta_v}{\delta_h}\right) \tag{10.5.17}$$

Since the damped natural frequency ω_0 is same as the critical speed of the shaft, the shaft will develop unstable whirl at speeds slightly greater than the critical speed, if the hysteresis damping coefficient is large compared to the viscous damping coefficient. On the other hand, if there is sufficient viscous damping, say $\delta_v = 2\delta_h$, the shaft will have no unstable whirl due to hysteresis, until about three times the critical speed.

Additional references may be used for a further study; instability due to oil film bearings [12-28], and internal hysteresis [29-33].

10.6 Instability due to Negative Cross-Coupled Stiffness

In section 8.1, we considered a Jeffcott rotor mounted on fluid film bearings. When one of the cross coupled stiffness coefficients is negative.

Instability Due to Fluid Film Forces 215

we have also shown by equation (8.1.22) that the two distinct critical speeds exist under a specific condition. In example 8.2, we considered such a rotor and calculated the unbalance response, even though there are no two distinct critical speeds. Under those conditions the system is actually unstable.

Consider the equations of motion for free vibration given by (8.1.27) again. We can write

$$z = Z \exp(ip^* t)$$
$$y = Y \exp(ip^* t) \tag{10.6.1}$$

where p^* is a complex frequency. Substituting the above in equation (8.1.27), we get

$$\begin{bmatrix} K_1^* - Mp^{*2} & K_{12}^* \\ K_{21}^* & K_2^* - Mp^{*2} \end{bmatrix} \begin{Bmatrix} Z \\ Y \end{Bmatrix} = 0 \tag{10.6.2}$$

For nontrivial values of Z and Y, we obtain

$$p^{*4} - p^{*2} \frac{K_1^* + K_2^*}{M} + \frac{K_1^* K_2^* - K_{21}^* K_{12}^*}{M} = 0 \tag{10.6.3}$$

The roots of the above equation are

$$p_{1,2}^{*2} = \frac{K_1^* + K_2^*}{2M} \pm \sqrt{\left[\frac{K_1^* - K_2^*}{2M}\right]^2 + \frac{K_{12}^* K_{21}^*}{M^2}} \tag{10.6.4}$$

Let the complex eigen values be expressed in the form

$$\lambda = m + in \tag{10.6.5}$$

Then we obtain

$$\begin{Bmatrix} Z \\ Y \end{Bmatrix} = \begin{Bmatrix} Z \\ Y \end{Bmatrix} e^{qt} \sin(pt + \theta) \tag{10.6.6}$$

where

$$p = (m^2 + n^2)^{1/4} \cos \frac{\phi}{2}$$
$$q = -(m^2 + n^2)^{1/4} \sin \frac{\phi}{2} \tag{10.6.7}$$
$$\phi = \tan^{-1}(n/m)$$

The stability of the system depends on the sign of q. If q is negative, the free response is damped and the system is stable. If q is positive, the free vibration response grows unbounded and the system is unstable. We denote q as stability index.

Example 10.3

Consider the rotor of example 10.1, once again, this time to determine the instability that may exist due to the negative cross-coupled stiffness coefficient. In example 8.2, for this rotor K_{21} is negative at 7500 rpm. The stiffness and damping coefficients for this bearing can be taken from

Figs. 7.5 and 7.6. These figures can be expressed as polynomial functions of Sommerfeld number, see equation 7.1.12, for obtaining the stiffness and damping coefficients.

$$CK_{zz}/W = 8.727 - 35.03S + 87.23S^2 - 120.1S^3 + 93.65S^4$$
$$- 41S^5 + 9.35S^6 - 0.863S^7$$
$$CK_{yy}/W = 1.472 + 3.01S - 4.244S^2 + 4.012S^3 - 2.67S^4$$
$$+ 1.11S^5 - 0.248S^6 + 0.023S^7$$
$$CK_{zy}/W = 5.338 - 9.65S + 20.62S^2 - 20.96S^3 + 11.37S^4$$
$$- 3.07S^5 + 0.324S^6$$
$$CK_{yz}/W = 0.493 - 2.117S + 0.0899S^2 - 3.656S^3 + 4.086S^4$$
$$- 1.555S^5 + 0.2S^6$$

(10.6.8)

$$\omega CC_{zz}/W = 8.44 - 11.61S + 16.63S^2 + 7.1S^3 - 30.78S^4$$
$$+ 24.87S^5 - 8.14S^6 + 0.96S^7$$
$$\omega CC_{yy}/W = 1.789 - 5.76S + 52.92S^2 - 143.16S^3 + 208.74S^4$$
$$- 171.55S^5 + 79.11S^6 - 18.96S^7 + 1.828S^8$$
$$\omega CC_{zy}/W = 1.941 + 0.672S - S^2 + 0.671S^3 - 0.206S^4$$
$$+ 0.0236S^5$$
$$\omega CC_{yz}/W = 1.852 + 1.264S - 2.295S^2 + 1.847S^3 - 0.625S^4$$
$$+ 0.075S^5$$

(10.6.9)

Case 1: Undamped rotor.

For the undamped rotor, Rao [34] found the eigen values to be complex conjugates under unstable conditions. Table 10.1 gives the natural frequencies and stability indices for different rotational speeds.

Table 10.1 Natural Frequencies and Stability Indices—Undamped Rotor

RPM	Major Frequency		Minor Frequency	
	p_1 — RPM	q_1 — 1/s	p_2 — RPM	q_2 — 1/s
1000	4860	0	6086	0
1200	5144	0	5795	0
1290	5471	8	5471	−8
1500	5484	52	5484	−52
2000	5553	90	5553	−90
3000	5774	129	5774	−129
4000	6000	154	6000	−154
5000	6199	171	6199	−171
6000	6368	180	6368	−180
7000	6509	184	6509	−184
8000	6626	187	6626	−187

Instability Due to Fluid Film Forces 217

The results are also plotted in Fig. 10.11. Under stable conditions, there are two distinct major and minor natural frequencies. From Fig. 10.11, it can be seen that the rotor becomes unstable at 1285 rpm. Here the two natural frequencies coalesce, one of them being unstable.

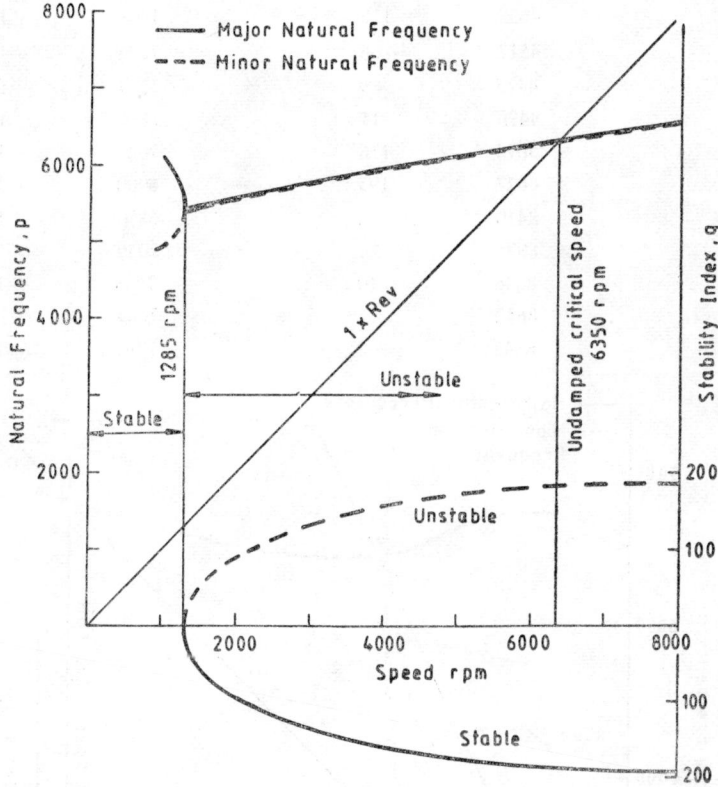

Fig. 10.11 Natural frequencies and corresponding stability indices for undamped case

The stability index initially increases with speed from 1285 rpm, but remains constant for higher speeds. The undamped critical speed from Fig. 10.11 is 6350 rpm. Super-synchronous excitation is required to make the system unstable for speeds below the critical speed 6350 rpm. For speeds above the critical speed, sub-synchronous excitation is required to make the system unstable.

Case 2. Damped Rotor.

Rao [34] found that, nominal values of shaft damping have no significant effect on the performance of the rotor, but the bearing damping has predominant influence on its dynamic behaviour. The results obtained by including bearing damping are given in Table 10.2 and plotted in Fig. 10.12.

Table 10.2 Natural Frequencies and Stability Indices—Damped Rotor

RPM	Major Frequency		Minor Frequency	
	p_1 — RPM	$q_1 - 1/s$	p_2 — RPM	$q_2 - 1/s$
1000	4287	−109	7483	−119
2000	4633	−116	7408	−140
2500	4517	−66	7276	−147
2700	4475	−9	7185	−149
2750	4496	11	7160	−150
3000	5074	126	7027	−154
3300	6627	195	6900	−159
4000	8410	92	6850	−168
5000	8625	26	7039	−172
6000	8046	10	7255	−176
7000	8642	5	7465	−165
8000	8643	2	7709	−156

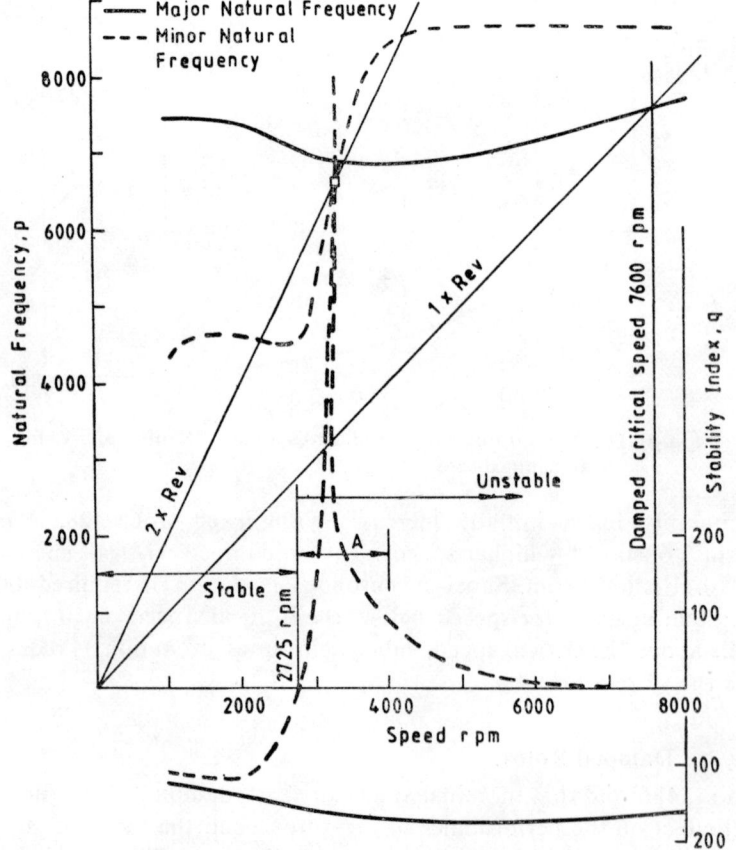

Fig. 10.12 Natural frequency and corresponding stability indices of the rotor with bearing damping coefficients included

The rotor becomes unstable at 2725 rpm, instead of 1285 rpm when damping in the bearing is accounted. The stable natural frequency, denoted major in Fig. 10.12, remains practically same around 7000 rpm. The damped critical speed with this frequency is 7600 rpm. The unstable natural frequency, denoted minor in Fig. 10.12, increases rapidly from around 4500 rpm to 8500 rpm, as soon as the instability condition is set in at 2725 rpm. In this range of rapid increase in the minor natural frequency 2725-4000 rpm, the instability index also increases rapidly until around 3300 rpm and then decreases rapidly becoming almost zero at 6000 rpm onwards.

From Fig. 10.12, it is interesting to note that the peak value of instability index occurs when the unstable natural frequency is twice the rotational speed. Thus the region A marked on Fig. 10.12, can be dangerous for operation, particularly around 3300 rpm, when the excitation is $2 \times$ rev, as commonly found in rotors with asymmetry such as generator rotors. Also see Fig. 9.19.

It should be noted here that the half frequency whirl instability, section 10.3, examples 10.1 and 10.2, imposes the condition that the whirl frequency is slightly less than half the rotational speed. In fact this type of instability cannot occur if the shaft whirls at a frequency higher than half the rotational frequency. The rotor whirl in the instability considered in Fig. 10.12 occurs at twice the rotational frequency and therefore called as supersynchronous whirl instability.

10.7 Orbital Analysis by Transfer Matrix Method

For common rotor dynamics problems, the linear analytical tools are adequate to study the behaviour of the rotor bearing system. Due to special conditions such as, the bearing vibration amplitudes becoming large and approaching bearing clearances, system response due to earthquakes and marine turbines under conditions of high intensity shock, nonlinearities will occur in the bearing. When strong nonlinearities are present in the bearing, linear theory can provide grossly erroneous predictions qualitatively as well as quantitatively. Unlike the linear theory, the bearing properties are to be obtained at each step by solving Reynolds lubrication equation at each journal bearing by adopting a suitable integration procedure. (This solution procedure for estimating the nonlinear bearing forces is out of the present scope of discussion here.) Quite often, the linear theory of prediction of instability thresholds is just not enough to a modern high speed or heavy duty machine designer. Beyond the instability threshold speed as predicted by linear theory, there is a fairly sizable safe zone of operation, which a designer may like to extract as much as possible in an attempt to increase the capacity and speed of the machine. The criterion for such operations can be laid by the allowable orbital amplitudes to determine the safe operating speeds of the machine. Such a design capability requires a good transient analysis of the rotor. For Jeffcott type of rotors, the analysis is somewhat simple by adopting a time marching technique, as considered in section 9.5.

In Chapter 4, the transfer matrix method for sample shafts, bending in x-z plane was described. Then in Chapter 5, this method was extended to determine the steady state unbalance response, by considering the shaft bending in x-z and x-y planes. Since the unbalance had cosine and sine excitation components, the steady state response vector was also split into two cosine and sine components. Rao and Bhaskara Sarma [35] developed a transfer matrix method with its state vector quantities as general functions of time, to determine the transient orbital response of rotors mounted on hydrodynamic bearings. This method is described here.

10.7.1 Field matrix

In Fig. 4.6, the relations for ith field are given. Denoting the state vector quantities as functions of time, we have from equations (4.3.4) and (4.3.5) and (4.3.1) and (4.3.2),

$$w_i^L(t) = w_{i-1}^R(t) - \theta_{i-1}^R(t)l_i - M_{y,i}^L(t)\frac{l_i^2}{2E_iI_i} + V_{z,i}^L(t)\frac{l_i^3}{3E_iI_i}$$

$$\theta_i^L(t) = \theta_{i-1}^R(t) + M_{y,i}^L(t)\frac{l_i}{E_iI_i} - V_{z,i}^L(t)\frac{l_i^2}{2E_iI_i}$$

$$M_{y,i}^L(t) = M_{y,i-1}^R(t) + V_{z,i}^L(t)l_i \qquad (10.7.1)$$

$$V_{z,i}^L(t) = V_{z,i-1}^R(t)$$

The above can be rewritten as

$$-w_i^L(t) = -w_{i-1}^R(t) + \theta_{i-1}^R(t)l_i + M_{y,i-1}^R(t)\frac{l_i^2}{2E_iI_i} + V_{z,i-1}^R(t)\frac{l_i^3}{6E_iI_i}$$

$$\theta_i^L(t) = \theta_{i-1}^R(t) + M_{y,i-1}^R(t)\frac{l_i}{E_iI_i} + V_{z,i-1}^R(t)\frac{l_i^2}{2E_iI_i}$$

$$M_{y,i}^L(t) = M_{y,i-1}^R(t) + V_{z,i-1}^R(t)l_i \qquad (10.7.2)$$

$$V_{z,i}^L(t) = V_{z,i-1}^R(t)$$

Differentiating these equations with respect to time, we have

$$-\dot{w}_i^L(t) = -\dot{w}_{i-1}^R(t) + \dot{\theta}_{i-1}^R(t)l_i + \dot{M}_{y,i-1}^R(t)\frac{l_i^2}{2E_iI_i} + \dot{V}_{z,i-1}^R(t)\frac{l_i^3}{6E_iI_i}$$

$$-\ddot{w}_i^L(t) = -\ddot{w}_{i-1}^R(t) + \ddot{\theta}_{i-1}^R(t)l_i + \ddot{M}_{y,i-1}^R(t)\frac{l_i^2}{2E_iI_i} + \ddot{V}_{z,i-1}^R(t)\frac{l_i^3}{6E_iI_i}$$

$$\qquad (10.7.3)$$

$$-\dddot{w}_i^L(t) = -\dddot{w}_{i-1}^R(t) + \dddot{\theta}_{i-1}^R(t)l_i + \dddot{M}_{y,i-1}^R(t)\frac{l_i^2}{2E_iI_i} + \dddot{V}_{z,i-1}^R(t)\frac{l_i^3}{6E_iI_i}$$

$$-\ddddot{w}_i^L(t) = -\ddddot{w}_{i-1}^R(t) + \ddddot{\theta}_{i-1}^R(t)l_i + \ddddot{M}_{y,i-1}^R(t)\frac{l_i^2}{2E_iI_i} + \ddddot{V}_{z,i-1}^R(t)\frac{l_i^3}{6E_iI_i}$$

Similarly,

$$\theta_i^L(t) = \theta_{i-1}^R(t) + \dot{M}_{y,i-1}^R(t)\frac{l_i}{E_iI_i} + \dot{V}_{z,i-1}^R(t)\frac{l_i^2}{2E_iI_i}$$

$$\ddot{\theta}_i^L(t) = \ddot{\theta}_{i-1}^R(t) + \ddot{M}_{y,i-1}^R(t)\frac{l_i}{E_iI_i} + \ddot{V}_{z,i-1}^R(t)\frac{l_i^2}{2E_iI_i}$$

$$\dddot{\theta}_i^L(t) = \dddot{\theta}_{i-1}^R(t) + \dddot{M}_{y,i-1}^R(t)\frac{l_i}{E_iI_i} + \dddot{V}_{z,i-1}^R(t)\frac{l_i^2}{2E_iI_i} \quad (10.7.4)$$

$$\ddddot{\theta}_i^L(t) = \ddddot{\theta}_{i-1}^R(t) + \ddddot{M}_{y,i-1}^R(t)\frac{l_i}{E_iI_i} + \ddddot{V}_{z,i-1}^R(t)\frac{l_i^2}{2E_iI_i}$$

$$\dot{M}_{y,i}^L(t) = \dot{M}_{y,i-1}^R(t) + \dot{V}_{z,i-1}^R(t)l_i$$

$$\ddot{M}_{y,i}^L(t) = \ddot{M}_{y,i-1}^R(t) + \ddot{V}_{z,i-1}^R(t)l_i$$

$$\dddot{M}_{y,i}^L(t) = \dddot{M}_{y,i-1}^R(t) + \dddot{V}_{z,i-1}^R(t)l_i \quad (10.7.5)$$

$$\ddddot{M}_{y,i}^L(t) = \ddddot{M}_{y,i-1}^R(t) + \ddddot{V}_{z,i-1}^R(t)l_i$$

$$\dot{V}_{z,i}^L(t) = \dot{V}_{z,i-1}^R(t)$$

$$\ddot{V}_{z,i}^L(t) = \ddot{V}_{z,i-1}^R(t)$$

$$\dddot{V}_{z,i}^L(t) = \dddot{V}_{z,i-1}^R(t) \quad (10.7.6)$$

$$\ddddot{V}_{z,i}^L(t) = \ddddot{V}_{z,i-1}^R(t)$$

The above relations can be expressed in matrix form as

$$\{S_{zt}\}_i^L = [F_t]\{S_{zt}\}_{i-1}^R \quad (10.7.6a)$$

In the above, the elements of $[F_t]$ are

$$F_{ti,i} = 1 \quad i = 1 \text{ to } 20$$

$$F_{ti,i+5} = 1 \quad i = 1 \text{ to } 5 \text{ and } 11 \text{ to } 15$$

$$F_{ti,i+5} = l/EI \quad i = 6 \text{ to } 10$$

$$F_{ti,i+10} = l^2/2EI \quad i = 1 \text{ to } 10$$

$$F_{ti,i+15} = l^3/6EI \quad i = 1 \text{ to } 5$$

and the rest $F_t = 0$.

The same relations are valid for x-y plane bending also, i.e.,

$$\{S_{yt}\}_i^L = [F_t]\{S_{yt}\}_{i-1}^R \quad (10.7.6b)$$

10.7.2 Point matrix

In Fig. 4.7, let us include modal damping force with ζ as damping ratio for nth natural mode ω_n, then

$$w_i^R(t) = w_i^L(t)$$

$$\theta_i^R(t) = \theta_i^L(t) \quad (10.7.7)$$

$$M_{y,i}^R(t) = M_{y,i}^L(t)$$

$$V_{z,i}^R(t) = V_{z,i}^L(t) + m_i\ddot{w}_i^L(t) + 2m_i\zeta\omega_n\dot{w}_i^L(t)$$

Differentiating the above and neglecting fifth order terms,

$$\dot{w}_i^R(t) = \dot{w}_i^L(t)$$
$$\ddot{w}_i^R(t) = \ddot{w}_i^L(t)$$
$$\dddot{w}_i^R(t) = \dddot{w}_i^L(t)$$
$$\ddddot{w}_i^R(t) = \ddddot{w}_i^L(t)$$
$$\dot{\theta}_i^R(t) = \dot{\theta}_i^L(t)$$
$$\ddot{\theta}_i^R(t) = \ddot{\theta}_i^L(t)$$
$$\dddot{\theta}_i^R(t) = \dddot{\theta}_i^L(t)$$
$$\ddddot{\theta}_i^R(t) = \ddddot{\theta}_i^L(t) \qquad (10.7.8)$$
$$\dot{M}_{y,i}^R(t) = \dot{M}_{y,i}^L(t)$$
$$\ddot{M}_{y,i}^R(t) = \ddot{M}_{y,i}^L(t)$$
$$\dddot{M}_{y,i}^R(t) = \dddot{M}_{y,i}^L(t)$$
$$\ddddot{M}_{y,i}^R(t) = \ddddot{M}_{y,i}^L(t)$$
$$\dot{V}_{z,i}^R(t) = \dot{V}_{z,i}^L(t) + m_i \dddot{w}_i^L(t) + 2m_i \zeta \omega_n \ddot{w}_i^L(t)$$
$$\ddot{V}_{z,i}^R(t) = \ddot{V}_{z,i}^L(t) + m_i \ddddot{w}_i^L(t) + 2m_i \zeta \omega_n \dddot{w}_i^L(t)$$
$$\dddot{V}_{z,i}^R(t) = \dddot{V}_{z,i}^L(t) + 2m_i \zeta \omega_n \ddddot{w}_i^L(t)$$
$$\ddddot{V}_{z,i}^R(t) = \ddddot{V}_{z,i}^L(t)$$

The above can be expressed as

$$\{S_{zt}\}_i^R = [P_t]\{S_{zt}\}_i^L \qquad (10.7.8a)$$

where

$$P_{ti,i} = 1, \; i = 1 \text{ to } 20$$
$$P_{t16,2} = P_{t17,3} = P_{t18,4} = P_{t19,5} = -2\zeta m \omega_n$$
$$P_{t16,3} = P_{t17,4} = P_{t18,5} = -m$$

Rest P_t are zero.

The relations for x-y plane can be derived in a similar manner.

$$\{S_{yt}\}_i^R = [P_t]\{S_{yt}\}_i^L \qquad (10.7.8b)$$

10.7.3 *Point matrix and unbalance*

Including the damping term as in section above, equations (5.2.1) and (5.2.2) relating to the unbalance mass in Fig. 5.4b can be written as follows.

$$V_{z,i}^R(t) = V_{z,i}^L(t) + m_i \ddot{w}_i^L(t) + 2m_i \zeta \omega_n \dot{w}_i^L(t) - \omega^2 u_{z,i} \cos \omega t + \omega^2 u_{y,i} \sin \omega t$$
$$V_{y,i}^R(t) = V_{y,i}^L(t) + m_i \ddot{v}_i^L(t) + 2m_i \zeta \omega_n \dot{v}_i^L(t) - \omega^2 u_{y,i} \cos \omega t + \omega^2 u_{z,i} \sin \omega t$$
$$(10.7.9)$$

The deflection, slope and moment are continuous and hence are same to the left and right of the station i and the relations can be directly taken from the previous section 10.7.2. Differentiating equation (10.7.9) with respect to time t and neglecting fifth order terms

$$\dot{V}_{z,i}^R(t) = \dot{V}_{z,i}^L(t) + m_i \dddot{w}_i^L(t) + 2m_i\zeta\omega_n \ddot{w}_i^L(t) + \omega^3 u_{z,i} \sin \omega t + \omega^3 u_{y,i} \cos \omega t$$

$$\ddot{V}_{z,i}^R(t) = \ddot{V}_{z,i}^L(t) + m_i \ddddot{w}_i^L(t) + 2m_i\zeta\omega_n \dddot{w}_i^L(t) + \omega^4 u_{z,i} \cos \omega t - \omega^4 u_{y,i} \sin \omega t$$

$$\dddot{V}_{z,i}^R(t) = \dddot{V}_{z,i}^L(t) + 2m_i\zeta\omega_n \ddddot{w}_i^L(t) - \omega^5 u_{z,i} \sin \omega t - \omega^5 u_{y,i} \cos \omega t$$

$$\ddddot{V}_{z,i}^R(t) = \ddddot{V}_{z,i}^L(t) - \omega^6 u_{z,i} \cos \omega t + \omega^6 u_{y,i} \sin \omega t \qquad (10.7.10)$$

Similarly for the x-y plane

$$\dot{V}_{y,i}^R(t) = \dot{V}_{y,i}^L(t) + m_i \dddot{v}_i^L(t) + 2m_i\zeta\omega_n \ddot{v}_i^L(t) + \omega^3 u_{y,i} \sin \omega t + \omega^3 u_{z,i} \cos \omega t$$

$$\ddot{V}_{y,i}^R(t) = \ddot{V}_{y,i}^L(t) + m_i \ddddot{v}_i^L(t) + 2m_i\zeta\omega_n \dddot{v}_i^L(t) + \omega^4 u_{y,i} \cos \omega t - \omega^4 u_{z,i} \sin \omega t$$

$$\dddot{V}_{y,i}^R(t) = \dddot{V}_{y,i}^L(t) + 2m_i\zeta\omega_n \ddddot{v}_i^L(t) - \omega^5 u_{y,i} \sin \omega t - \omega^5 u_{z,i} \cos \omega t$$

$$\ddddot{V}_{y,i}^R(t) = \ddddot{V}_{y,i}^L(t) - \omega^6 u_{y,i} \cos \omega t + \omega^6 u_{z,i} \sin \omega t \qquad (10.7.11)$$

To account for the unbalance relations, one more column is added to the matrix $[P_t]$ containing $\{m_{zt}\}$ and $\{m_{yt}\}$ for x-z and x-y planes respectively, i.e.,

$$\{S_t\}_i^R = \begin{Bmatrix} \{S_{zt}\} \\ \{S_{yt}\} \\ 1 \end{Bmatrix}_i^R = \begin{bmatrix} [P_t] & 0 & \{m_{zt}\} \\ 0 & [P_t] & \{m_{yt}\} \\ 0 & 0 & 1 \end{bmatrix} \{S_t\}_i^L = [\bar{P}_t]\{S_t\}_i^L \qquad (10.7.12)$$

where

$$m_{zt,16} = \omega^2(-u_z \cos \omega t + u_y \sin \omega t)$$
$$m_{zt,17} = \omega^3(u_z \sin \omega t + u_y \cos \omega t)$$
$$m_{zt,18} = \omega^4(u_z \cos \omega t - u_y \sin \omega t)$$
$$m_{zt,19} = \omega^5(-u_z \sin \omega t - u_y \cos \omega t)$$
$$m_{zt,20} = \omega^6(-u_z \cos \omega t + u_y \sin \omega t)$$
$$m_{yt,16} = \omega^2(u_y \cos \omega t + u_z \sin \omega t)$$
$$m_{yt,17} = \omega^3(-u_y \sin \omega t + u_z \cos \omega t)$$
$$m_{yt,18} = \omega^4(-u_y \cos \omega t - u_z \sin \omega t)$$
$$m_{yt,19} = \omega^5(u_y \sin \omega t - u_z \cos \omega t)$$
$$m_{yt,20} = \omega^6(u_y \cos \omega t + u_z \sin \omega t)$$

Rest $m_{zt} = m_{yt} = 0$

Similarly, the modified field matrix is given by

$$\{S_t\}_i^L = \begin{bmatrix} [F_t] & 0 & 0 \\ 0 & [F_t] & 0 \\ 0 & 0 & 1 \end{bmatrix} \{S_t\}_{i-1}^R = [\bar{F}_t]\{S_t\}_{i-1}^R \qquad (10.7.13)$$

10.7.4 Bearing matrix

The shear force relations across a bearing station are, see Fig. 8.12

$$V_{z,i}^R(t) = V_{z,i}^L(t) + K_{zz}w_i^L(t) + K_{zy}v_i^L(t) + C_{zz}\dot{w}_i^L(t) + C_{zy}\dot{v}_i^L$$
$$+ m_i\ddot{w}_i^L(t) + F_z'(w, \dot{w}, v, \dot{v})$$

$$V_{y,i}^R(t) = V_{y,i}^L(t) + K_{yy}v_i^L(t) + K_{yz}w_i^L(t) + C_{yy}\dot{v}_i^L(t) + C_{yz}\dot{w}_i^L(t)$$
$$+ m_i\ddot{v}_i^L(t) + F_y'(w, \dot{w}, v, \dot{v}) \qquad (10.7.14)$$

In the above F_z' and F_y' represent the nonlinear part of the bearing forces. The bearing forces can be approximated as

$$\begin{Bmatrix} F_z \\ F_y \end{Bmatrix} = \begin{bmatrix} K_{zz}^{(1)} & K_{zy}^{(1)} \\ K_{yz}^{(1)} & K_{yy}^{(1)} \end{bmatrix} \begin{Bmatrix} w_0(t) \\ v_0(t) \end{Bmatrix} + \begin{bmatrix} C_{zz}^{(1)} & C_{zy}^{(1)} \\ C_{yz}^{(1)} & C_{yy}^{(1)} \end{bmatrix} \begin{Bmatrix} \dot{w}_0(t) \\ \dot{v}_0(t) \end{Bmatrix}$$
$$+ \begin{bmatrix} K_{zz}^{(2)} & K_{zy}^{(2)} \\ K_{yz}^{(2)} & K_{yy}^{(2)} \end{bmatrix} \begin{Bmatrix} w_0^2(t) \\ v_0^2(t) \end{Bmatrix} + \begin{bmatrix} C_{zz}^{(2)} & C_{zy}^{(2)} \\ C_{yz}^{(2)} & C_{yy}^{(2)} \end{bmatrix} \begin{Bmatrix} \dot{w}_0^2(t) \\ \dot{v}_0^2(t) \end{Bmatrix} \qquad (10.7.15)$$

Here superscript 1 represents linear bearing force terms and superscript 2 the second order nonlinear coefficients of the bearing forces. The linear terms are expressed in the bearing matrix and the secondary nonlinear terms are expressed in the form of F_z' and F_y' in the force column of the bearing matrix. The bearing transfer relation is

$$\{S_t\}_i^R = [B_t]\ \{S_t\}_i^L \qquad (10.7.16)$$

where the nonzero elements of the bearing matrix terms are

$$b_{r,s} = 1, \quad r = s = 1, 2, \ldots, 41$$

$$b_{16,1} = b_{17,2} = b_{18,3} = b_{19,4} = b_{20,5} = -K_{zz}^{(1)}$$

$$b_{16,2} = b_{17,3} = b_{18,4} = b_{19,5} = -C_{zz}^{(1)}$$

$$b_{16,3} = b_{17,4} = b_{18,5} = -m$$

$$b_{16,21} = b_{17,22} = b_{18,23} = b_{19,24} = b_{20,25} = K_{zy}^{(1)}$$

$$b_{16,22} = b_{17,23} = b_{18,24} = b_{19,25} = C_{zy}^{(1)}$$

$$b_{36,21} = b_{37,22} = b_{38,23} = b_{39,24} = b_{40,25} = -K_{yy}^{(1)}$$

$$b_{36,1} = b_{37,2} = b_{38,3} = b_{39,4} = b_{40,5} = K_{yz}^{(1)}$$

$$b_{36,22} = b_{37,23} = b_{38,24} = b_{39,25} = -C_{yy}^{(1)}$$

$$b_{36,2} = b_{37,3} = b_{38,4} = b_{39,5} = C_{yz}^{(1)}$$

$$b_{36,3} = b_{37,4} = b_{38,5} = -m$$

$$b_{16,41} = F_z', b_{17,41} = \dot{F}_z', b_{18,41} = \ddot{F}_z', b_{19,41} = \dddot{F}_z', b_{20,41} = \ddddot{F}_z'$$

$$b_{36,41} = F_y', b_{37,41} = \dot{F}_y', b_{38,41} = \ddot{F}_y', b_{39,41} = \dddot{F}_p, b_{40,41} = \ddddot{F}_y'$$

10.7.5 Method of solution

The transfer matrix for the ith station $[T]_i = [\bar{F}_t]_i [\bar{P}_t]_i$ can be obtained and the overall transfer matrix of an n stationed system is given by

$$\{S_t\}_n = [T]_n [T]_{n-1} \cdots [T]_2 [T]_1 \{S_t\}_0 = [T] \{S_t\}_0 \qquad (10.7.17)$$

where $[T]$ is the overall transfer matrix of size 41×41.

For free ends of the rotor, the bending moments and shear forces are zero and hence equation (10.7.17) can be reduced to

$$\begin{bmatrix} t_{11,1} & t_{11,2} & \cdots & t_{11,10} & t_{11,21} & \cdots & t_{11,30} \\ t_{12,1} & t_{12,2} & \cdots & t_{12,10} & t_{12,21} & \cdots & t_{12,30} \\ \cdots & \cdots & & \cdots & \cdots & & \cdots \\ t_{15,1} & t_{15,2} & \cdots & t_{15,10} & t_{15,21} & \cdots & t_{15,30} \\ t_{16,1} & t_{16,2} & \cdots & t_{16,10} & t_{16,21} & \cdots & t_{16,30} \\ \cdots & \cdots & & \cdots & \cdots & & \cdots \\ t_{20,1} & t_{20,2} & \cdots & t_{20,10} & t_{20,21} & \cdots & t_{20,30} \\ t_{31,1} & t_{31,2} & \cdots & t_{31,10} & t_{31,21} & \cdots & t_{31,30} \\ \cdots & \cdots & & \cdots & \cdots & & \cdots \\ t_{35,1} & t_{35,2} & \cdots & t_{35,10} & t_{35,21} & \cdots & t_{35,30} \\ t_{36,1} & t_{36,2} & \cdots & t_{36,10} & t_{36,21} & \cdots & t_{36,30} \\ \cdots & \cdots & & \cdots & \cdots & & \cdots \\ t_{40,1} & t_{40,2} & \cdots & t_{40,10} & t_{40,21} & \cdots & t_{40,30} \end{bmatrix} \begin{Bmatrix} w(t) \\ \dot{w}(t) \\ \cdots \\ \dddot{w}(t) \\ \theta(t) \\ \cdots \\ \ddddot{\theta}(t) \\ v(t) \\ \cdots \\ \ddddot{v}(t) \\ \phi(t) \\ \cdots \\ \ddddot{\phi}(t) \end{Bmatrix} = \begin{Bmatrix} -t_{11,41} \\ -t_{12,41} \\ \cdots \\ -t_{15,41} \\ -t_{16,41} \\ \cdots \\ -t_{20,41} \\ -t_{31,41} \\ \cdots \\ -t_{35,41} \\ -t_{36,41} \\ \cdots \\ -t_{40,41} \end{Bmatrix}$$

(10.7.18)

Equations (10.7.18) contain fourth order coupled differential equations with the dependent variables $w(t)$, $\theta(t)$, $v(t)$ and $\phi(t)$. The first, sixth, eleventh and sixteenth rows are rewritten below, as the rest of the rows are respectively their derivatives.

$$[\bar{M}]\{\ddot{U}\} + [\bar{C}]\{\dot{U}\} + [\bar{K}]\{U\} = \{\bar{F}\} \qquad (10.7.19)$$

where

$$[\bar{M}] = \begin{bmatrix} t_{11,3} & t_{11,8} & t_{11,23} & t_{11,28} \\ t_{16,3} & t_{16,8} & t_{16,23} & t_{16,28} \\ t_{31,3} & t_{31,8} & t_{31,23} & t_{31,28} \\ t_{36,3} & t_{36,8} & t_{36,23} & t_{36,28} \end{bmatrix}$$

$$[\bar{C}] = \begin{bmatrix} t_{11,2} & t_{11,7} & t_{11,22} & t_{11,27} \\ t_{16,2} & t_{16,7} & t_{16,22} & t_{16,27} \\ t_{31,2} & t_{31,7} & t_{31,22} & t_{31,27} \\ t_{36,2} & t_{36,7} & t_{36,22} & t_{36,27} \end{bmatrix}$$

$$[\bar{K}] = \begin{bmatrix} t_{11,1} & t_{11,6} & t_{11,21} & t_{11,26} \\ t_{16,1} & t_{16,6} & t_{16,21} & t_{16,26} \\ t_{31,1} & t_{31,6} & t_{31,21} & t_{31,26} \\ t_{36,1} & t_{36,6} & t_{36,21} & t_{36,26} \end{bmatrix}$$

$$\{\bar{F}\} = \begin{Bmatrix} -t_{11,41} \\ -t_{16,41} \\ -t_{31,41} \\ -t_{36,41} \end{Bmatrix} \quad \{U\} = \begin{Bmatrix} -w(t) \\ \theta(t) \\ v(t) \\ \phi(t) \end{Bmatrix}$$

The above equation (10.7.19) can be solved by the time marching technique described in section 9.5.2.

Example 10.4

As a simple example for illustration, consider an overhung rotor shown in Fig. 10.13. The distance between the supports is assumed very small compared to the overhung portion and hence it can be treated as a cantilever with only two stations, one at the fixed end 0, and the other at

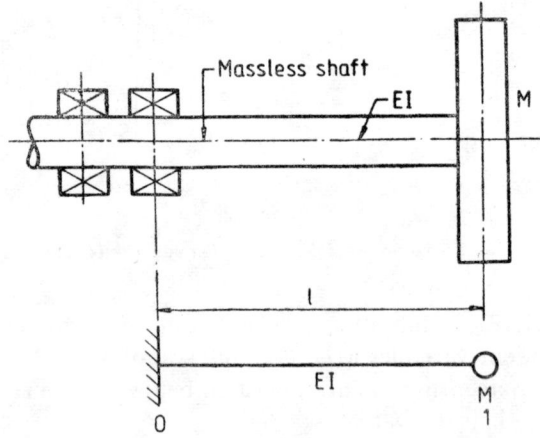

Fig. 10.13 Cantilever rotor

the free end 1. The unbalance in the rotor is considered to be only u_z. i.e., $u_y = 0$. Since the rotor is symmetric $\{s_z\}$ vector is sufficient to describe the problem. Then

$$\begin{Bmatrix} \{S_{zt}\} \\ 1 \end{Bmatrix}_1 = \begin{bmatrix} [P_t][F_t] & \{m_{zt}\} \\ 0 & 1 \end{bmatrix}_1 \begin{Bmatrix} \{S_{zt}\} \\ 1 \end{Bmatrix}_0 \quad (10.7.20)$$

From the elements of $[F_t]$ and $[P_t]$ given in equations (10.7.6a) and (10.7.8a) and using

$$\{S_{zwt}\} = \begin{Bmatrix} -w(t) \\ -\dot{w}(t) \\ -\ddot{w}(t) \\ -\dddot{w}(t) \\ -\ddddot{w}(t) \end{Bmatrix} \quad \text{etc.}$$

we can write

$$\begin{Bmatrix} \{S_{zwt}\} \\ \{S_{z\theta t}\} \\ \{S_{zMyt}\} \\ \{S_{zVzt}\} \\ 1 \end{Bmatrix}_1 = \begin{bmatrix} [I] & [I] & \dfrac{l^2}{2EI}[I] & \dfrac{l^3}{6EI}[I] & 0 \\ 0 & [I] & \dfrac{l}{EI}[I] & \dfrac{l^2}{2EI}[I] & 0 \\ 0 & 0 & [I] & l[I] & 0 \\ [\widetilde{P}] & l[\widetilde{P}] & \dfrac{l^2}{2EI}[\widetilde{P}] & \left(\dfrac{l^3}{6EI}[\widetilde{P}]+[I]\right) & \{m_{zvt}\} \\ 0 & 0 & 0 & 0 & 1 \end{bmatrix} \begin{Bmatrix} \{S_{zwt}\} \\ \{S_{z\theta t}\} \\ \{S_{zMyt}\} \\ \{S_{zVzt}\} \\ 1 \end{Bmatrix}_0$$

(10.7.21)

where $\{m_{zvt}\}$ have the last five terms of $\{m_{zt}\}$ in equation (10.7.12), and

$$[\widetilde{P}] = \begin{bmatrix} 0 & -C & -M & 0 & 0 \\ 0 & 0 & -C & -M & 0 \\ 0 & 0 & 0 & -C & -M \\ 0 & 0 & 0 & 0 & -C \\ 0 & 0 & 0 & 0 & 0 \end{bmatrix} \quad (10.7.22)$$

The boundary conditions are

$$\{S_{zwt}\}_0 = 0$$
$$\{S_{z\theta t}\}_0 = 0 \quad (10.7.23)$$

and

$$\{S_{zMyt}\}_1 = 0$$
$$\{S_{zVzt}\}_1 = 0 \quad (10.7.24)$$

Hence equations (10.7.19) reduce to

$$\begin{bmatrix} [I] & l[I] \\ \dfrac{l^2}{2EI}[\widetilde{P}] & \dfrac{l^3}{6EI}[\widetilde{P}]+[I] \end{bmatrix} \begin{Bmatrix} \{S_{zMyt}\} \\ S_{zVzt} \end{Bmatrix}_0 = \begin{Bmatrix} 0 \\ -\{m_{zVt}\} \end{Bmatrix} \quad (10.7.25)$$

i.e.,

$$\{S_{zMyt}\}_0 = -l\{S_{zVzt}\}_0$$

$$\dfrac{l^2}{2EI}[\widetilde{P}]\{S_{zMyt}\}_0 + \left[\dfrac{l^3}{6EI}[\widetilde{P}]+[I]\right]\{S_{zVzt}\}_0 = -\{m_{zVt}\} \quad (10.7.26)$$

We can eliminate $\{S_{zMyt}\}_0$ from the above to get

$$\dfrac{-l^3}{3EI}[\widetilde{P}]\{S_{zVzt}\}_0 + \{S_{zVzt}\}_0 = -\{m_{zvt}\} \quad (10.7.27)$$

Substituting for $[\widetilde{P}]$ from equation (10.7.22) in the above equation and neglecting damping terms, the following equations and their derivatives are obtained.

$$\dfrac{Ml^3}{3EI}\ddot{V}_{z0} + V_{z0} = u_z\omega^2 \cos\omega t$$

$$M_{y0} = lV_{z0} \quad (0.7.28)$$

The steady state solution of the above is

$$V_{z0} = \frac{u_z \omega^2}{1 - \frac{\omega^2}{\omega_n^2}} \cos \omega t \qquad (10.7.29)$$

where

$$\omega_n^2 = \sqrt{\frac{3EI}{Ml^3}}$$

Substituting V_{z0} and M_{y0} in equation (10.7.21), the deflection at the free end is given by

$$w_1 = \frac{u_z \frac{\omega^2}{\omega_n^2}}{1 - \frac{\omega^2}{\omega_n^2}} \cos \omega t \qquad (10.7.30)$$

which is the closed form conventional steady state solution. If we considered the transient solution of equation (10.7.28), we can likewise determine the transient solution for w_1.

Example 10.5

Consider a 50 kg rotor shown in Fig. 10.14, having a rigid bearing critical speed of 6800 rpm. The unbalance is 0.5 kg mm with an eccentri-

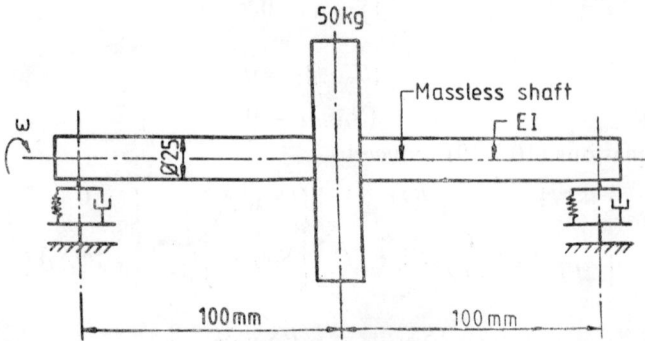

Fig. 10.14 Simply supported rotor

city of 0.01 mm. Let the bearings be isotropic and undamped with $K_{zz} = K_{yy} = 0.174 \times 10^5$ N/mm. To obtain a quick convergency, shaft damping $\zeta = 0.1$ may be assumed.

Figure 10.15 gives the orbital response at the bearing stations and midspan at 3000 rpm. The rotor attains the steady state solution in around three cycles. At the midspan, the whirl orbit is circular and synchronous with the shaft speed. The steady state whirl amplitude is $r_u/a = 0.5$. It is found that the response of the two bearings is identical and again synchronous with the speed of rotation. The whirl amplitude at the bearings is $r_u/a = 0.21$. Also the angular locations of the deflected form of the rotor at any instant of time are same for all three locations.

Instability Due to Fluid Film Forces 229

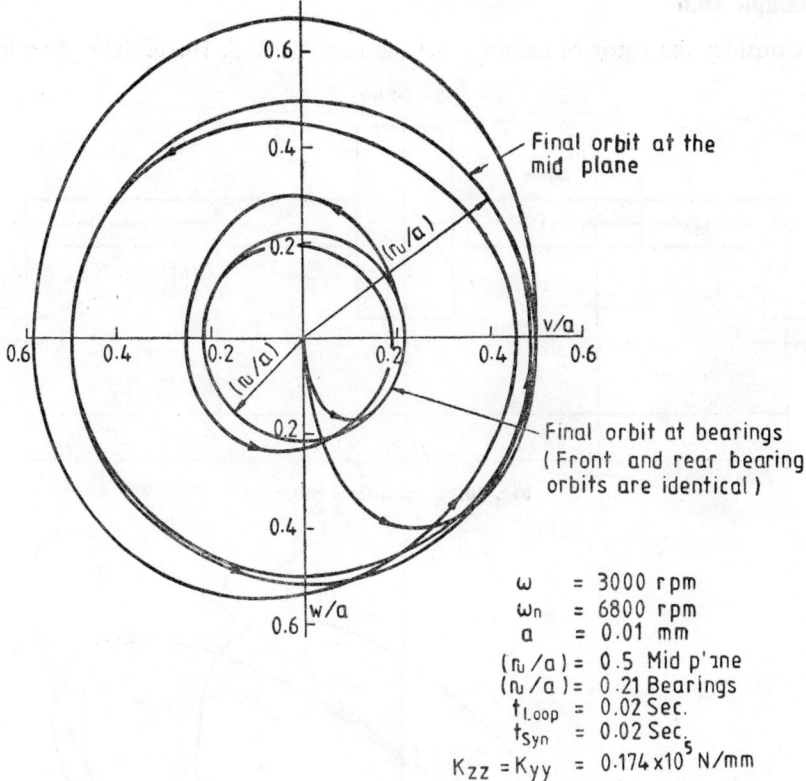

Fig. 10.15 Transient whirl orbit of rotor on isotropic flexible bearings at 3000 rpm

The steady state solution can be directly obtained from section 8.1.1. The shaft stiffness

$$K = M\omega_n^2 = 0.253 \times 10^5 \text{ N/mm}$$

$$K_z = \frac{2K_{zz}K}{2K_{zz} + K} = 0.1465 \times 10^5 \text{ N/mm}$$

$$\omega_{n\,\text{eff}} = \sqrt{\frac{K_z}{M}} = 541.4 \text{ rad/sec}$$

$$\frac{\omega}{\omega_{n\,\text{eff}}} = 0.5805$$

$$\frac{r_u}{a} = \frac{\frac{\omega^2}{\omega_{n\,\text{eff}}^2}}{1 - \frac{\omega^2}{\omega_{n\,\text{eff}}^2}} = 0.506$$

$$\frac{r_{u0}}{a} = \frac{K}{2K_{zz} + K} \frac{r_u}{a} = 0.213$$

The steady state response obtained by the transient analysis agrees with the above.

Example 10.6

Consider the rotor of example 8.1, shown in Fig. 10.16. The bearing

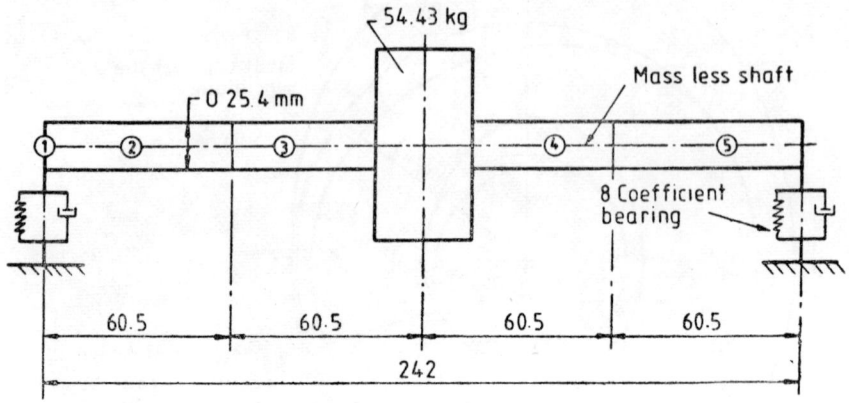

Fig. 10.16 Gunter's rotor

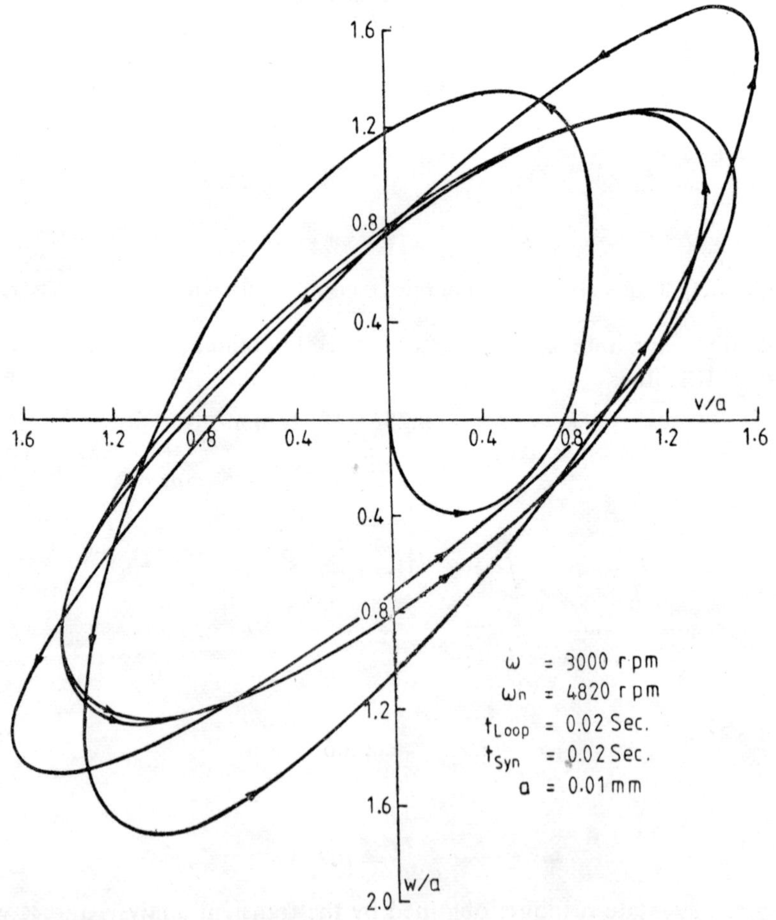

ω = 3000 rpm
ω_n = 4820 rpm
t_{Loop} = 0.02 Sec.
t_{Syn} = 0.02 Sec.
a = 0.01 mm

Fig. 10.17 Transient whirl orbit of Gunter's rotor on flexible bearings at 3000 rpm

Instability Due to Fluid Film Forces 231

properties at 3000 rpm are

$$K_{zz} = 0.175 \times 10^5 \text{ N/mm} \quad K_{zy} = 0.1 \times 10^5 \text{ N/mm}$$

$$K_{yz} = 0.1 \times 10^5 \text{ N/mm} \quad K_{yy} = 0.175 \times 10^5 \text{ N/mm}$$

An unbalance of 0.544 kg mm is assumed. A shaft damping ratio $\zeta = 0.1$ may be assumed for quick convergence.

The transient unbalance response for this rotor at the midplane is shown in Fig. 10.17. The steady state amplitude is 0.0185 mm and the time taken for one complete cycle is equal to the synchronous time 0.02 sec. Figure 10.18 shows the deflected form of the shaft at $t = 0.0788$ sec after the start and it can be seen that the response and orientation of the orbits at both the bearings is same.

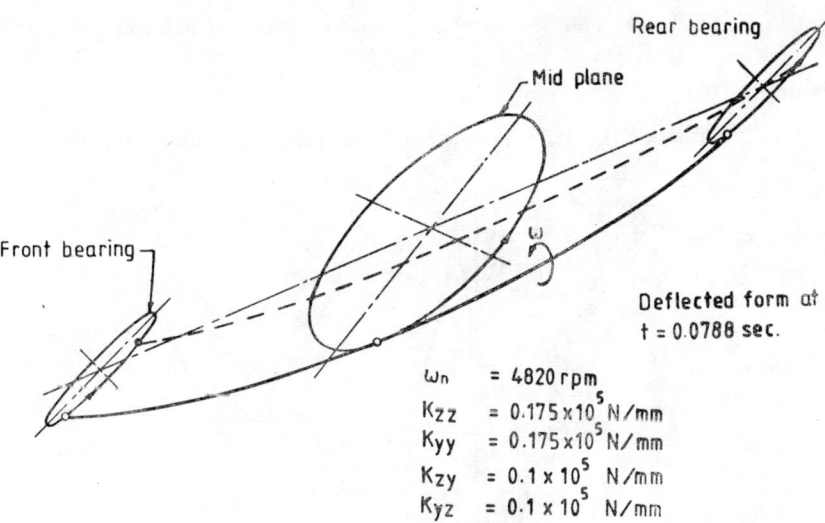

Fig. 10.18 Deflected form of Gunter's rotor at 3000 rpm

Example 10.7

Consider the 6 MW rotor of example 8.7, where the steady state solution was obtained by the transfer matrix method and Jeffcott model. The same rotor is analyzed here to find the transient response at 4000 rpm.

The transient whirl orbit at the central plane of the rotor is given in Fig. 10.19. The whirl is found to be always synchronous. For clarity, the whirl orbits are shown between 0 sec and 0.135 sec and then from 0.192 sec till the steady state is attained. The steady state solution is 1.8, which agrees with the value in Fig. 8.24. The final steady state orbit in Fig. 10.19 is obtained after 17 cycles which is considerably smaller than the transient values.

232 *Rotor Dynamics*

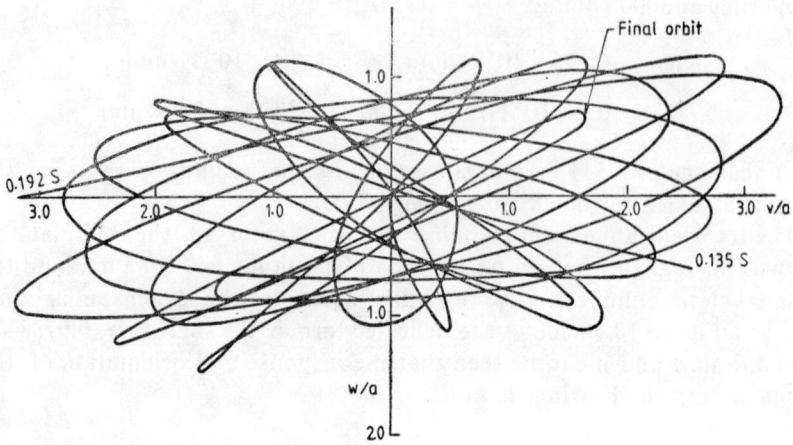

Fig. 10.19   Transient whirl orbit of a 6 MW rotor at 4000 rpm

Example 10.8

Finally consider the Jeffcott rotor of example 10.5, mounted on non-

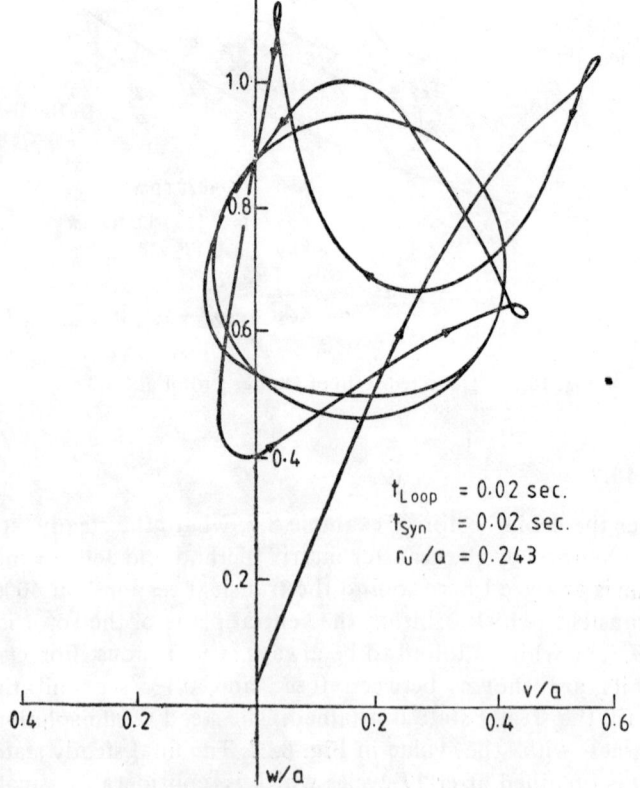

Fig. 10.20   Transient whirl orbit of rotor on nonlinear bearings at 3000 rpm

linear bearings with the following properties at 3000 rpm,

$$F_z = K_{zz}w(1 + w + w^2) + C_{zz}\dot{w}(1 + \dot{w} + \dot{w}^2)$$

$$F_y = K_{yy}v(1 + v + v^2) + C_{yy}\dot{v}(1 + \dot{v} + \dot{v}^2)$$

$$K_{zz} = 0.146 \times 10^5 \text{ N/mm}, K_{yy} = 0.101 \times 10^5 \text{ N/mm}$$

$$C_{zz} = C_{yy} = 70 \text{ N sec/mm}$$

Figure 10.20 shows the rotor response obtained due to the above non-linear bearing forces. The rotor amplitude shoots up to a high value of $w/a > 1$, then takes up a backward whirl and finally converges to a circular orbit whose radius is $r_u/a = 0.23$.

For a further study, references may be made to the additional literature [36 to 50] given at the end of the chapter.

References

1. Newkirk, B.L. and Taylor, H.D. Oil film whirl—An investigation of disturbances on oil films in journal bearings, General Electric review, Vol. 28, 1925, p. 559.
2. Robertson, D. Whirling of journal in sleeve bearings, Phil. Mag., Vol. 15, 1933, p. 113.
3. Pinkus, O., and Sternlicht, B. Theory of hydrodynamic lubrication, McGraw-Hill Book Co., 1961.
4. Morrison, D. and Peterson, A.N. Criteria for unstable oil whirl of flexible rotor, Proc. I. Mech. E., Vol. 179, Pt. 3j, 1964-65, p. 45.
5. Rieger, N.F. and Thomas, C.B. Some recent computer studies on the stability of rotors in fluid film bearings, Proc. IUTAM Symp. Dynamics of Rotors, 1974, p. 436.
6. Lund, J.W. Rotor bearing dynamic design technology, Part III: Design handbook for fluid film bearings, Mechanical Technology Inc., AFAPL-Tr-65-45, 1965.
7. Rao, J.S., Raju, R.J. and Reddy, K.B.V. Experimental investigation on oil whip of flexible rotors, Tribology, 1970, p. 100.
8. Rao, J.S. Instability of rotors in fluid film bearings, J. Vib. Acoust. Stress Rel. Des., ASME, Vol. 105, 1983, p. 274.
9. Lund, J.W. Computer programs for unbalance response and stability, Rotor bearing dynamics design technology, Part V, AFAPL-Tr-65-45 1965.
10. Den Hartog, J.P. Mechanical Vibrations, McGraw-Hill Book Co., 1956.
11. Tondl, A. Some problems in rotor dynamics, Publishing House of Czechoslovakian Academy of Sciences, Prague, 1965.
12. Hori, Y. Theory of oil whip. J. Appld. Mech., ASME, 1959, p. 189.
13. Gunter, E.J. Dynamic stability of rotor bearing systems, NASA report SP-113, 1966.
14. Gladwell, G.M.L. and Stammers, C.W. On the stability of an unsymmetrical rigid rotor supported in unsymmetrical bearings, J. of Sound and Vib., Vol. 3, 1966, p. 211.
15. Black, H.F. and Mcternam, A.J. Vibrations of a rotating asymmetric shaft supported in asymmetric bearings, J. of Mech. Engng. Sci., Vol. 10, 1968, p. 252.
16. Gunter, E.J. and DeChowdhury, P. Dynamic stability of flexible rotor bearing systems, Report No. ME-4040-104-700, Res. Lab. for Engng. Sci., Univ. of Virginia, Charlottesville, 1970.
17. Iwatsubo, T. et. al. Vibrations of asymmetric rotors suppoted by asymmetric bearings, Ingenuir Archiv, Vol. 42, 1973, p. 416.

18. Dostal, M. et al. Stability control of flexible shafts on oil film bearings, J. of Sound and Vib., Vol. 35, 1974, p. 361.
19. Lund, J. W. Stability and damped critical speeds of a flexible rotor in fluid film bearings, J. of Engng. for Indus., ASME, 1974, p. 509.
20. Black, H.F. Calculation of forced whirling and stability of centrifugal pump rotor systems, J. of Engng. for Indus., ASME, 1974, p. 1076.
21. Kirk, R.G. DeChowdhury, P. and Gunter, E.J. The effect of support flexibility on the stability of rotors mounted in plain cylindrical bearings, Proc. IUTAM Symp. Dynamics of Rotors, 1974, p. 244.
22. Mohan, S. and Hahn, E.J. Design of squeeze film damper supports for rigid rotors, J. of Engng. for Indus., ASME, 1974, p. 976.
23. Cunningham, R.E., Fleming, D.P. and Gunter, E.J. Design of a squeeze film damper for a multimass flexible rotor, J. of Engng. for Indus., ASME, 1975, p. 1383.
24. Bansal, P.N. and Kirk, R.G. Stability and damped critical speeds of rotor bearing systems, J. of Engng. for Indus., ASME, 1975, p. 1325.
25. Myrick, S.T. and Rylander, H.G. Analysis of flexible rotor whirl and whip using a realistic hydrodynamic journal bearing model, J. of Engng. for Indus., 1976, p. 1135.
26. Shapiro, W. Fluid film bearing characteristics and their influence on rotor vibrations, Tech. Report 32 TR 75-2 (revised) Franklin Institute Research Laboratories, 1976.
27. Someya, T. An investigation into the stability of rotating shaft supported by journal bearings, 4 reports, JSME, Vol. 42, 1976, pp. 2345, 2355, 2363, 2373.
28. Iwatsubo, T. Stability of rotor systems having asymmetric elements, Ingenuir Archiv, Vol. 47, 1978, p. 293.
29. Kimball, A.L. Internal friction theory of shaft whirling, General Electric review, Vol. 27, 1924, p. 244.
30. Ehric, F.F. Shaft whirl induced by rotor internal damping, J. of Appld. Mech., ASME, 1964, p. 279.
31. Ehric, F.F. Identification and avoidance of instabilities and self-excited vibrations in rotating machinery, ASME 72-DE-21.
32. Begg, I.C. Friction induced rotor whirl—a study in stability, J. of Engng. for Indus, ASME, 1974, p. 451.
33. Zorzi, E.S. and Nelson, H.D. Finite element simulation of rotor bearing systems with internal damping, J. of Engng. for Indus., ASME, 1977, p. 71.
34. Rao, J.S. Instability of rotors mounted in fluid film bearings with a negative cross-coupled stiffness coefficient, Mechanism and Machine Theory, v. 20, No. 3, 1985, p. 181.
35. Rao, J.S. and Bhaskara Sarma, K.V. Transient analysis of rotors by transfer matrix method, 11th Biennial conf. Vibration and Noise, ASME, Rotating Machinery Dynamics, DE-vol. 2, 1987, p. 545.
36. Castelli, V. and McCabe, J.T. Transient dynamics of a tilting pad gas bearing system, J. Lub. Tech. ASME, 1967, p. 299.
37. Sweet, J. and Genin, J. Nonlinear rotor bearing behaviour, Intl. J. Nonlinear Mechanics, v. 7, 1972, p. 407.
38. Licht, L. The dynamic characteristics of a turbogenerator simulator supported on gas lubricated foil bearings—part 3, J. Lub. Tech. ASME, v. 94, 1972, p. 211.
39. Tondl, A. Notes on the identification of subharmonic resonance of rotors, J. Sound Vib., v. 31, No. 1, 1973, p. 119.
40. Kirk, R.G. and Gunter, E.J. Transient response of rotor bearing systems, J. Engng. Indus. ASME, v. 96, 1974, p. 682.
41. Hibner, D.H. Dynamic response of viscous damped multishaft jet engines, J. Aircraft, v. 12, 1975, p. 305.
42. Pons, A. Study of some effects of hydrodynamic bearings nonlinear behaviour on rotating machine operation, Intl. conf. Rotor Dynamics Problems in Power plants,

IFToMM, Sept. 1982, Rome, p. 397.
43. Holmes, A.G. The vibration of multirotor systems supported upon oil film bearings, Ph.D. Thesis, London University, 1978.
44. Adams, M.L. Nonlinear dynamics of multibearing flexible rotors, J. Sound Vib., v. 71, No. 1, 1980, p. 129.
45. Nelson, H.D. and Meacham, W.L. Transient analysis of rotor-bearing systems using component mode synthesis, ASME 81-GT-110.
46. Gallardo, V.C., Black, G. and Stallone, M.J. Blade loss transient dynamic analysis, Vols. I, II and III, General Electric's Final reports and program documentation for the TETRA computer code, NASA Lewis Research Center, NAS3-22053, June 1981.
47. Adams, M.L. Protect against large rotor unbalance, Power, v. 125, No. 7, 1982, p. 52.
48. Adams, M.L., Padovan, J. and Fertis, D.G. Engine dynamics analysis with general nonlinear finite element codes, part I—overall approach and development for bearing damper element, J. Engng. Power, ASME, v. 104, No. 3, 1982, p. 586.
49. Adams, M.L. and Payandesh, J. Self excited vibration of statically unloaded pads in tilting pad journal bearings, J. Lub. Tech., ASME, v. 105, No. 3, 1983, p. 377.
50. Adams, M.L. and McCloskey, T.H. Large unbalance vibrations in steam turbine generator sets. Proc. Vibrations in Rotating Machinery Conf. I. Mech. E., 1984, p. 491.

Chapter 11

Instability in Torsional Vibration

Instabilities in torsional vibration can occur, because of variable stiffness or variable inertia of the system. Variable stiffness occurs, when the driven load is coupled to the driver through coupler rods, as in the case of an electric side rod locomotive discussed by Den Hartog [1]. Variable inertia of a system occurs in a reciprocating engine mechanism. In Chapter 3, the mass moment of inertia of an engine cylinder is taken as a constant value corresponding to the average inertia of the engine. This is not true, however, the results of linear analysis with constant inertia are reliable, if the mass of the reciprocating parts is small compared to the rotating parts. In some marine engines, the mass of reciprocating parts is quite large, and instabilities can occur as discussed by Goldsborough [2]. These two cases are discussed here.

11.1 Variable Stiffness in Torsional Vibrations

Figure 11.1 shows schematically an electric motor driving the wheels of a locomotive through side rods. The two side rods on the left and the right of the motor are 90° offset, so that there is no dead centre for the drive. Because of friction, the wheels are locked to the rail under normal driving conditions and they do not participate in the torsional vibration. We can consider them as fixed. The motor has a constant mass moment of inertia, being a pure rotating part, but the two side rods contribute to variable stiffness.

Consider first, one of the side rods at the dead centre position. The motor inertia can now have a free oscillation in torsion as the coupler rod can freely oscillate about the pin on the wheel. Hence there is no resistance for the motor rotor in torsional oscillation. When the coupler rod is 90° away from this position, large resistance is offered by the rod, for torsional oscillation of the motor rotor. Thus we have two positions of maximum resistance and two positions of zero resistance as shown in Fig. 11.1b, in one revolution of the gear. The spring constant for the second coupler rod is 90° away from the first rod and it is represented by the dotted curve. The total torsional stiffness is obtained by summing these two curves, which has an average value K and an oscillatory or variable component ΔK with a frequency ω_k equal to 4ω, where ω is the rotational speed of the wheels. Hence the equivalent model of the system, referred to the

Instability in Torsional Vibration 231

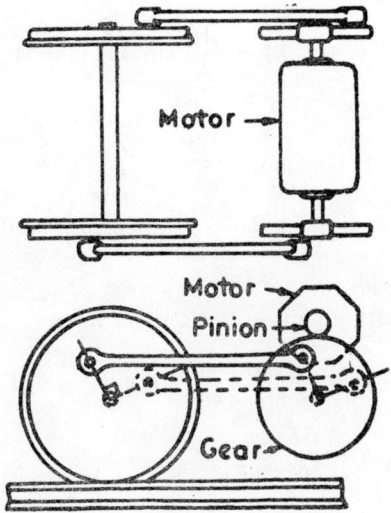

(a) Schematic layout of the drive of a side rod locomotive

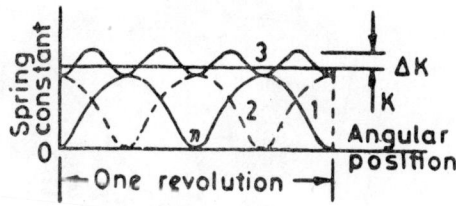

(b) Torsional stiffness as a function of angular position

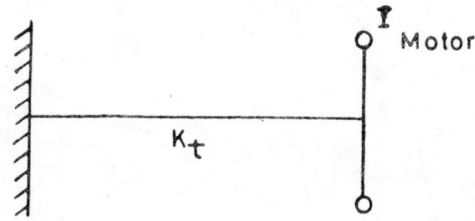

$K_t = K + \Delta K \sin 4\omega t$

(c) Torsional vibration model for the side rod locomotive

Fig. 11.1 Torsional system with variable stiffness

speed of the wheels is as shown in Fig. 11.1c. The equation of motion for free vibrations of the system is

$$I\ddot{\theta} + (K + \Delta K \sin 4\omega t)\theta = 0 \qquad (11.1.1)$$

i.e.,

$$\ddot{\theta} + \omega_n^2\left(1 + \frac{\Delta K}{K}\sin 4\omega t\right)\theta = 0 \qquad (11.1.2)$$

Let

$$\tau = 2\omega t \qquad (11.1.3)$$

then, equation (11.1.2) becomes

$$\theta'' + \left(\frac{\omega_n^2}{4\omega^2} + \frac{\Delta K}{K}\frac{\omega_n^2}{4\omega^2}\sin 2\tau\right)\theta = 0 \qquad (11.1.4)$$

where ( )' denotes $d/d\tau$.

Comparing this with equation (9.1.2), we have

$$a = \frac{\omega_n^2}{4\omega^2} = 4\frac{\omega_n^2}{\omega_k^2}; \text{ since } \omega_k = 4\omega$$

$$q = \frac{\Delta K}{K}\frac{\omega_n^2}{4\omega_k^2} \qquad (11.1.5)$$

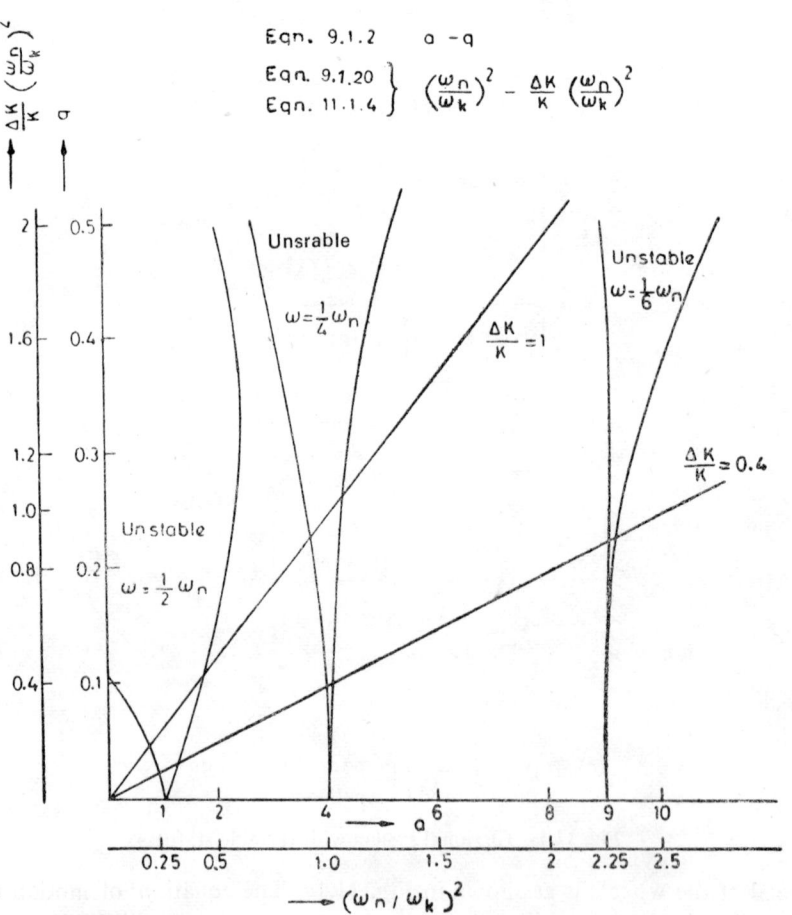

Fig. 11.2 Mathieu equation—Strutt diagram instabilities in torsional vibration of a system with variable stiffness

Instability in Torsional Vibration 239

Hence the instability conditions are given by Fig. 9.1. The important thing to notice is, the major instability condition corresponds to

$$\frac{\omega_n}{\omega_k} = \frac{1}{2}$$

i.e.,

$$\omega = \tfrac{1}{2}\omega_n \qquad (11.1.6)$$

unlike in the case of asymmetric shafts in bending vibration, where the condition is $\omega = \omega_n$. In fact the torsional system under consideration is not unstable at $\omega = \omega_n$. The predominant instability occurs at a speed equal to half the natural frequency, besides the others at $\omega_n/4$, $\omega_n/6$, ... The instability diagram for torsional vibrations is shown in Fig. 11.2. We may conclude from this, that the severity of torsional instability at half the critical speed is very high compared to the severity of bending vibration instability that may occur at half critical speed, in a shaft of the dissimilar moment of area.

11.2 System with Variable Inertia

In Fig. 11.3, a single cylinder engine with a large fly wheel is shown. The mass moment of inertia of the engine varies with time, because of the

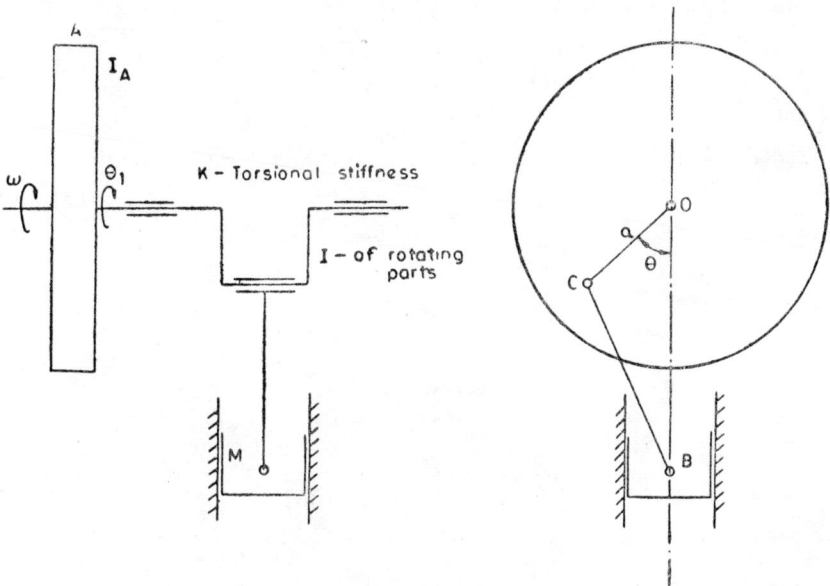

Fig. 11.3 Variable inertia system

motion of the piston. In linear analysis of torsional vibrations, we take the average value of mass moment of inertia for each cylinder of the engine assuming the change in inertia is small compared to the average value of inertia. However, for engines with heavy pistons, as in some marine

240 Rotor Dynamics

applications, this assumption is not valid. In practice, large amplitudes of vibration have been observed at speeds about one half of the torsional critical speed of the linear system.

For simplicity, let us assume simple harmonic motion of the piston, whose mass is denoted as M. If a is crank length, the displacement of the piston is

$$x = a \cos \theta$$
$$\dot{x} = -a \sin \theta \, \dot{\theta} \qquad (11.2.1)$$

The square of linear velocity of the piston is therefore

$$\dot{x}^2 = \tfrac{1}{2} a^2 \dot{\theta}^2 (1 - \cos 2\theta) \qquad (11.2.2)$$

Assuming I_A, the mass moment of inertia of the flywheel to be large compared to I, the mass moment of inertia of the rotating parts of the engine, we can ignore the kinetic energy of the flywheel in the oscillatory motion, by considering a node at this point. Hence the kinetic energy is

$$T = \tfrac{1}{2} I \dot{\theta}^2 + \tfrac{1}{2} M a^2 \dot{\theta}^2 \tfrac{1}{2}(1 - \cos 2\theta) + \tfrac{1}{2} I_A \omega^2 \qquad (11.2.3)$$

The strain energy in the shaft is

$$V = \tfrac{1}{2} K (\theta - \theta_1)^2 \qquad (11.2.4)$$

Using Lagrange's equation

$$\frac{d}{dt}\left(\frac{\partial T}{\partial \dot{\theta}}\right) - \frac{\partial T}{\partial \theta} + \frac{\partial V}{\partial \theta} = 0 \qquad (11.2.5)$$

we get

$$\ddot{\theta}(I + \tfrac{1}{2} Ma^2 - \tfrac{1}{2} Ma^2 \cos 2\theta) + \tfrac{1}{2} Ma^2 \dot{\theta}^2 \sin 2\theta + K(\theta - \theta_1) = 0 \qquad (11.2.6)$$

Writing

$$\theta_1 = \omega t$$
$$\theta = \omega t + \gamma$$
$$\tau = \omega t$$
$$\epsilon = \frac{\tfrac{1}{2} Ma^2}{I + \tfrac{1}{2} Ma^2}$$
$$\frac{1}{r^2} = \frac{\omega_n^2}{\omega^2} = \frac{K}{\omega^2 (I + \tfrac{1}{2} Ma^2)} \qquad (11.2.7)$$

we get

$$\gamma''(1 - \epsilon \cos 2\tau) + 2\epsilon \gamma' \sin 2\tau + \gamma \left(\frac{1}{r^2} + 2\epsilon \cos 2\tau\right) = -\epsilon \sin 2\tau \qquad (11.2.8)$$

While arriving at equation (11.2.8), higher order terms like $\epsilon \gamma \gamma''$, $\epsilon \gamma'^2$ ... are neglected [2, 3]. We notice that equation (11.2.8) is neither Mathieu equation nor Hill equation [5]. Hence we adopt a numerical procedure to find the response γ, for a given ϵ as a function of r.

Equation (11.2.8) can be in general expressed as

$$\gamma'' = f(\tau, \gamma, \gamma') \tag{11.2.9}$$

Solution of the above equation can be written as

$$\gamma = F(\tau) \tag{11.2.10}$$

and can be represented as shown in Fig. 11.4a. We use modified Euler's method [6] as follows to determine $F(\tau)$. Since we are interested in the stability of a free vibratory solution, we can start with, at time $t = 0$,

$$\gamma_0 = 1; \qquad \gamma'_0 = 0 \tag{11.2.11}$$

Assuming γ to be linear for a small interval of τ

$$\Delta \gamma \approx \Delta \tau \tan \alpha = \left\{\frac{d\gamma}{d\tau}\right\}_0 \Delta \tau \tag{11.2.12}$$

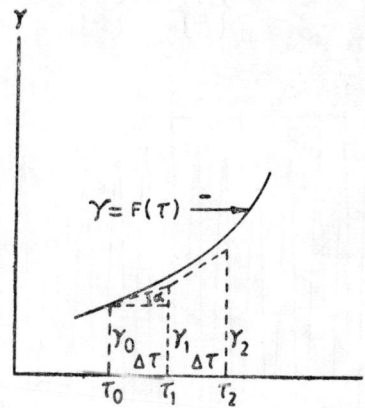

(a) Solution $F(\tau)$ in $\gamma \sim \tau$ plane

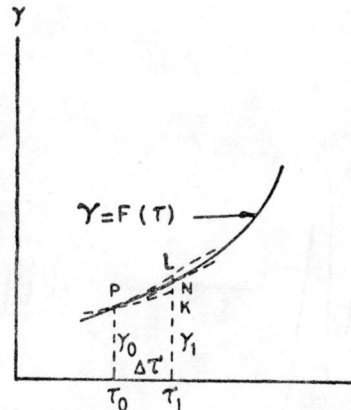

(b) Modification of solution at $\tau = \tau_1$

Fig. 11.4 Solution by modified Euler method

242 *Rotor Dynamics*

Hence

$$\gamma_1 \approx \gamma_0 + \left\{\frac{d\gamma}{d\tau}\right\}_0 \Delta\tau \tag{11.2.13}$$

In general

$$\gamma_i \approx \gamma_{i-1} + \left\{\frac{d\gamma}{d\tau}\right\}_{i-1} \Delta\tau; \quad i = 1, 2, \ldots \tag{11.2.14}$$

The values of $d\gamma/d\tau$ are evaluated from

$$\left\{\frac{d\gamma}{d\tau}\right\}_i \approx \left\{\frac{d\gamma}{d\tau}\right\}_{i-1} + \left\{\frac{d^2\gamma}{d\tau^2}\right\}_{i-1} \Delta\tau \tag{11.2.15}$$

From Fig. 11.4a, it is clear that there will be a cumulative error as τ increases, in estimating γ, even if $\Delta\tau$ is very small. We can improve the accuracy of the solution, by a modification of the above method, as illustrated in Fig. 11.4b, which gives

$$\gamma_1^{(2)} \approx \gamma_0 + \frac{\left\{\frac{d\gamma}{d\tau}\right\}_0 + \left\{\frac{d\gamma}{d\tau}\right\}_1^{(1)}}{2}\Delta\tau \tag{11.2.16}$$

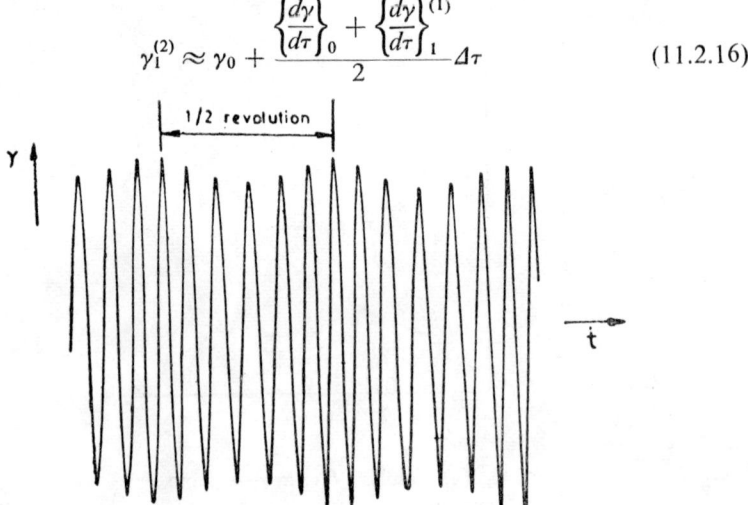

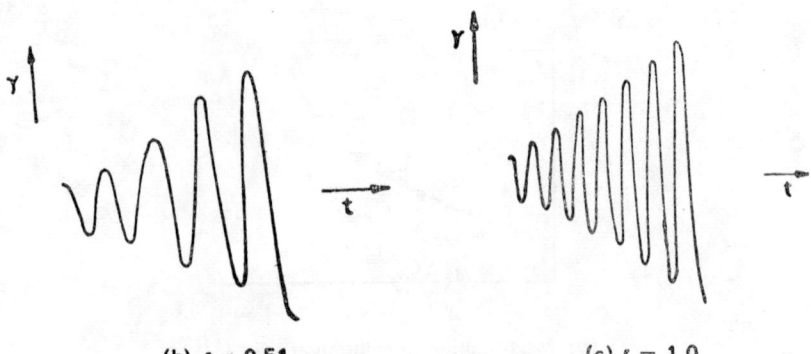

(b) $r = 0.51$ (c) $r = 1.0$

Fig. 11.5 Theoretical wave forms of $\gamma - t$, $\epsilon = 0.3$

Instability in Torsional Vibration 243

The superscripts in the above denote the iteration number and the iterations can be carried on until the required accuracy is obtained.

$$\gamma_1^{(j)} \approx \gamma_0 + \frac{\left\{\dfrac{d\gamma}{d\tau}\right\}_0 + \left\{\dfrac{d\gamma}{d\tau}\right\}_1^{(j-1)}}{2} \Delta\tau \qquad (11.2.17)$$

For ith step, therefore, we get the following set of equations:

$$\gamma_i''^{(j-1)} = F(\tau_i, \gamma_i^{(j-1)}, \gamma_i'^{(j-1)})$$
$$\gamma_i'^{(j)} = \gamma_{i-1}' + \tfrac{1}{2}(\gamma_{i-1}'' + \gamma_i''^{(j-1)})\Delta\tau$$
$$\gamma_i^{(j)} = \gamma_{i-1} + \tfrac{1}{2}(\gamma_{i-1}' + \gamma_i'^{(j-1)})\Delta\tau \qquad (11.2.18)$$

The above equations can be programmed to determine $\gamma = F(\tau)$, for a given ϵ and r.

Typical response curves for $r = 0.1$, 0.51 and 1.0 are shown in Fig. 11.5 for $\epsilon = 0.3$. The free vibration solution for $r = 0.1$ is bounded, whereas for $r = 0.51$ and 1.0, it is unbounded. Hence the system is unstable at $r = 0.51$ and 1.0, corresponding to rotational speeds of $\tfrac{1}{2}\omega_n$ and ω_n of the linear system.

The actual stability boundaries can be plotted by varying r for a given ϵ

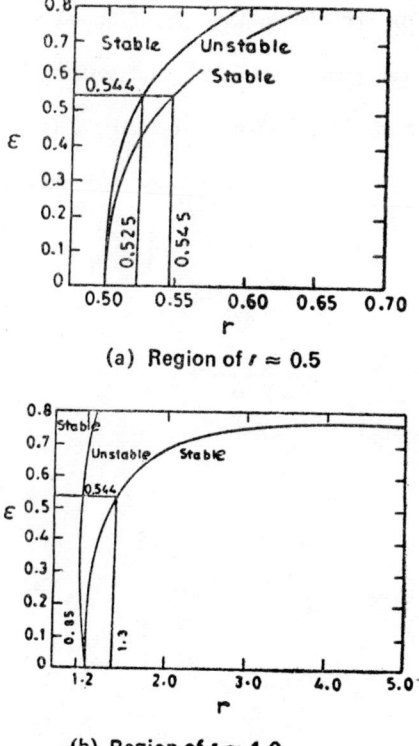

(a) Region of $r \approx 0.5$

(b) Region of $r \approx 1.0$

Fig. 11.6 Stability diagram of variable inertia system

and noting the values, where the solution is unstable. Such stability charts are shown in Fig. 11.6.

Example 11.1

Consider a typical marine engine which has the following data:

$$\tfrac{1}{2}Ma^2 = 5774 \text{ kg m}^2$$

$$I + \tfrac{1}{2}Ma^2 = 10614 \text{ kg m}^2$$

$$\omega_n = 682 \text{ rpm}$$

From Fig. 11.6, it can be seen that the engine will run roughly at speeds between 358-370 rpm and 580-890 rpm. The effect of damping on this engine has been studied by Pasricha and Carnegie [4].

References

1. Den Hartog, J.P. Mechanical vibrations, McGraw-Hill Book Co., 1956.
2. Goldsborough, G.R. Torsional vibrations in reciprocating engine shafts, Proc. Royal Soc., Vol. 109, 1925, p. 99.
3. Carnegie, W.D., Rao, J.S. and Pasricha, M.S. A theoretical study of the effects of variable inertia on the torsional vibrations of a single cylinder engine system, Instn. of Marine Engrs., London, August 1971.
4. Pasricha, M.S. and Carnegie, W.D. Effects of variable inertia on the damped torsional vibrations of diesel engine systems, J. of Sound and Vib., Vol. 46, 1976, p. 339.
5. Hayashi, T. Nonlinear oscillations in physical systems, McGraw-Hill Book Co., 1965.
6. Scarborough, J.B. Numerical mathematical Analysis, Oxford Book Co., Calcutta, 1964.

ABRO dynamic balancing machine showing a 3 tonne SO_2 compressor rotor being balanced to below 1 micron level (*Courtesy*: ABRO Balancing Machines, Delhi)

Chapter 12

Balancing of Rotors

The unbalance in the rotors considered before will not only cause rotor vibrations, but also transmit rotating forces to the bearings and to the foundation structure. The forces thus transmitted may cause damage to the machine parts and its foundation. In some cases, if the transmitted force is large enough, it might affect even the neighbouring machines and structures. Thus, it is necessary to remove the unbalance of a rotor, to as large an extent as possible, for its smooth running.

If we have a single disc of mass M, with eccentricity a, the correction mass m required at a radial distance r is

$$m = \frac{a}{r} M$$

The unbalance mr is usually measured in inch ounces or centimetre grams. The location where the balance mass is to be attached or removed can be determined simply by an experiment of statics. The rotor can be placed on two parallel horizontal knife edges and allowed to roll freely until the heavy spot comes to lowest position. The balance mass can be applied at the top of the disc tentatively and allowed to roll again and repeat the process of adding the required mass until the rotor is in indifferent equilibrium. Such a process is called static balancing and is valid, if the rotor has only one disc or the balancing is required in one plane.

If the unbalance is distributed along the length of the rotor, as it is in many cases, then a static balancing procedure cannot be used to determine the correction masses. This is because the centre of gravity of the rotor may be brought to the centre line of the shaft, but there could be an unbalanced moment left in the rotor, generating equal and opposite rotating reactions. This unbalance can be removed by dynamic tests only and hence called dynamic unbalance. This procedure is simple if the rotor is ideal or rigid. If the shaft deflects, and the deflection changes with speed, as it does in the vicinity of critical speeds, the problem of balancing is complicated. With the increase in size of machinery, the rotors of modern machines are made flexible and they run above the first critical speed and in some cases even beyond the second critical speed. These rotors need special flexible rotor balancing procedures.

12.1 Classification of Rotors

For the purpose of balancing, five classes of rotors are recognised [1].

Class 1—Rigid rotors. A rotor is considered rigid, when it can be corrected in any two arbitrary planes and after correction, its unbalance does not significantly change at any speed up to the maximum operating speed and when running under conditions which approximate closely to the final supporting system. A rotor not satisfying this condition is a flexible rotor and they are classified further as under.

Class 2—Quasi flexible rotors. Rotors that cannot be considered rigid but can be balanced adequately in a low speed balancing machine.

Class 3—Flexible rotors. Rotors that cannot be balanced in a low speed balancing machine and that require balancing by some special flexible rotor balancing technique.

Class 4—Flexible attachment rotors. Rotors that could fall in categories 1 or 2 but have in addition one or more components that are themselves flexible or are flexibly attached.

Class 5—Single speed flexible rotors. Rotors that could fall into category 3, but for some reason, e.g., economy, are balanced only for one speed of operation.

Table 12.1 gives examples of typical rotor configurations for each of the above classification [4]. For class 2, quasi-flexible rotors, the illustrations give the reasons why rotors can often be satisfactorily balanced at low speeds, as rigid rotors, even though they are normally flexible.

12.2 Rigid Rotor Classification and Balancing Criteria

Consider a rotor with the unbalance distribution represented by several unbalance masses in different planes of the rotor along its axis, located at different angular locations in each plane, as shown in Fig. 12.1. Let two convenient planes be arbitrarily chosen and denote them as left and right planes L and R respectively. The ith unbalance force, can be resolved into a force $F_{i'}^L$, and moment M_i as shown in Fig. 12.1b. The moment can be further split into two equal and opposite forces in left and right planes as F_i', as denoted in Fig. 12.1c. Thus, the ith unbalance force is split into two forces $F_i^L = F_{i'}^L - F_i'$ and $F_i^R = F_i'$ in the left and right planes which are arbitrarily chosen. We can repeat this process for all unbalance masses and obtain their components in left and right planes as shown in Fig. 12.1d. Combining the unbalance forces in these two planes, we arrive at two unbalance forces $F_L = \sum F_i^L$; $F_R = \sum F_i^R$. Thus, for a rigid rotor the unbalance distribution can be considered as equivalent to two unbalance masses in any two arbitrary planes and hence the definition of class 1 rotors in section 12.1.

Balancing of Rotors 249

Table 12.1 Classification of rotors

Class of Rotor	Description	Example
Class 1 Rotors *Rigid Rotors*	A rotor is considered rigid when its unbalance can be corrected in any two (arbitrarily selected) planes and, after that correction, its unbalance does not significantly change at any speed up to maximum service speed	gear wheel
Class 2 Rotors *Quasi-Flex Rotors*	A rotor that cannot be considered rigid but that can be balanced in a low speed balancing machine	

Rotors where the axial distribution of unbalance is known

Class 2A	A rotor with a single transverse plane of unbalance, e.g. single mass on a light shaft whose unbalance can be neglected	Shaft with grinding wheel
Class 2B	A rotor with two axial planes of unbalance, e.g., two masses on a light shaft whose unbalance can be neglected	Shaft with grinding wheel and pulley
Class 2C	A rotor with more than two transverse planes of unbalance	Jet engine compressor rotor
Class 2D	A rotor with uniformly distributed unbalance	Printing press roller
Class 2E	A rotor consisting of a rigid mass of significant axial length supported by a flexible shaft whose unbalance can be neglected	Computer memory drum

250 *Rotor Dynamics*

Table 12.1 (Cont.)

Class of Rotor	Description	Example
Rotors where axial distribution of unbalance is not known		
Class 2F	A symmetrical rotor, with two end correction planes, whose maximum speed does not significantly approach second critical speed, whose service speed range does not contain first critical speed and with controlled initial unbalance	5-stage centrifugal pump
Class 2G	A symmetrical rotor with two end correction planes and a central correction plane whose maximum speed does not significantly approach second critical speed; and with a controlled initial unbalance	Multi-stage pump impeller
Class 2H	An unsymmetrical rotor with controlled initial unbalance treated in a similar manner to class 2F rotor	I.P. steam turbine rotor
Class 3 Rotors *Flexible Rotors*	A rotor that cannot be balanced in a low speed balancing machine and that requires high speed balancing techniques	Generator rotor
Class 4 Rotors *Special Flexible Rotors*	A rotor that could fall into classes 1, 2 or 3, but has in addition one or more components that are themselves flexible or are flexibly attached	Rotor with centrifugal switch
Class 5 Rotors *Single Speed Flexible Rotors*	A rotor that could fall into class 3 but for some reason, e.g. economy balanced only for a single service speed	High speed motor

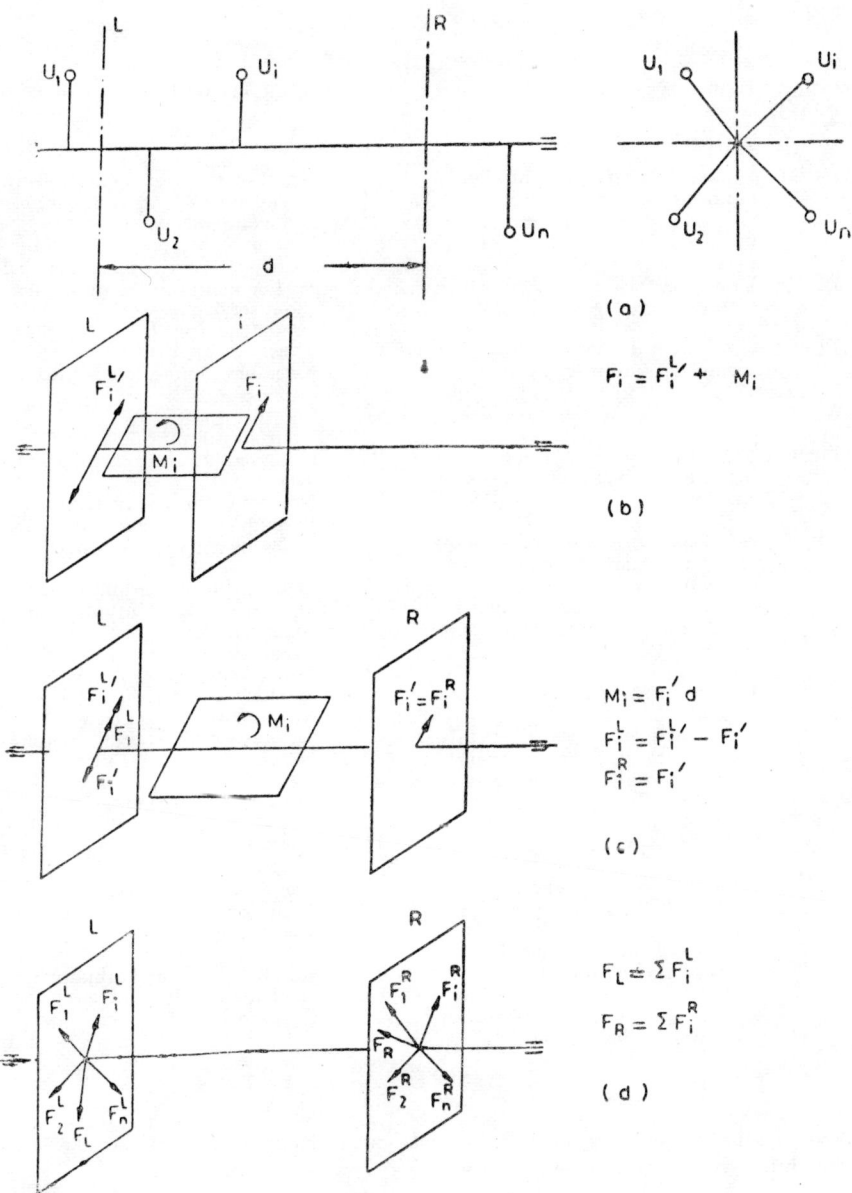

Fig. 12.1 Rigid rotor unbalance in two planes

Table 12.2 gives the classification of rigid rotor types with the unbalance quality for each classification. The quality of balance required for different groups is given in Table 12.3, and for each grade indicated in Table 12.2, the accepted residual unbalance is given in Fig. 12.2.

Table 12.2 Classification of rigid rotor types

Balance Quality Grade G	Rotor Types—General Examples
G 4000	Crankshaft-drives (2) of rigidly mounted slow marine diesel engines with uneven number of cylinders (3).
G 1600	Crankshaft-drives of rigidly mounted large two-cycle engines.
G 630	Crankshaft-drives of rigidly mounted large four-cycle engines. Crankshaft-drives of elastically mounted marine diesel engines.
G 250	Crankshaft-drives of rigidly mounted fast four-cylinder diesel engines (3).
G 100	Crankshaft-drives of fast diesel engines with six and more cylinders (3). Complete engines (gasoline or diesel) for cars, trucks and locomotives (4).
G 40	Car wheels, wheel rims, wheel sets, drive shafts. Crankshaft-drives of elastically mounted fast four-cycle engines (gasoline or diesel) with six and more cylinders (3). Crankshaft-drives for engines of cars, trucks and locomotives.
G 16	Drive shafts (propeller shafts, cardan shafts) with special requirements. Parts of crushing machinery. Parts of agricultural machinery. Individual components of engines, (gasoline or diesel) for cars, trucks and locomotives.
	Crankshaft-drives of engines with six and more cylinders under special requirements.
G 6.3	Parts of process plant machines. Marine main turbine gears (merchant service). Centrifuge drums. Fans. Assembled aircraft gas turbine rotors. Fly wheels. Pump impellers. Machine-tool and general machinery parts. Normal electrical armatures. Individual components of engines under special requirements.
G 2.5	Gas and steam turbines, including marine main turbines (merchant service). Rigid turbo-generator rotors. Turbo-compressors. Machine-tool drives. Medium and large electrical armatures with special requirements. Small electrical armatures. Turbine-driven pumps.
G 1	Tape recorder and phonograph drives. Grinding-machine drives. Small electrical armatures with special requirements.
G 0.4	Spindles, discs, and armatures of precision grinders. Gyroscopes.

Notes (1) The quality grade number represents the maximum permissible circular velocity of the centre of gravity $a\omega$ in mm/sec.

(2) A crankshaft drive is an assembly which includes the crankshaft, a flywheel, clutch, pulley, vibration damper, rotating portion of connecting rod etc.

(3) For the purposes of this recommendation, slow diesel engines are those with a piston velocity of less than 10 m per sec.; fast diesel engines are those with a piston velocity of greater than 10 m per sec.

(4) In complete engines the rotor mass comprises the sum of all masses belonging to the crankshaft-drive.

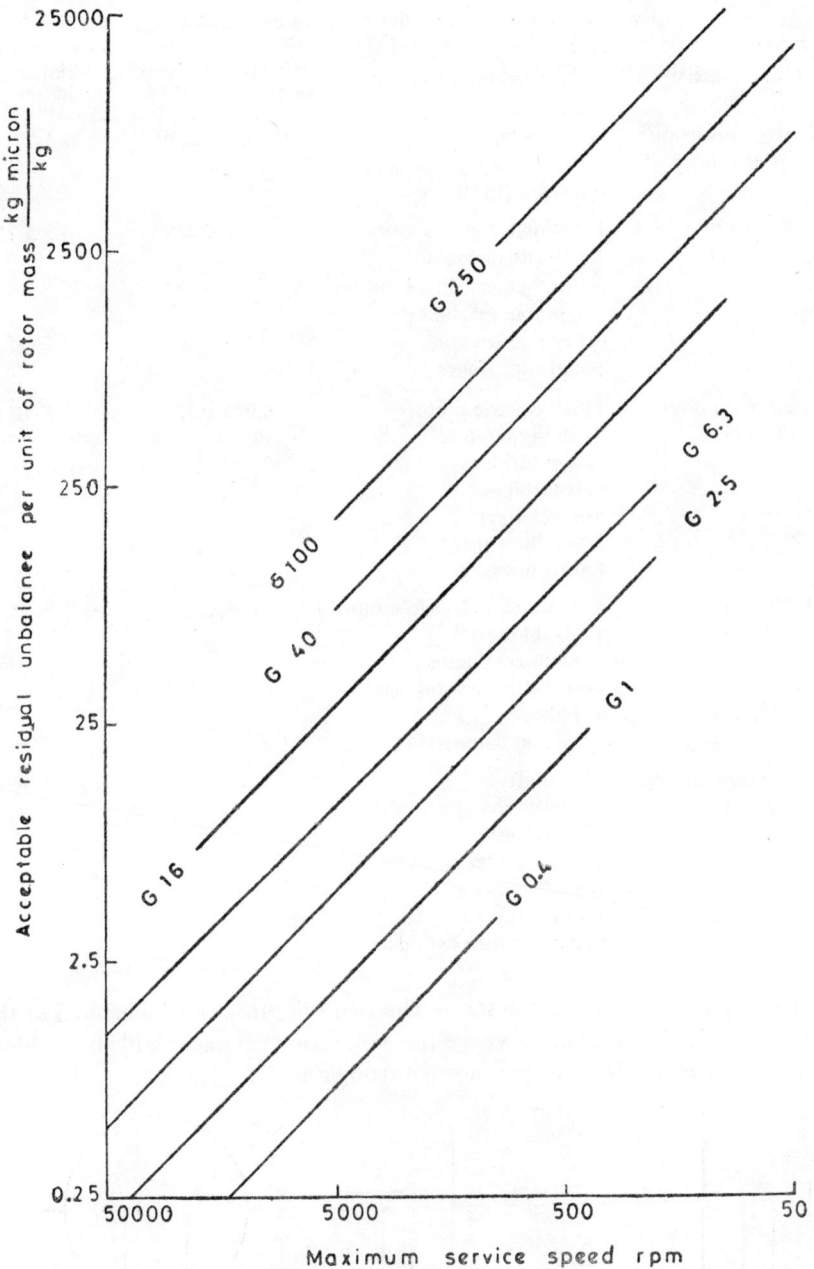

Fig. 12.2 Accepted residual unbalance for rigid rotors

12.3 Balancing of Rigid Rotors

A very simple procedure to balance a rigid rotor is described by Den Hartog [6] and Macduff and Curreri [7], using a cradle balancing machine. The rotor is placed in the two bearings of a cradle C shown in Fig. 12.3. The cradle is placed on two springs and can be fulcrumed about F_1 or F_2

Table 12.3 Quality of balance required by machine type

Quality group	Machine type	Residual unbalance kg mm per 100 kg	Balance microns
A. High precision balancing	Gyro rotors Centrifuges Superfinish grinders	0.0025–0.0120	0.2–1
B. Precision balancing	Ultra high speed motors Small gas turbines Superchargers, Precision grinding machines Jet engine rotors Small centrifuges	0.005–0.025	0.5–1.75
C. High quality balancing	Small electric motors Turbo-generators Steam turbines Gas turbines Superchargers (medium speed) Centrifuges	0.025–0.12	1.75–10
D. Good quality balancing	Commercial electric motors Fans, blowers Centrifugal pumps Four cylinder crank shafts Flywheels Torsional dampers	0.05–0.25	5–25
E. Average quality balancing	Line shafts Propeller shafts Gear trains 1 to 3 cylinder engines Large flywheels Farm machines Paper machine cylinders	0.15–0.8	20–100

to form a simple vibrating system. The two fulcrums can be located at the two chosen balance planes, where the correction masses will be added. The rotor can be driven by a motor through a belt.

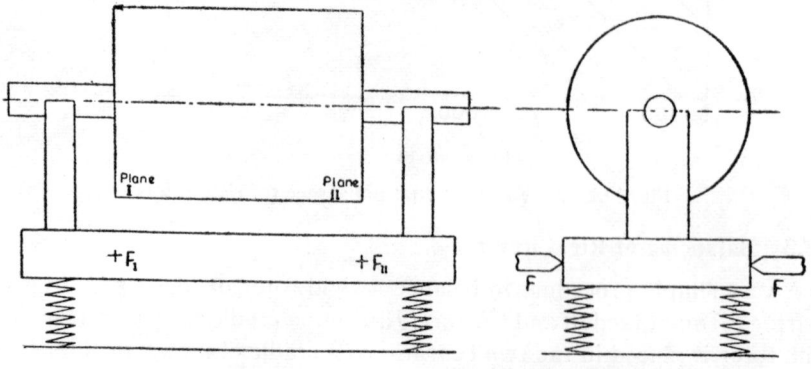

Fig. 12.3 Schematic diagram of a cradle balancing machine

Balancing of Rotors

If the spring system is such that the natural frequency of the system is in the range of motor speed, the phase angle or the location of the mass in either plane can be determined as follows.

Fulcrum the cradle in plane I, by fixing F_1 and releasing F_2. Run the rotor to resonance, observing the maximum amplitude to the right of fulcrum F_2. This vibration is due to all the unbalance in plane II, since the unbalance in plane I has no moment about F_1. Use a trial mass at a chosen location and determine the amplitude of vibration. Make a plot of this amplitude for different locations of the same trial mass. This plot looks as shown in Fig. 12.4. The trial mass for correction is added at the

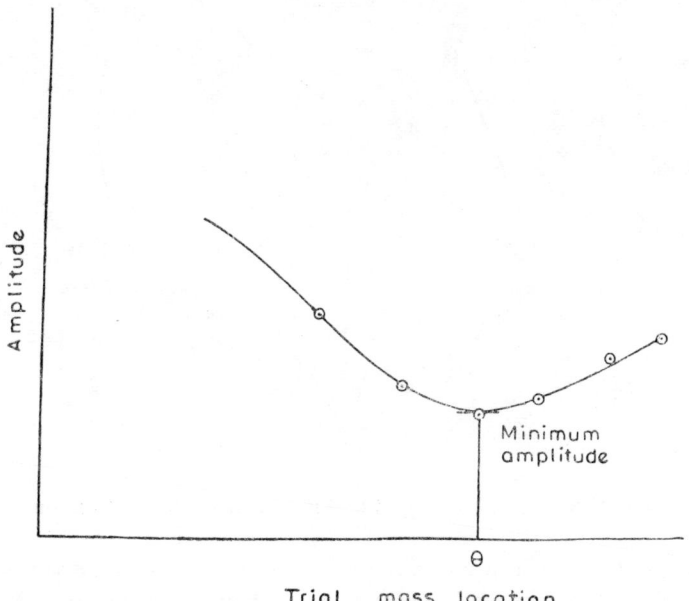

Fig. 12.4 A simple trial to determine correct location of trial mass

location where the amplitude of vibration is minimum. Increase or decrease the trial mass at the same location, until the desired level of balance is achieved. In determining this location, the rotor may be run at a speed well below resonance, and a mark is made on the rotor, by holding a chalk very close to the surface in plane II. This will be obviously the heavy spot of the rotor where the mass is to be added. However in practice, it is difficult to determine this spot accurately and a plot as given in Fig. 12.4, around this spot is necessary. This procedure is also tedious and sometimes may be time consuming.

A procedure to determine the correction mass and location can be laid as follows, based on four observations of amplitude: (1) without any addition to the rotor, (2) with a trial mass at $\theta = 0°$, (3) with the same trial mass at $\theta = 180°$ and (4) with the same trial mass at $\theta = \pm 90°$, where θ is measured from a conveniently chosen location.

256 Rotor Dynamics

Figure 12.5 illustrates the procedure to determine the correction mass. OA is the amplitude measured with trial run (1). OB is the amplitude measured in trial run (2) by the addition of a trial mass w_t at 0°. Hence

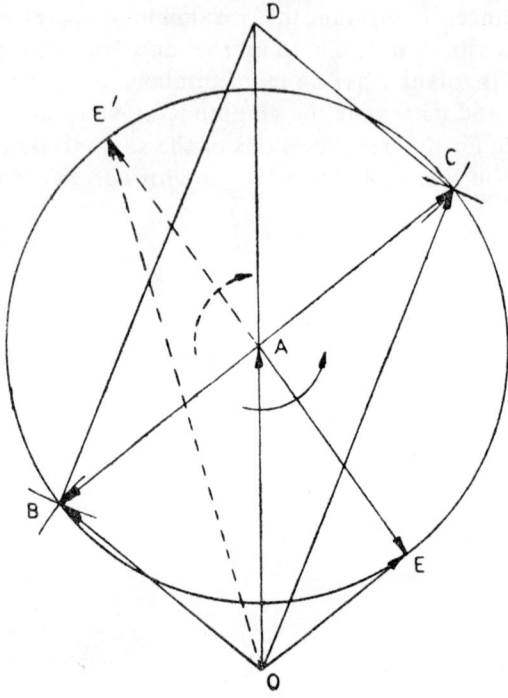

Fig. 12.5 Determination of trial mass magnitude and location

the vector AB will represent the effect of trial mass w_t. At this stage we should remember that we know the chosen location of 0° on the rotor, but we have no idea of location of vector OA on the rotor. If OC is the vibration measured in trial run (3), with the trial mass at 180°, then $AB = AC$ with 180° phase difference between them as shown in Fig. 12.5. However, we only know the magnitudes OA, OB and OC from three test runs and we have to lay down a procedure to arrive at the diagram in Fig. 12.5. This is quite simple. Erect a line OAD equal to $2 \cdot OA$. With O as centre and OB and OC as radii draw arcs of circles and with D as centre draw arcs with radii OC and OB respectively to intersect the previous arcs. Construct the parallelogram $OBDC$. Now AB represents 0° position and AC 180° position on the rotor. The angular measurement may be clockwise or counterclockwise and is determined from the fourth observation. The fourth observation could be either OE or OE'. If the value observed is in the vicinity of OE, then the angle is to be measured counterclockwise, however it will be clockwise if OE' is the reading observed in the test. The fourth run also checks the validity of the linearity used in the balancing procedure. The magnitude of the trial mass is determined in proportion of AB and AO and the location from the angle $O\hat{A}B$, and the direction obtained from Fig. 12.5. The test is repeated by making the cradle fulcrumed at F_{II} and measurements made in plane I.

The cradle balancing procedure just described is very time consuming and also restricts the mass and size of the rotor. Modern balancing machines use amplitude and phase measurements in two planes for balancing a rotor. The machines are either soft support or hard support machines. In soft support machines, the resonant frequency of the rotor support system is low and the rotor runs at a speed above the resonance of the support system. Vibratory amplitudes are measured, which are then converted to forces. In hard support machines, the support natural frequency is very high and they measure the rotor unbalance forces directly, independent of the rotor mass and configuration.

The balancing procedure is based on influence coefficients measurement which is made as follows. We choose two convenient planes L and R and the two measurement planes a and b can be chosen as bearing locations. Let L_1 and R_1 be the initial readings of vibration levels measured with phase angles γ_1 and δ_1 respectively. The phase angles are measured with the same reference during the test and their relative location with respect to rotor is initially unknown. Figure 12.6a shows these values L_1 and R_1.

In the second run, place a trial mass T_R at a convenient location in plane R and let the observations be L_2 and R_2 at angles γ_2 and δ_2 respectively in the left and right planes of measurement a and b. The difference between R_2 and R_1 will be the effect of trial mass in right plane R on the measurement made in plane b. We can denote this as an influence coefficient $\bar{\alpha}_{bR}$. Similarly we have the influence coefficient $\bar{\alpha}_{aR}$. These are indicated in Fig. 12.6b.

$$\bar{\alpha}_{bR} = (\bar{R}_2 - \bar{R}_1)/\bar{T}_R$$
$$\bar{\alpha}_{aR} = (\bar{L}_2 - \bar{L}_1)/\bar{T}_R \qquad (12.3.1)$$

We remove the trial mass now from plane R and place $\bar{T}_L$ in plane L and repeat the test to obtain the measured values in Fig. 12.6c. This gives the following influence coefficients:

$$\bar{\alpha}_{bL} = (\bar{R}_3 - \bar{R}_1)/\bar{T}_L$$
$$\bar{\alpha}_{aL} = (\bar{L}_3 - \bar{L}_1)/\bar{T}_L \qquad (12.3.2)$$

Let the correct balance masses be $\bar{w}_R$ and $\bar{w}_L$. Since the original unbalance is R_1 and L_1 as measured in right and left planes, we can write

$$-\bar{R}_1 = \bar{w}_R \bar{\alpha}_{bR} + \bar{w}_L \bar{\alpha}_{bL}$$
$$-\bar{L}_1 = \bar{w}_R \bar{\alpha}_{aR} + \bar{w}_L \bar{\alpha}_{aL} \qquad (12.3.3)$$

Hence

$$\begin{Bmatrix}\bar{R}_1\\\bar{L}_1\end{Bmatrix} = -\begin{bmatrix}\bar{\alpha}_{bR} & \bar{\alpha}_{bL}\\\bar{\alpha}_{aR} & \bar{\alpha}_{aL}\end{bmatrix}\begin{Bmatrix}\bar{w}_R\\\bar{w}_L\end{Bmatrix} \qquad (12.3.4)$$

258 *Rotor Dynamics*

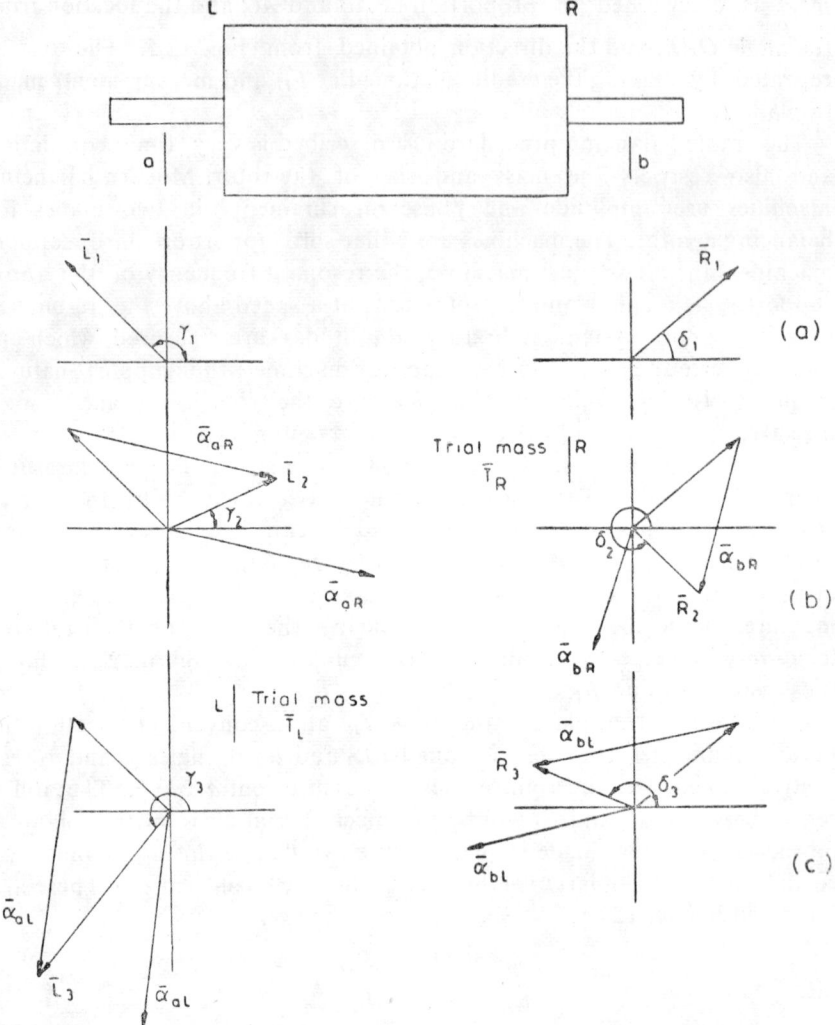

Fig. 12.6 Balancing measurements and influence coefficients for a rigid rotor

$$\bar{w}_R = \frac{\bar{L}_1 \bar{\alpha}_{bL} - \bar{R}_1 \bar{\alpha}_{aL}}{\bar{\alpha}_{bR} \bar{\alpha}_{aL} - \bar{\alpha}_{aR} \bar{\alpha}_{bL}} \qquad (12.3.5)$$

$$\bar{w}_L = \frac{\bar{R}_1 \bar{\alpha}_{aR} - \bar{L}_1 \bar{\alpha}_{bR}}{\bar{\alpha}_{bR} \bar{\alpha}_{aL} - \bar{\alpha}_{dR} \bar{\alpha}_{bL}} \qquad (12.3.6)$$

These can be calculated by either a graphical method of multiplication/division and addition/subtraction of vectors [7] or by a programmable calculator [8]. Commercial balancing machines incorporate electronic means of plane separation to simplify the balancing procedures [9].

Example 12.1

Consider the set of following observations [10]. The original unbalance vectors as measured for a turbine rotor are:

$$\overline{L}_1 = 9.144 \text{ microns peak to peak at } 90° = 9.144i$$
$$\overline{R}_1 = 10.16 \text{ microns peak to peak at } 45° = 7.188 + 7.188i$$

The trial mass in right plane:

$$\overline{T}_R = 6.8 \text{ gm at } 22\tfrac{1}{2}° = 6.294 + 2.608i$$
$$\overline{L}_2 = 5.08 \text{ microns peak to peak at } 27° = 4.521 + 2.311i$$
$$\overline{R}_2 = 6.35 \text{ microns peak to peak at } 99° = -.9906 + 6.274i$$

The trial mass in left plane:

$$\overline{T}_L = 6.8 \text{ gm at } 36° = 5.500 + 3.997i$$
$$\overline{L}_3 = 9.40 \text{ microns peak to peak at } 0° = 9.40$$
$$\overline{R}_3 = 30.5 \text{ microns peak to peak at } 99° = -4.775 + 30.1i$$

Then, from equations (12.3.1),

$$\overline{a}_{bR} = -1.160 + 0.336i$$
$$\overline{a}_{aR} = 0.229 - 1.181i$$

and from equations (12.3.2)

$$\overline{a}_{bL} = 0.558 + 3.760i$$
$$\overline{a}_{aL} = 0.327 - 1.900i$$

From equations (12.3.5) and (12.3.6), we have

$$\overline{w}_R = 10.94 + 1.56i$$
$$\overline{w}_L = -2.81 - 1.90i$$

Thus, the correction masses in the right and left planes are 11.06 gm at 8.1° and 3.40 gm at 214° respectively.

12.4 Balancing of Flexible Rotors

As long as the rotor experiences no deformations, that is, it remains as a rigid body, the balancing procedure discussed earlier is effective. Once the rotor bends, while approaching a critical speed, the bent centre line whirls around and additional centrifugal forces are set up and the rigid rotor balance becomes ineffective. Rieger and Badgley [11] gave a situation where rigid rotor balancing worsens bending mode whirl amplitude as shown in Fig. 12.7.

For class 3 rotors, it is a common practice to balance the individual components on mandrels and then stack balance the component assemblies one after another until a balanced rotor is formed. If necessary the rotor is balanced again as a rigid rotor at a speed below half the first critical speed. Before balancing the rotor for each critical speed, the rotor should be stabilised in the middle speed range, until the temperature in all parts of the rotor is equalised and unbalance readings are repeatable.

260 *Rotor Dynamics*

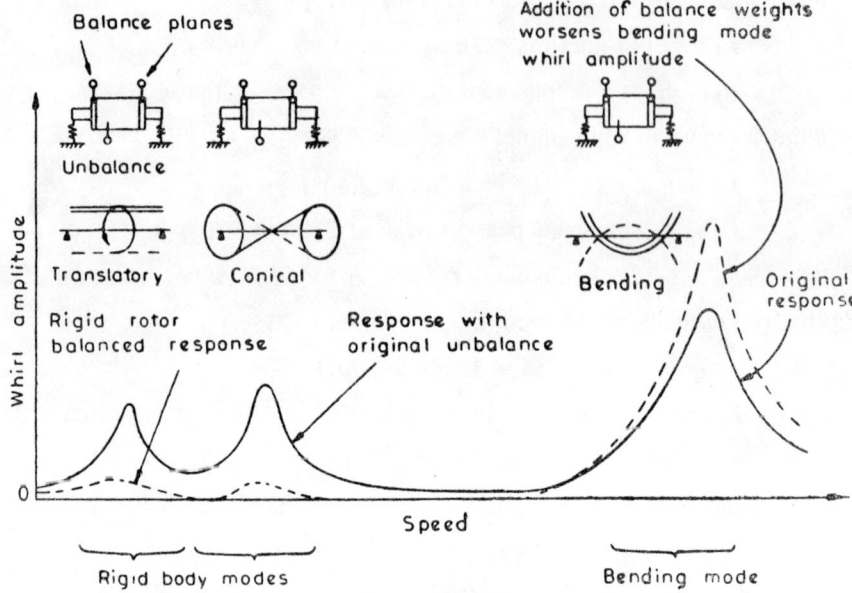

Fig. 12.7 Effect of two-plane balance at rigid body modes and at bending mode

Two different techniques are commonly employed, the modal balancing technique developed by Bishop, Gladwell and Parkinson [12, 13] and Moore and Dodd [14]; the influence coefficient method developed by Tessarzik, Badgley and Rieger [11, 15]. There are several papers that have appeared since then and reference may be made to the monograph by Rieger [5].

12.4.1 *Modal balancing theory*

For convenience, assume all the unbalance to be distributed in the plane *y-x*. This allows decoupling of *y* and *z* modes and the discussion will be limited to only *y-x* plane. The deflected shape of the rotor at any speed can be expressed in modal series as

$$y(x) = \sum_{i=1,2,\ldots} \phi_i Y_i(x) \qquad (12.4.1)$$

where $Y_i(x)$ are mode shapes and ϕ_i are arbitrary parameters. If $y(x)$ is measured at a given speed, ϕ_i can be determined by using the orthogonality principle as

$$\int_0^L y(x) m(x) Y_i(x)\, dx = \int_0^L \phi_i Y_i^2(x) m(x)\, dx = \phi_i M_i$$

or

$$\phi_i = \frac{1}{M_i} \int_0^L y(x) m(x) Y_i(x)\, dx \qquad (12.4.2)$$

In the above $m(x)$ is mass distribution of the rotor and M_i is generalised mass in *i*th mode.

The mass eccentricity in the rotor is expressed in the modal series as follows:

$$a(x) = \sum_{i=1,2,\ldots} \alpha_i Y_i(x) \qquad (12.4.3)$$

Therefore, the unbalance distribution of the rotor (Fig. 12.8), can be expressed as

$$U(x) = \sum_{i=1,2,\ldots} \lambda_i Y_i(x) \qquad (12.4.4)$$

To determine α_i in equation (12.4.3), we use the principle of virtual work. The virtual displacement of the rotor, at the given rotational speed is

$$\delta y = \sum \delta \phi_i Y_i(x) \qquad (12.4.5)$$

The work done by the centrifugal forces during ith mode virtual displacement is

$$W_{ce} = \int_0^L m(x)\omega^2[y(x) + a(x)]\delta\phi_i Y_i(x)\,dx \qquad (12.4.6)$$

Using equations (12.4.1) and (12.4.3) in the above, we get by using orthogonality principle, the following:

$$W_{ce} = \delta\phi_i M_i \omega^2 (\phi_i + \alpha_i) \qquad (12.4.7)$$

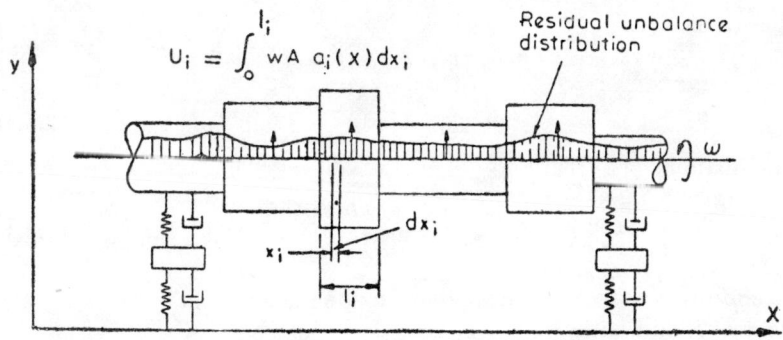

Fig. 12.8 Distribution of residual unbalance in rotor

The strain energy stored in the shaft is

$$V = \int_0^L \tfrac{1}{2} EI(x)\{\sum \phi_i Y_i''(x)\}^2\,dx = \tfrac{1}{2} \sum_{i=1,2,\ldots} \phi_i^2 K_i \qquad (12.4.8)$$

In equation (12.4.8), K_i is ith mode generalised stiffness of the rotor. Work done by the elastic force in the virtual displacement of ith mode is

$$W_{el} = -\frac{\partial V}{\partial \phi_i}\delta\phi_i = -\delta\phi_i\,\phi_i K_i \qquad (12.4.9)$$

Since $W_{ce} + W_{el} = 0$, we get

$$\phi_i = \frac{\omega^2 M_i \alpha_i}{K_i - \omega^2 M_i} \qquad (12.4.10)$$

From equation (12.4.10), we obtain

$$\alpha_i = \frac{1 - \dfrac{\omega^2}{\omega_{in}^2}}{\dfrac{\omega^2}{\omega_{in}^2}} \phi_i \qquad (12.4.11)$$

λ_i in equation (12.4.4) can be evaluated by using orthogonality principle.

$$\lambda_i = \frac{1}{M_i} \int_0^L m(x) U(x) Y_i(x)\, dx \qquad (12.4.12)$$

In the above, $U(x)$ is evaluated from Fig. 12.8, using $a(x)$ from equation (12.4.3) with the help of equations (12.4.11) and (12.4.2).

Let $P(x)$ be the balance distribution required in the ith mode. We can express $P(x)$ in modal series as

$$P(x) = \sum_{i=1,2,\ldots} \beta_i Y_i(x) \qquad (12.4.13)$$

Then

$$\beta_i = \frac{1}{M_i} \int_0^L P(x) m(x) Y_i(x)\, dx \qquad (12.4.14)$$

For balance condition,

$$P(x) + U(x) = 0 \qquad (12.4.15)$$

In the ith mode,

$$\beta_i Y_i(x) + \lambda_i Y_i(x) = 0$$

i.e.,

$$\beta_i = -\lambda_i \qquad (12.4.16)$$

From equations (12.4.12) and (12.4.14), we get

$$\int_0^L P(x) m(x) Y_i(x)\, dx = -\int_0^L m(x) U(x) Y_i(x)\, dx \qquad (12.4.17)$$

If we choose to correct the unbalance by one mass w_c at radius a_c at a plane $x = c$, and letting

$$U(x) \approx m(x) a(x) \qquad (12.4.18)$$

we get

$$w_c a_c m(c) Y_i(c) = -\int_0^L m^2(x) a(x) Y_i(x)\, dx \qquad (12.4.19)$$

Therefore

$$w_c = \frac{-\int_0^L m^2(x) a(x) Y_i(x)\, dx}{a_c m(c) Y_i(c)} \qquad (12.4.20)$$

Hence, to balance ith mode, we need the following data:

Mass distribution $m(x)$.
Critical speed ω_{in}.
Mode shape $Y_i(x)$.
Measured mode shape near critical speed $y(x)$.

Balancing of Rotors 263

Then, we use equation (12.4.2) to determine ϕ_i, equation (12.4.11) to determine α_i, equation (12.4.3) to determine $a(x)$ and equation (12.4.20) to determine w_c.

In practice the theory described as above becomes complicated and except for mass distribution $m(x)$ which can be well determined from the geometry of the rotor, the rest of the quantities cannot be accurately determined. The interaction between the two planes of bending during whirl also plays an important role.

A practical procedure to balance the rotor by modal correction, masses equal in number to the flexible mode shapes, N, known as N plane method is as follows.

Run the rotor in a suitable hard bearing balancing machine, to a safe speed approaching the first critical speed and record the bearing vibrations or forces. Choose an appropriate location for the trial mass. For the first critical speed, this should be roughly in the middle for a symmetrical rotor in its axial distribution of mass. Record the readings at the same speed as before.

Using the above two readings, the correct mass and location can be determined. With this correction mass, it should be possible to run the rotor through the first critical speed without appreciable vibration.

Next, run the rotor to a safe speed approaching the second critical speed, if the operating speed is near second critical or above the second critical speed. Note the readings. Add a pair of trial masses 180° apart in two planes, without affecting the first mode. Note the readings at the same speed near the second critical. From the two readings we can determine the correction masses required.

Instead of N plane correction, Kellenberger [16] suggests that the rotor should be corrected in $N + 2$ planes, so as not to disturb the rigid body balance. A guidance for the location of trial masses, for a rotor with symmetric distribution of mass along its axis, is shown in Fig. 12.9, for the rigid body modes and flexural modes.

12.4.2 *Influence coefficient method*

This method is an extension of the two plane correction used for rigid rotors described earlier. We choose p number of convenient balancing planes and q number of measurement planes for amplitude and phase as shown in Fig. 12.10. We can reasonably assume that all the unbalance is located in the p planes. As a result of this, we can write the following:

$$\{\bar{v}\}_{q \times 1} = [\bar{\alpha}]_{q \times p} \{\bar{U}\}_{p \times 1} \qquad (12.4.21)$$

where $\{\bar{v}\}$ is a complex vector containing the amplitudes and phase angles measured, expressed in real and imaginary components of size $q \times 1$.

$\{\bar{U}\}$ is a complex vector containing the unbalance magnitude and location expressed in real and imaginary components of size $p \times 1$, and $[\bar{\alpha}]$ is

264 *Rotor Dynamics*

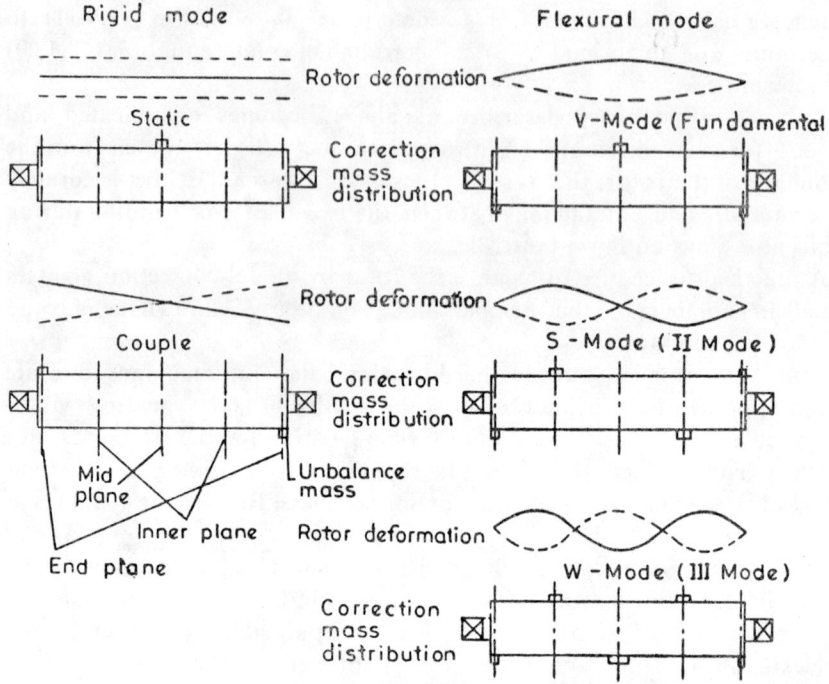

Fig. 12.9 Rotor mode shapes and correction masses

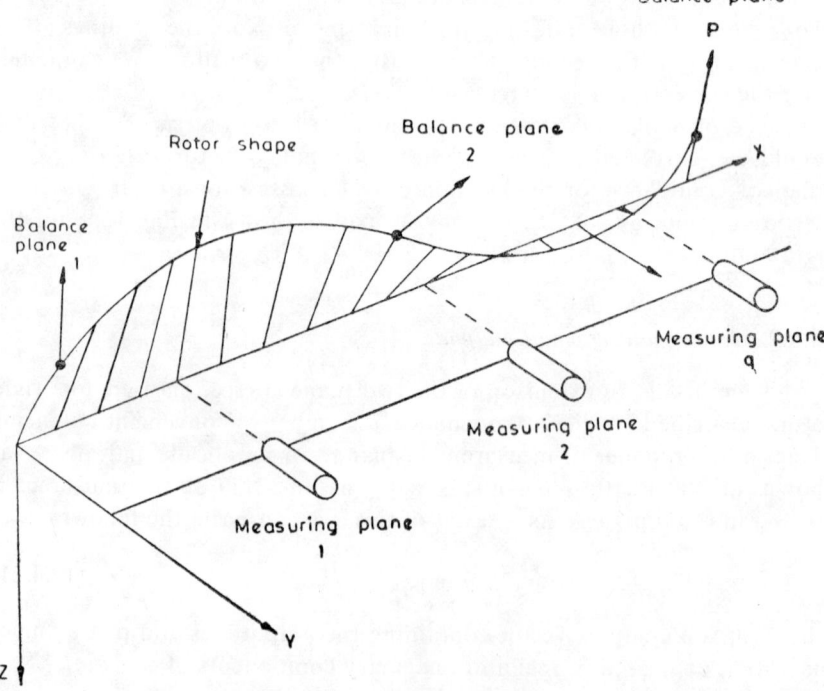

Fig. 12.10 Balancing planes "p" and measuring planes "q" for influence coefficient method

Balancing of Rotors   265

the influence coefficient matrix relating the response $\{\bar{v}\}$ and unbalance $\{\bar{U}\}$, of size $q \times p$.

Equation (12.4.21) is a more general expression of equation (12.3.3) used in rigid rotor balancing. The complex elements of the influence coefficient matrix are to be evaluated by using the observations recorded at a suitable number of speeds, say n. We select a suitable number of speeds $\omega_1, \omega_2, \ldots, \omega_n$ in the vicinity of critical speed which we wish to balance. Measure and record the amplitude and phase of the vibration signal at speed ω_1, in each of the q planes. Repeat the test for speeds $\omega_2, \omega_3, \ldots, \omega_n$. These observations are related to the unknown initial unbalance of the rotor in the p planes, through the influence coefficient matrix as given below.

$$\begin{Bmatrix} v_1^1 \\ v_2^1 \\ \vdots \\ v_q^1 \\ \hline v_1^2 \\ v_2^2 \\ \hline \vdots \\ v_q^2 \\ \hline \vdots \\ v_1^n \\ v_2^n \\ \vdots \\ v_q^n \end{Bmatrix} = \begin{bmatrix} \alpha_{11}^1 & \alpha_{12}^1 & \alpha_{13}^1 & \cdots & \alpha_{1p}^1 \\ \alpha_{21}^1 & \alpha_{22}^1 & \alpha_{23}^1 & \cdots & \alpha_{2p}^1 \\ \vdots & & & & \\ \alpha_{q1}^1 & \alpha_{q2}^1 & \alpha_{q3}^1 & \cdots & \alpha_{qp}^1 \\ \hline \alpha_{11}^2 & \alpha_{12}^2 & \alpha_{13}^2 & \cdots & \alpha_{1p}^2 \\ \alpha_{21}^2 & \alpha_{22}^2 & \alpha_{23}^2 & \cdots & \alpha_{2p}^2 \\ \hline & & & & \\ \alpha_{q1}^2 & \alpha_{q2}^2 & \alpha_q^2 & \cdots & \alpha_{qp}^2 \\ \hline & & \text{etc.} & & \\ \alpha_{11}^n & \alpha_{12}^n & \alpha_{13}^n & \cdots & \alpha_{1p}^n \\ \alpha_{21}^n & \alpha_{22}^n & \alpha_{23}^n & \cdots & \alpha_{2p}^n \\ \vdots & & & & \\ l^n & \alpha_{q2}^n & \alpha_{q3}^n & \cdots & \alpha_{qp}^n \end{bmatrix} \begin{Bmatrix} U_1 \\ U_2 \\ \vdots \\ U_p \end{Bmatrix} \quad (12.4.22)$$

From equation (12.4.22) we observe that

$$nq = p \quad (12.4.23)$$

for $[\bar{\alpha}]$ to be a square matrix.

The influence coefficient matrix elements are evaluated by adding trial masses in each balance plane p and measuring the response in qth planes. Let a trial mass T be added in plane l, for speed ω_1 at an angle, say 0° defined in the rotor. Then

$$\begin{Bmatrix} v_{11}^1 \\ v_{21}^1 \\ \vdots \\ v_{q1}^1 \end{Bmatrix} = \begin{bmatrix} \alpha_{11}^1 & \alpha_{12}^1 & \alpha_{13}^1 & \cdots & \alpha_{1p}^1 \\ \alpha_{21}^1 & \alpha_{22}^1 & \alpha_{23}^1 & \cdots & \alpha_{2p}^1 \\ \vdots & & & & \\ \alpha_{q1}^1 & \alpha_{q2}^1 & \alpha_{q3}^1 & \cdots & \alpha_{qp}^1 \end{bmatrix} \begin{Bmatrix} U_1 + T \\ U_2 \\ \vdots \\ U_p \end{Bmatrix} \quad (12.4.24)$$

Subtracting the equations corresponding to speed ω_1 in equation (12.4.23), from equations (12.4.24), we get

$$\alpha_{11}^1 = \frac{v_{11}^1 - v_1^1}{T}; \qquad \alpha_{21}^1 = \frac{v_{21}^1 - v_2^1}{T}$$

etc.

$$\alpha_{i1}^1 = \frac{v_{i1}^1 - v_i^1}{T} \tag{12.4.25}$$

The trial mass is removed from plane 1 and placed in plane 2 and the test repeated to determine $\alpha_{12}^1, \alpha_{22}^1, \ldots, \alpha_{q2}^1$, by observing the response $v_{12}^1, v_{22}^1, \ldots, v_{q2}^1$, i.e.,

$$\alpha_{i2}^1 = \frac{v_{i2}^1 - v_i^1}{T} \tag{12.4.26}$$

etc.

$$\alpha_{ij}^1 = \frac{v_{ij}^1 - v_i^1}{T} \tag{12.4.27}$$

For kth speed of rotation, in general,

$$\alpha_{ij}^k = \frac{v_{ij}^k - v_i^k}{T}$$

where,

$$i = 1, 2, \ldots; \quad j = 1, 2, \ldots, p; \quad k = 1, 2, \ldots, n \tag{12.4.28}$$

Thus, the influence coefficient matrix is completely determined. The determination of unbalance moments in the p planes is then a simple matter of inversion of influence coefficient matrix and writing equation (12.4.22) as

$$\{\bar{U}\} = [\bar{\alpha}]^{-1}\{\bar{v}\} \tag{12.4.29}$$

Generally the measurements are made in two planes where the bearings are located. Hence, if $q = 2$, the number of balancing speeds, keeping in mind the relation (12.4.23), are

$$n = \tfrac{1}{2}(p + 1) \quad \text{for } p \text{ odd}$$

$$n = \tfrac{1}{2}p \qquad \text{for } p \text{ even} \tag{12.4.30}$$

When p is chosen odd, one line in equation (12.4.22) is deleted to keep the influence coefficient matrix as a square one.

Alternatively, a least square principle can be used [17], to utilise as much data as possible that can be obtained. The influence coefficient method has been tested by several authors, Lund, Tonnesen, Tessarzik, Badgley and Anderson [17–19], in the laboratory and good balance condition achieved. Badgley and Rieger [20] have demonstrated the method by analytical calculations.

Example 12.2

Figure 12.11 shows the configuration of the rotor of a gas turbine engine, considered by Badgley and Rieger. An unbalance of 0.072 gm cm, is intro-

Balancing of Rotors 267

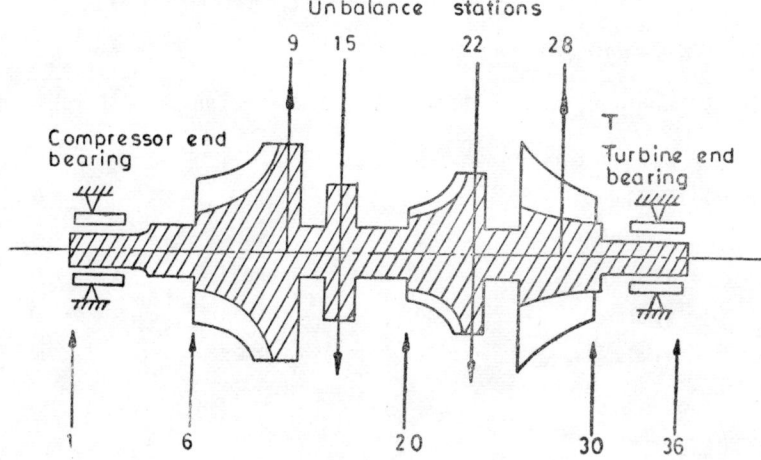

Fig. 12.11 Locations of unbalance masses (stations 9, 15, 22, 28 and balancing planes 1, 6, 20, 30, 36)

duced in each of the planes 9, 15, 22, and 28 as shown. Due to these unbalances, the damped mode shapes calculated for the first flexural mode are given in Figs. 12.12 and 12.13. The rigid body mode shapes are not shown. The rotor is to operate at 75000 rpm, in between the two critical speeds 72200 and 85800 rpm. The calculations made according to the influence coefficient method described are given in Table 12.4.

The unbalance whirl amplitudes of the rotor with the original unbalance in planes 9, 15, 22, and 28 only, with the original unbalance together with correction masses in planes 1, 6, 20, 30, and 36 (five plane balancing),

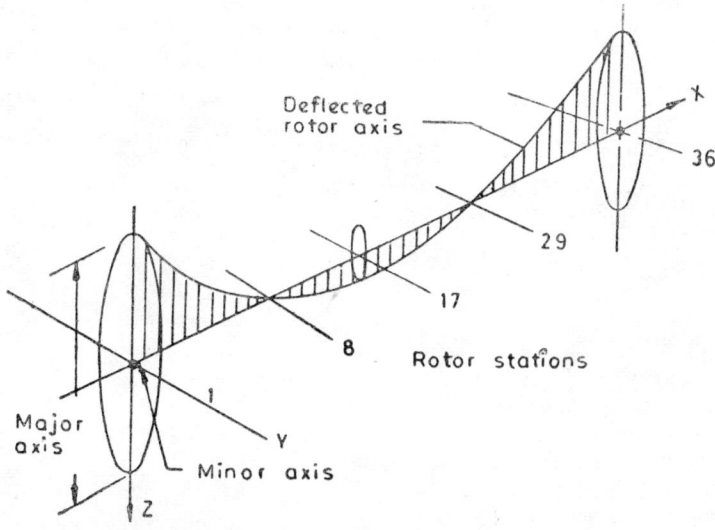

Fig. 12.12 Damped mode shape for critical speed at 72,200 rpm

268 Rotor Dynamics

Table 12.4 Calculation of balance correction masses [20]

Identification of data acquisition runs

Run No.	Run type	Plane No.	Trial mass, gm cm	Location, deg.
1	Base line data	—	—	—
2	Trial mass	1	0.072	0
3	Trial mass	6	0.072	0
4	Trial mass	20	0.072	0
5	Trial mass	30	0.072	0
6	Trial mass	36	0.072	0

Calculated unbalance vibration data

Balancing speed-rpm	Run No.	Compressor-End		Turbine-End	
		Amplitude microns	Phase angle deg.	Amplitude microns	Phase angle deg.
72,200	1	0.15	354.0	1.2	353.6
	2	0.25	3.1	0.69	1.3
	3	0.05	29.9	0.04	3.5
	4	0.04	347.1	0.03	197.1
	5	0.33	349.6	0.07	231.2
	6	0.64	350.8	0.12	312.9
85,500	1	3.69	271.7	3.18	272.6
	2	12.91	270.6	11.09	273.7
	3	5.38	269.7	4.62	274.5
	4	0.87	273.8	0.75	270.6
	5	5.06	273.6	4.37	270.8
	6	11.30	272.7	9.74	271.6
75,000	1	0.49	0.0	0.42	0.0
	2	1.36	0.0	1.65	0.0
	3	0.47	0.0	0.74	0.0
	4	0.09	0.0	0.02	0.0
	5	0.78	0.0	0.37	0.0
	6	0.17	0.0	1.03	0.0

Calculated balance correction masses:
Plane 1: 0.001584 gm cm 28.3 deg.
Plane 6: 0.031252 gm cm 183.2 deg.
Plane 20: 0.072873 gm cm 1.3 deg.
Plane 30: 0.053719 gm cm 181.7 deg.
Plane 36: 0.006985 gm cm 3.2 deg.

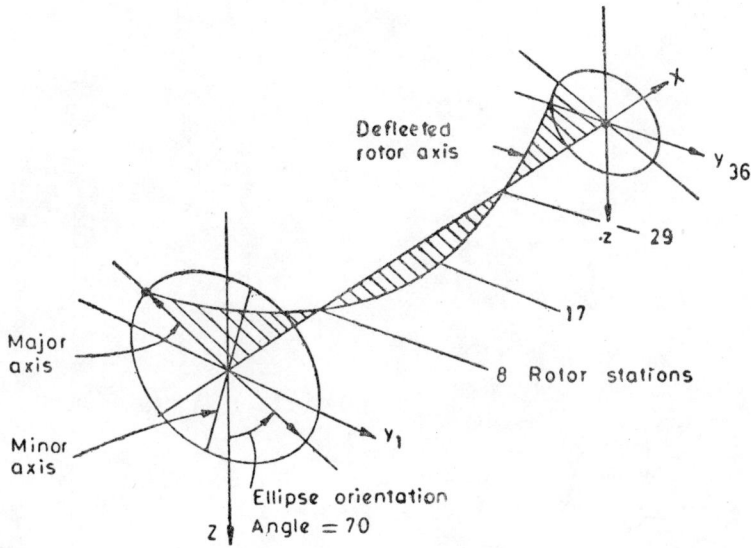

Fig. 12.13 Damped mode shape for critical speed at 85,800 rpm

are shown in Fig. 12.14. Rieger and Badgley [11], originally calculated the response with four plane correction and this amplitude response is also given in Fig. 12.14. There is a substantial improvement of the balance condition of the rotor in the entire region, and in particular around the three balancing speeds for five plane balancing. Hence, the rotor should be balanced at as many speeds as possible, with as many number of planes possible for balance correction.

Badgley and Rieger [20] also calculated the effect of introducing an error in the balance, in both size and location; the corresponding response is shown in Fig. 12.15. This shows that the rotor is very sensitive for errors in balancing and utmost care should be taken in adopting the total balancing procedure. Zhou and Rieger [21] studied several balancing methods and have shown that influence coefficient method gives high quality balance compared to others.

12.5 Balance Criteria for Flexible Rotors

Table 12.5 gives the suggested balance quality criteria for the five classes of rotors defined earlier [22]. Four qualities of precision are used:
A—Precision quality.
B—Commercially acceptable.
C—Needs attention at next overhaul.
D—Immediate attention to the rotor is needed.
The quality is expressed in terms of effective pedestal vibration velocity at once per rev frequency in mm/sec rms. A correction factor is suggested for these velocity values, depending on how they are obtained. If the

270 *Rotor Dynamics*

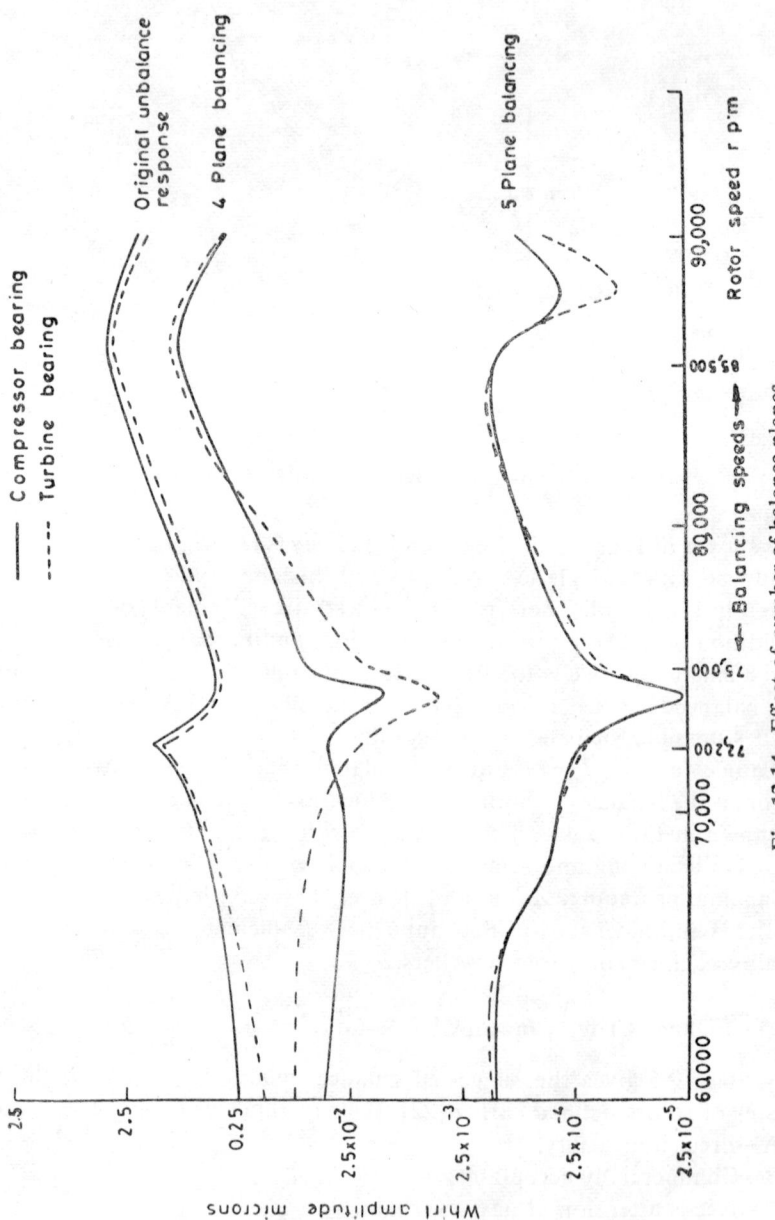

Fig. 12.14 Effect of number of balance planes

Balancing of Rotors 271

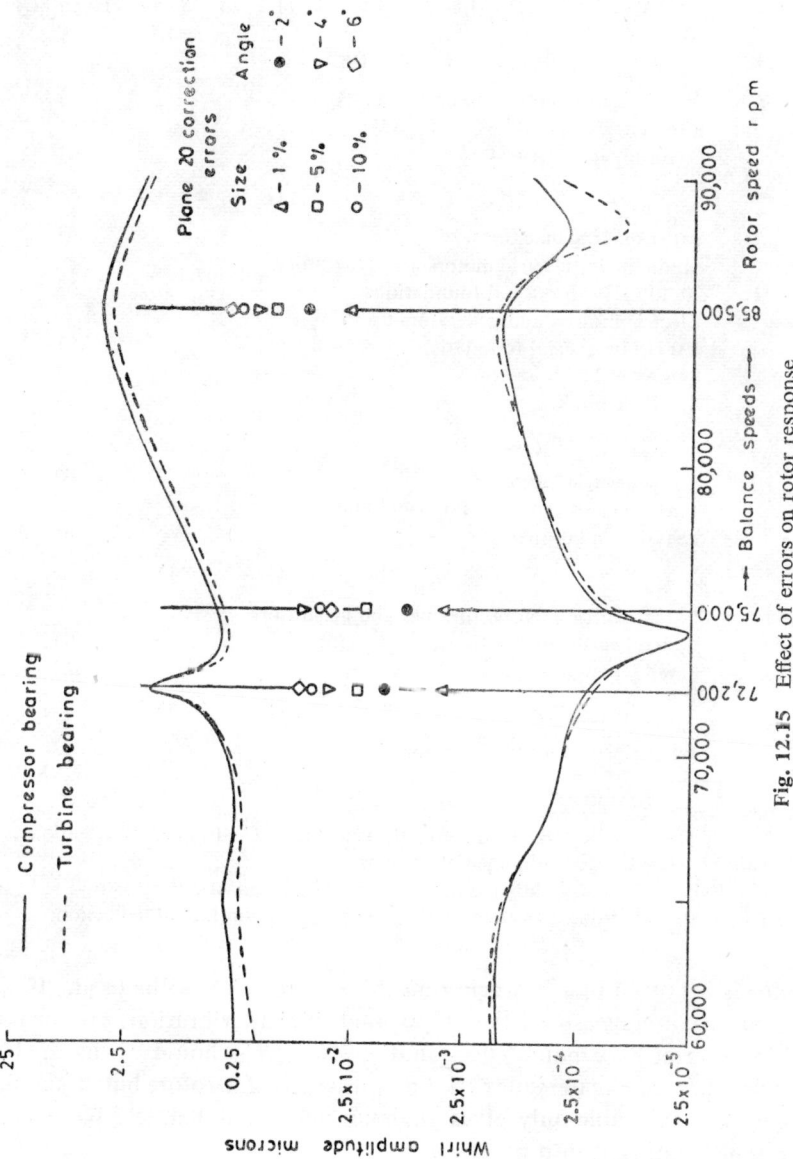

Fig. 12.15 Effect of errors on rotor response

Table 12.5 Balance Quality Criteria for Flexible Rotors

Rotor Category	Ranges of effective pedestal vibration velocity at once-per-rev frequency V_1 (mm/sec) rms												Correction Factor†		
	0.28	0.45	0.71	1.12	1.8	2.8	4.5	7.1	11.2	18	28	45	71	C_1	C_2	C_3
I	\multicolumn{13}{l}{A B C D*}			
	\multicolumn{13}{l}{Small electric motors up to 20 HP.}	.63		
	\multicolumn{13}{l}{Superchargers.}	.63	2	
	\multicolumn{13}{l}{Gyroscopes.}	.63	2	
II	\multicolumn{13}{l}{A B C D}			
	\multicolumn{13}{l}{Paper making machines.}	.63		
	\multicolumn{13}{l}{Medium size electric motors and generators, 20–100 HP on normal foundations.}	.63	4	
	\multicolumn{13}{l}{Electric motors and generators up to 400 HP on special foundations.}	.63	4	20
	\multicolumn{13}{l}{Pumps and compressors.}	.63	8	15
	\multicolumn{13}{l}{Small turbines.}	.63	4	8
III	\multicolumn{13}{l}{A B C D}			
	\multicolumn{13}{l}{Large electric motors.}	.63	5	
	\multicolumn{13}{l}{Turbines and generators on rigid and heavy foundations.}	.63	5	20
IV	\multicolumn{13}{l}{A B C D}			
	\multicolumn{13}{l}{Large electric motors, turbines and generators on lightweight foundations.}	.63	3	10
	\multicolumn{13}{l}{Small jet engines.}	.63		
V	\multicolumn{13}{l}{A B C D}			
	\multicolumn{13}{l}{Jet engines larger than category IV.}	.63	2	10

*Quality bands.
A—Precision quality; B—Commercially acceptable; C—In need of attention at next overhaul and D—In need of immediate attention.

†C_1—Balanced in a balancing machine; C_2—Shaft vibrations measured at bearings and C_3—Shaft vibrations measured at location of maximum lateral deflection.

rotor is balanced in a balancing machine factor C_1 is to be used. If shaft vibrations are measured then C_2 is used. If shaft vibrations are measured at location of maximum deflection, then factor C_3 should be used. Using Table 12.5, one can judge the performance of a rotor, but it should be noted that this table only gives guidance and caution should be exercised in implementing it into practice.

References

1. ISO Draft document TC 108/WG 6, 1967.
2. ISO 1925 Balancing-Vocabulary, 1974.

3. ISO 1940 Balance quality of rotating rigid bodies, 1973.
4. Aspects of flexible rotor balancing, Schenck Trebel Corporation, Deer Park, New York, 1976.
5. Rieger, N.F. Vibrations of rotating machinery, Part 1: Rotor bearing dynamics, The Vibration Institute, Clarendon Hills, Illinois, 1981.
6. Den Hartog, J.P. Mechanical Vibrations, McGraw-Hill Book Co., 1956.
7. Macduff, J.N. and Curreri, J.R. Vibration Control, McGraw-Hill Book Co., 1958.
8. Bruel and Kjaer, Balancing Kit type 9500, Naeurum, Denmark.
9. Mcquery, D.E. Understanding balancing machines, American Machinist Special report, 656, June 11, 1973.
10. Jackson, C. Two plane field balance hybrid system-vectors and orbits, Trim balance, Case History 1, The Vibration Institute, Clarendon Hills, Illinois.
11. Rieger, N.F. and Badgley, R.H. Flexible rotor balancing of a high speed gas turbine engine, SAE publication 720741.
12. Bishop, R.E.D. and Gladwell, G.M.L. The vibration and balancing of an unbalanced flexible rotor, J. of Mech. Engng. Sci., Vol. 1, 1959, p. 66.
13. Bishop, R.E.D. and Parkinson, A.G. On the use of balancing machines for flexible rotors, ASME 71-Vibr-73.
14. Moore, L.S. and Dodd, E.G. Mass balancing of large flexible rotors, GEC Journal, Vol. 31, 1964, p. 74.
15. Tessarzik, J.M., Badgley, R.H. and Anderson, W.J. Flexible rotor balancing by the exact point speed influence coefficient method, J. of Engng. for Indus., ASME, 1972, p. 148.
16. Kellenberger, W. Should a flexible rotor be balanced in N or $N+2$ planes? J. of Engng. for Indus., ASME, 1972, p. 548.
17. Tessarzik, J.M. and Badgley, R.H. Experimental evaluation of the exact point speed and least squares procedures for flexible rotor balancing by the influence coefficient method, J. of Engng. for Indus., ASME, 1974, p. 633.
18. Lund, J.W. and Tonnesen, J. Analysis and experiments in multiplane balancing of flexible rotors, J. of Engng. for Indus., ASME, 1972, p. 233.
19. Tessarzik, J.M., Badgley, R.H. and Fleming, D.P. Experimental evaluation of multiplane-multispeed rotor balancing through multiple critical speeds, J. of Engng. for Indus., ASME, 1976, p. 988.
20. Badgley, R.H. and Rieger, N.F. The effects of multiple balancing on flexible rotor whirl amplitudes, SAE 730102.
21. Zhou, S. and Rieger, N.F. Comparison of effectiveness of several balancing methods for flexible rotors, IFToMM Int. Conf. Rotor Dynamic Problems in Power Plants, 1982, p. 305.
22. ISO Draft document TC/108/SCI/N18, 1976.

A turbine on test bed (*Courtesy*, BHEL, Hyderabad)

Chapter 13

Condition Monitoring Using Vibration Measurements

The internal condition of a machine can be judged by a number of indicators such as:

Oil and coolant temperatures.
Oil contamination and leakage analysis.
Vibration and noise measurements.
Thermal imaging etc.

In the present context, we are obviously concerned with vibration measurements and their analysis in determining the performance of a rotating or reciprocating machine. Vibration measurements of the external surface of a machine contain a great deal of information on the internal processes and has become an established method of judging the machine condition. Since the rotating and reciprocating parts cause vibrations and that they are transmitted to the structure of the machine, foundation etc., we can predict the condition of the machine and avoid any failure or damage that may occur. The practice of taking measurements of vibration and judgement of the health of the machine from these measurements has been practised in industry, right from the beginning of industrial revolution with steam engines. With the advent of high capacity, light weight and high speed machinery, condition monitoring for maintenance of the machinery has become more important and sophisticated with modern electronic measurement instruments. As an example consider the report of John Hubbell [1], on the trials of Surface Effect Ship 100 tons SES-100 B, thus:
"Sitting behind the skipper, Lieutenant Harold Hall, in test director Charles Lester of Bell Aerospace Textron, builder of the SES-100 B. Behind Lester is John Roberts, also of Bell, he is studying an oscillogram showing vibration levels in our power plants' ten gear boxes. At the rear of the cabin, concentrating on some computers, is Electrician Mate Second class Eric Houston. He is operating the Data Acquisition system (DAS), monitoring some 350 readings, such as wave height sensors, temperature of each engine at its inlet and exhaust points. If any of these detects a serious problem, one that can't be solved from a control

panel, Engineerman Second class Dennis Fillhoelter is our super maintenance man and trouble shooter.

I race about the Gulf. Through my ear phones I hear men calmly exchanging information. John Roberts reports a rising vibration level in one engine. Hall makes a minuscule speed adjustment and the vibration level drops."

Condition monitoring based on vibration signature analysis is thus a custom built system for effective product quality assurance or machinery fault diagnosis. It normally includes:

1. Listing of all critical locations of a machine for monitoring the signatures
2. Identifying vibration generating mechanism and establishing acceptance levels of machinery vibration and its spectral characteristics
3. Determining the machine condition and normal vibration level and/or spectrum
4. Selecting the interval for periodic vibration checks
5. Starting from a simple data recording system which correlates the vibration signature with machine condition.

The vibration signature is generally analysed in the frequency domain, to obtain a display of signature amplitude versus frequency. The total energy in a vibration or sound emission is distributed over a range of frequencies consisting of different fundamental frequencies and their harmonics. The signal is resolved into its constituent frequencies, which identify the major sources of vibration and noise. The signals are filtered electronically into frequency bands which may be octave, fractional octave or narrow bandwidth. Such an analysis is then used in studying the condition of machine [2].

13.1 Vibration Generating Mechanisms

In the previous chapters, the sources of vibration and their effects have been studied as related to rotating shafts. The defects in a machine rotor, which may be due to a design fault, or developed during the life of the rotor, have distinct characteristics. Depending on these characteristics, one can identify the source of the trouble. They are listed below [3].

13.1.1 *Unbalance in the rotor*

Rotor unbalance is the major problem in most of rotating machines. The causes of this unbalance are given in Table 13.1. Pure unbalance vibration always occurs at the rotor speed, i.e., one/rev. Some specific unbalances created during the running of the machine give rise to additional characteristics, such as:

1. A blade or vane broken off—one/rev frequency signal is accompanied by possible process pulsations.
2. Loose bolt or component slip—one/rev frequency signal is accompanied by magnitude and phase changes.

Table 13.1 Typical causes of rotor unbalance

Cause of unbalance	Observable signs	Frequency of vibration
Disk or component eccentric on shaft.	Detectable runout on slow rotation, c.g. runs to bottom on knife edges.	One/rev
Dimensional inaccuracies	Measurable lack of symmetry.	One/rev
Eccentric machining or forming inaccuracies.	Detectable runout	One/rev
Oblique angled component.	Detectable angular runout. Measure with dial gauge on knife edges.	One/rev
Bent shaft. Distorted assembly. Stress relaxation with time.	Detectable runout on slow rotation, often heavy vibration during rotation.	One/rev
Section of blade or vane broken off.	Observable. Bearing vibration during operation.	One/rev. Possible process pulsations.
Eccentric accumulation of process dirt on blade.	Bearing vibration	One/rev
Differential thermal expansion	Shaft bends and throws c.g. out. Source of heavy vibration.	One/rev
Non-homogeneous component structure. Sub-surface voids in casting.	Rotor machined concentric. Bearing vibration during operation. c.g. runs to bottom on knife edges.	One/rev.
Non-uniform process erosion.	Bearing vibration.	One/rev
Loose bolt or component slip.	Vibration reappears after balancing due to component angular movement.	One/rev. Possible magnitude and phase changes.
Trapped fluid inside rotor, possible condensing/vaporizing with process cycle.	Vibration reappears after balancing. Apparent c.g. angular movement occurs.	One/rev. Possible magnitude and phase changes.
Ball bearing wear.	Bearing vibration. Eccentric orbit with possible multi-loops.	One, two, or higher per rev.

3. Ball bearing wear—one/rev frequency accompanied by high frequency signal corresponding to bearing vibration frequency.

Figure 13.1 shows a plot of frequency vs speed for common rotor vibration problems, as in a Campbell diagram, showing different possible per rev excitations. Table 13.2 gives a summary of these characteristics.

13.1.2 *Reciprocating unbalance*

Because of the inherent principles of design of a reciprocating machine, the gas force and inertia force of rotating and reciprocating parts are periodic in nature with several harmonics of the engine speed. The reci-

278 *Rotor Dynamics*

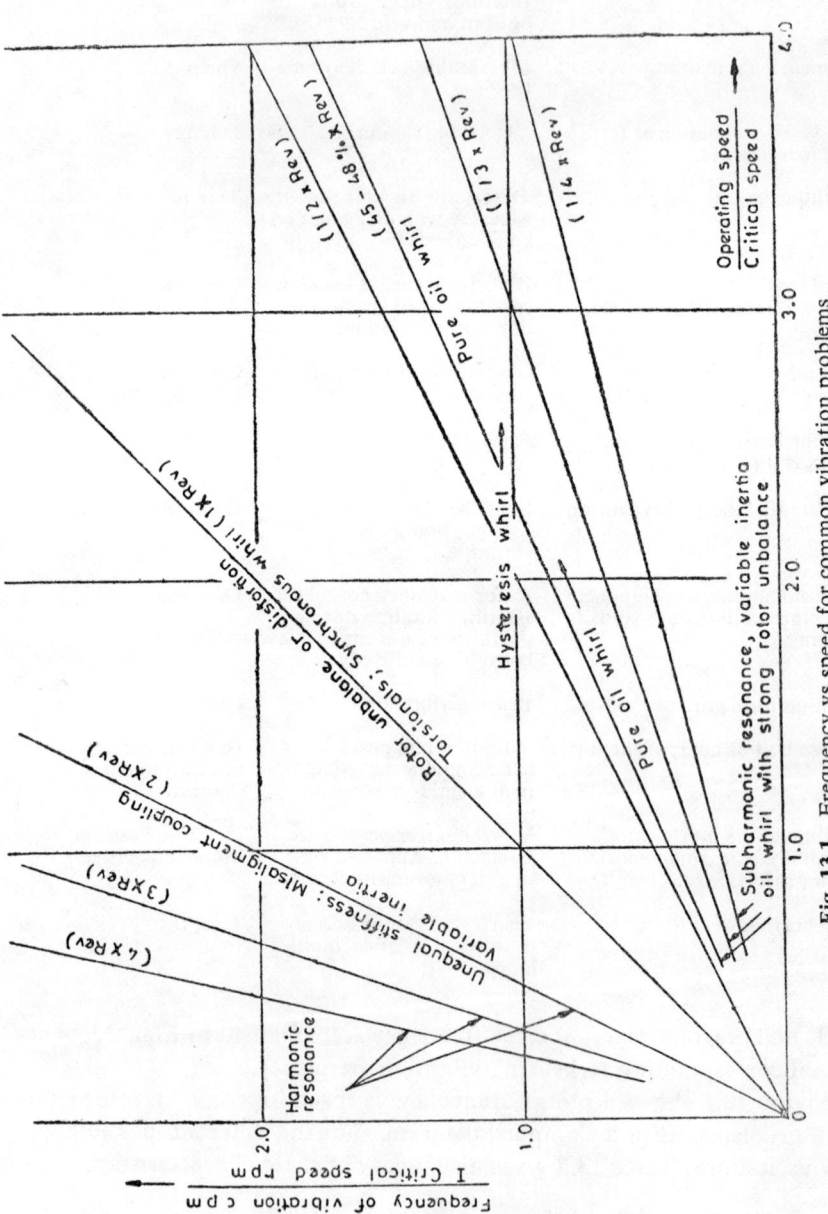

Fig. 13.1 Frequency vs speed for common vibration problems

Condition Monitoring Using Vibration Measurements 279

Table 13.2 Vibration characteristics of some machine parts

Fault	Dominant frequency	Direction/Position
1. Rotating unbalance	Rotating speed, n	R
2. Reciprocating unbalance	n, $2n$, $3n$, $n/2$, $n/3$, etc.	R B
3. Bent shaft or coupling	n	R, B
4. Misalignment	$2n$, sometimes n, $3n$, $4n$.	R, A B
5. Mechanical looseness	n, $2n$, $3n$, etc.	R, A B
6. Loose rotor components, Hysteresis	Rotor critical speed and/or higher harmonics	R B
7. Loose journal bearing	$n/2$, $n/3$	R, B
8. Oil film whirl	0.4 to $0.5n$ and harmonics	R B
9. Damaged hydrodynamic bearing	Change in higher frequencies as debris cause rubbing	R B
10. Asymmetric shaft	n, $2n$, $3n$ etc.	R, B

R—Radial direction; A—Axial direction and B—Bearing housing.

procating parts cannot be balanced exactly by a rotating mass and therefore single cylinder engines have lateral and torsional vibrations of the crankshaft with harmonics of the order of $\frac{1}{2}$, 1, $1\frac{1}{2}$ etc., of the engine speed. Multicylinder engines reduce the amount of reciprocating unbalance. The vibration levels allowed for reciprocating machines are high and heavy reciprocating machines run at very low speeds. The reciprocating unbalance does not change significantly with the period of running, rather the changes that may occur due to wear of the piston and other parts, are small compared to to the existing unbalance left in the machine. The change in the unbalance in a rotating machine that may occur due to differential thermal expansion, process dirt accumulation or wear, is very important.

13.1.3 *Misalignment*

Coupling misalignment causes friction and deflection forces, which cause the rotor bearing system to deflect, creating secondary phenomena such as harmonic resonances. They can become very severe. Frequencies are characteristically 2×rev, but 1×rev and sometimes 3×rev or 4×rev are also observed.

13.1.4 *Mechanical looseness*

This is one of the most vexing problems. Loose rotor components such as disks, sleeves, thrust collars etc., cause rotor internal friction problems. The frequency of vibration is always the rotor critical speed and hence easy to detect. The situation however becomes difficult, at speeds around and above twice the critical speed of the rotor, because this gets mixed up with rotor half frequency whirl due to bearing instability. The oil whirl generally locks in at the critical speed, thus creating a problem of identification between the two, the hysteresis and oil whirl.

Mechanical loose components like bolts, give rise to $1 \times$ rev and harmonic frequency signals due to secondary phenomena. The amplitude and phase continually change.

Loose assembly of bearings give rise to subharmonic response and the typical frequency response is $\frac{1}{2} \times$ rev and $\frac{1}{3} \times$ rev. This could be mistaken for oil film whirl, particularly in the region of twice the rotor critical speed, around the threshold of instability.

13.1.5 Oil film whirl

The frequency of response at the onset of the whirl is always a little less than $\frac{1}{2} \times$ rev, in the region of 0.4 to $0.5 \times$ rev. Generally it occurs around 0.45–$0.48 \times$ rev. The frequency gets locked at the critical speed in the unstable region, if the whip is severe. If the bearing is damaged, high frequency response will be observed, due to the rubbing of debris.

13.1.6 Asymmetric shaft

The response of an asymmetric shaft has several harmonics and the frequencies observed are $1 \times$ rev., $2 \times$ rev, $3 \times$ rev and sometimes even higher harmonics, if the asymmetry is predominant.

13.1.7 Bending Criticals and Resonance

Bending critical of the rotor occurs when the rotational speed is equal to its lateral vibration resonance frequency. Unlike a resonance frequency, the critical speed cannot be cured by addition of damping, which in fact may be harmful, causing internal hysteresis. Bending critical speed can be easily detected by synchronous whirl conditions and fairly large amplitude at the rotor speed. Because of the oil film, the critical speed region occurs over a relatively wide range and is often accompanied by backward whirl despite the excitation in the forward direction. The resonance of the structure, support and auxiliaries cause fairly large amplitudes of vibration at the rotor speed and this occurs over a narrow range of speed of operation. Such resonances can be cured by the addition of damping.

13.1.8 Torsional criticals

Torsional criticals are to be detected by torsional vibration measuring instruments. Unless there is a gear box in the transmission unit, it is difficult to detect these criticals by vibrometers or accelerometers used for conventional non-angular vibratory measurements. Torsional criticals are predominant in reciprocating machinery, as there is a large excitation torque. In rotating machinery, they are not easily excited, unless there is a disturbance arising out of gear transmission units. Reciprocating machines have several criticals, some of them major, between the starting and operating speeds.

The above discussion illustrates some guidelines for detecting possible faults in a machine. In practice, the problem is more complicated and experience is an important factor in detecting and diagnosing a fault. Figure 13.1 gives somewhat an overall picture of frequency of vibration versus operating speed for some common vibration problems. They are also represented in the form of a list in Table 13.2.

13.2 Condition Monitoring

Condition monitoring may be carried out by simple measurements, by octave band analysis or by a narrow band analysis of the measured signals. Each is briefly explained below and any method can be adopted depending on the need.

13.2.1 Condition monitoring by simple measurements

A simple broad band overall level measurement scheme is shown in Fig. 13.2. Measurements can be made periodically on suitable predetermined locations, e.g., bearings or casing and recorded in a table or

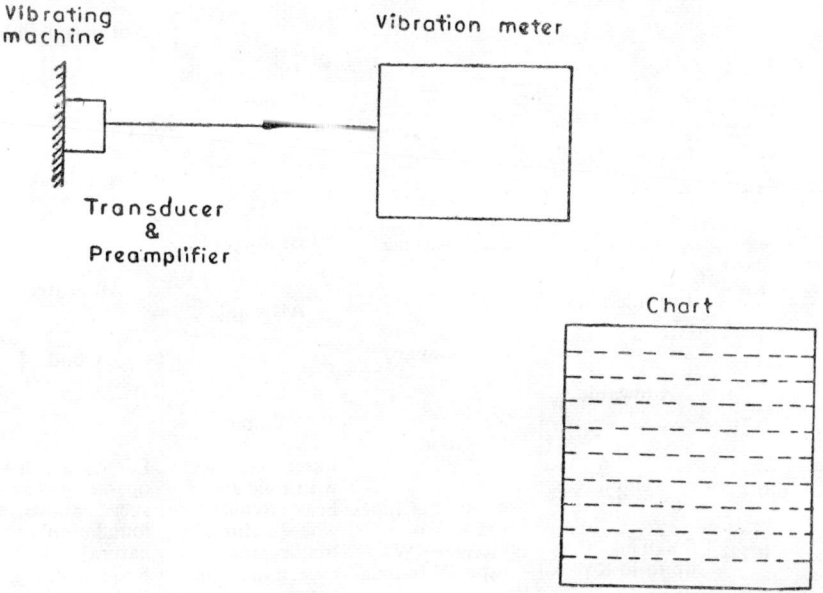

Fig. 13.2 Condition monitoring by simple manual measurement

precalibrated charts. Overall limits can be fixed based on international standards or as recommended by the manufacturer based on the previous experience. Since shear strain energy is a satisfactory criterion of failure under fatigue of ductile materials, rms velocity which relates to the energy, is adopted in some standards, e.g., VDI 2056/1964 [4], which was developed from the preliminary work of Rathbone [5]. Here the machines are classified into four groups:

K—small machines up to 15 KW.

M—medium machines up to 75 KW or up to 300 KW on special foundations.

G—large machines with speeds below foundation natural frequency.

T—large machines with operating speeds above the foundation natural frequency like turbomachines.

The velocity range is divided into four categories:

(i) Not permissible.

(ii) Just tolerable.

(iii) Allowable.

(iv) Good.

The chart is given in Fig. 13.3.

RMS Velocity (mm/sec)	GROUP K	GROUP M	GROUP G	GROUP T
45	Not permissible	Not permissible	Not permissible	Not permissible
28				
16				
11.2				Just tolerable
7.1				
4.5	Just tolerable	Just tolerable	Just tolerable	Allowable
2.8			Allowable	
1.6		Allowable		Good
1.12	Allowable			
0.71		Good	Good	
0.45	Good	Medium machines 15–75 KW or up to 300 KW on special foundations	Large machines with rigid and heavy foundations whose natural frequencies exceed machine speeds	Large machines operating at speeds above foundation's natural frequency, e.g. turbo-machines
0.28	Small machines up to 15 KW			
0.16				

Fig. 13.3 Vibration criterion chart VDI 2056/1964

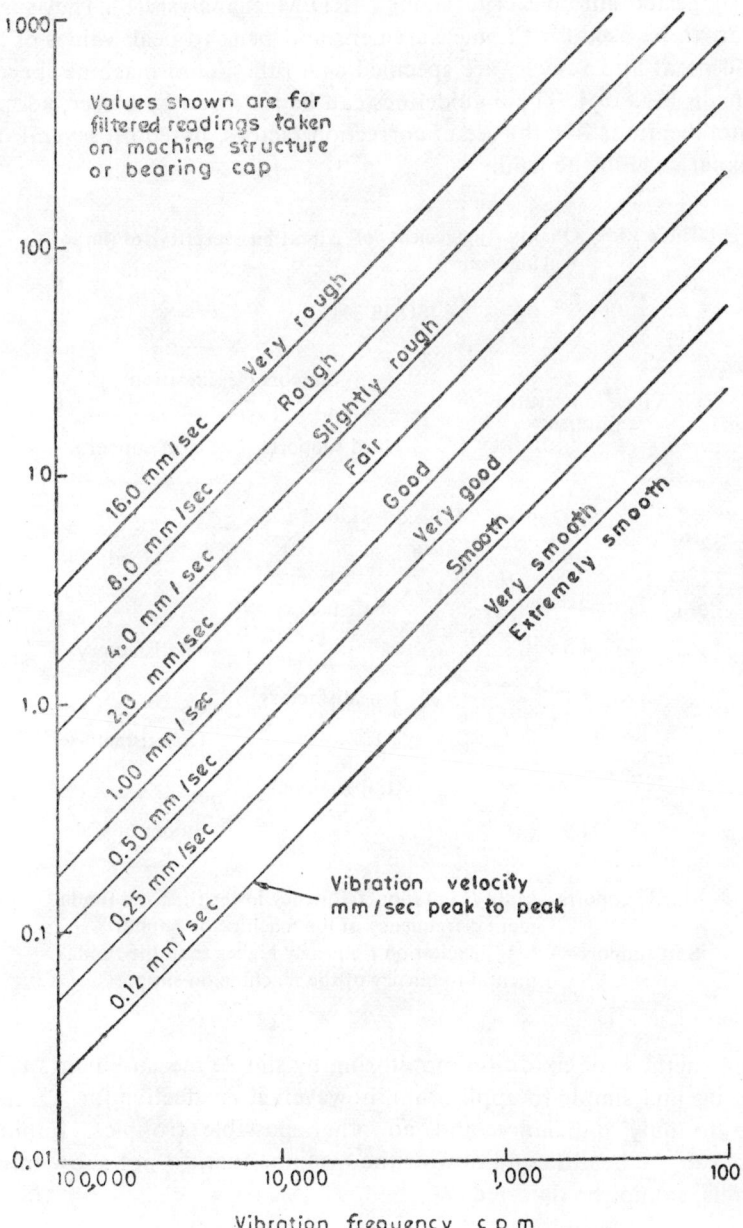

Fig. 13.4 General machinery vibration severity chart

ISO/IS2732 standard [6] is similar to the German standard [4], whereas ISO standard for large machines [7] is classified for hard and soft supports separately. This is given in Table 13.3. Some instrument manufacturers have suggested different criteria, e.g., IRD Mechanalysis [8]. They suggest a filter to be used for the measurement and peak to peak values of both displacement and velocity are specified as a function of machine speed as shown in Fig. 13.4. These guidelines can be followed by a user, adopting any modifications like the use of correction factors, based on several tests and experience in the field.

Table 13.3 Quality judgement of vibration severity of large machines

ISO/IS 3945

Vibration severity (mm/sec)	Support classification	
	Hard supports	Soft supports
0.45	Good	
1.8		Good
2.8	Satisfactory	
4.5		Satisfactory
7.1	Unsatisfactory	
11.2		Unsatisfactory
18.0	Impermissible	
71.0		Impermissible

Hard supports—Main excitation frequency lower than the fundamental frequency of the machine on supports.
Soft supports— Main excitation frequency higher than the fundamental frequency of the machine on supports.

This method of condition monitoring by simple measurements is least expensive and simple in application, however, it is effective for machines prone to only unbalance and no other possible troubles. Important changes in the health/condition of the machine that do not show in overall levels, cannot be detected.

13.2.2 *Condition monitoring by octave band analysis*

The above procedure can be improved by utilising an octave or 1/3 octave filter set to determine vibration levels in an octave band, as shown in Fig. 13.5. The manually recorded spectra can be compared with a reference spectrum either recorded earlier or the one which is assessed as ideal performance of the machine by previous experience, containing known components of the machine such as speed, its harmonics etc., as shown in Fig. 13.6. The most likely cause of an impending trouble can be detected, when the vibration inside an octave band containing a known component has increased. If the machine has, however, a large number of trouble prone components, it is difficult to make precise diagnosis.

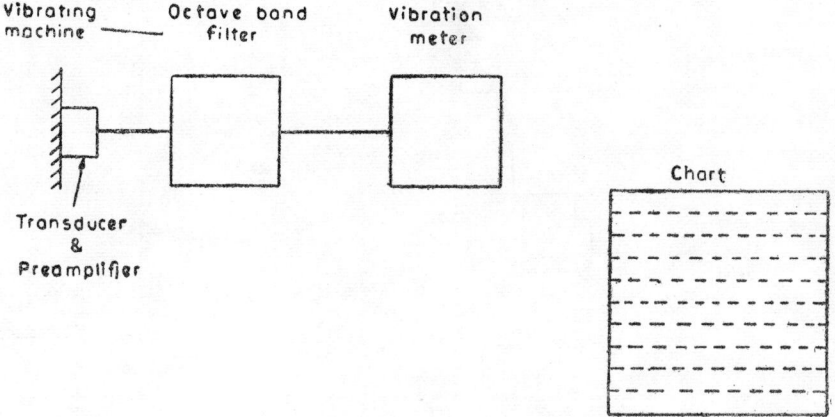

Fig. 13.5 Condition monitoring by octave band analysis

13.2.3 *Condition monitoring by narrow band analysis*

A high level condition monitoring can be achieved by using a narrow band frequency analyser with a level recorder as shown in Fig. 13.7. A tape recorder can be used in the field, if measurements are made in situ and analysed in the laboratory.

Some cases of diagnosis of the condition of machine using narrow band analysis are given below.

Figure 13.8 gives the spectrum of a vibration signature taken on a new reciprocating machine developed by M/s KG Khosla Compressors [9], which runs at 750 rpm. The study revealed at an early stage that the second and fourth harmonic components of response is fairly large, though within allowable limits.

An interesting case of hysteresis causing resonance is given by Erskine and Reeves [10]. Figure 13.9 gives the spectrum analysis of vibration signature taken from a boiler feed water pump, driven by a motor at 50 cps. The signatures for normal running and under large vibrations are shown together. The critical speed for the rotor predicted is 60 cps. The signature under abnormal running shows a high peak at 8 times the rotational speed and significant peaks at 60 and 120 cps, thus showing there is a rub in the system which has caused the resonance (instability).

286 Rotor Dynamics

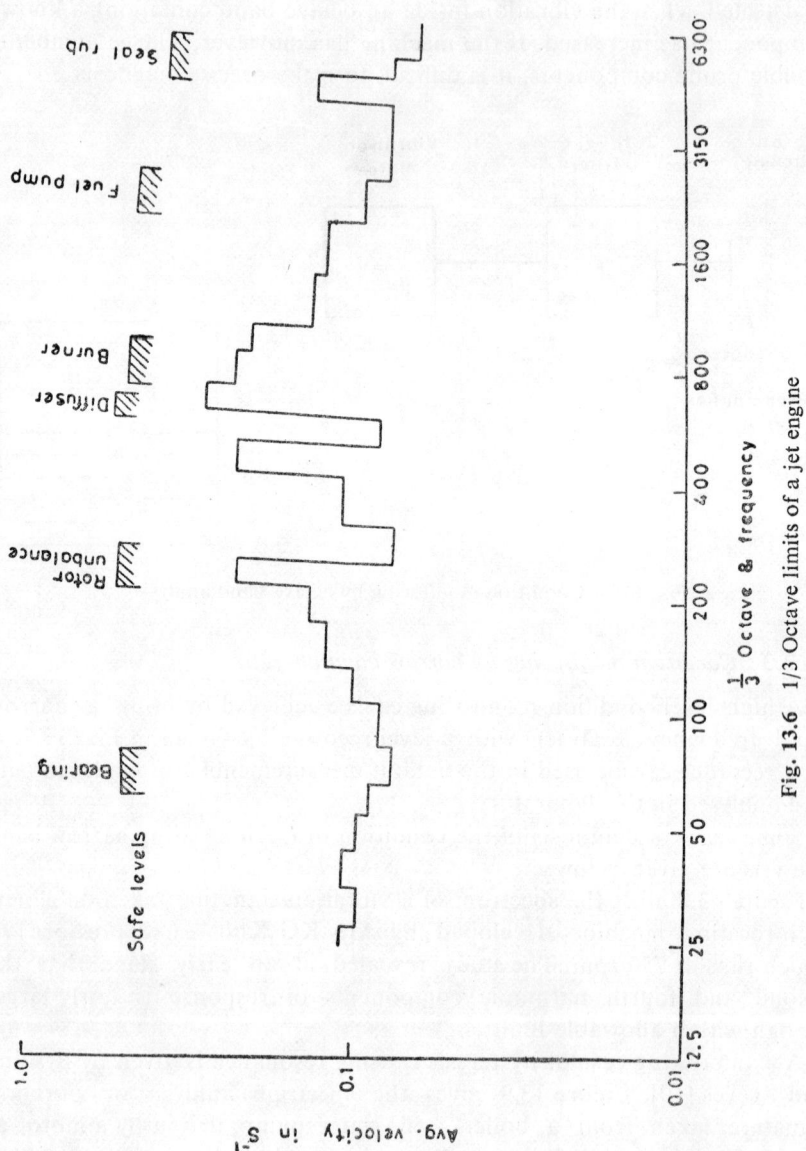

Fig. 13.6 1/3 Octave limits of a jet engine

The instabilities due to oil film whirl were detected on a 12.1 MW turbine driving a centrifugal compressor by Erskine and Reeves [10]. The

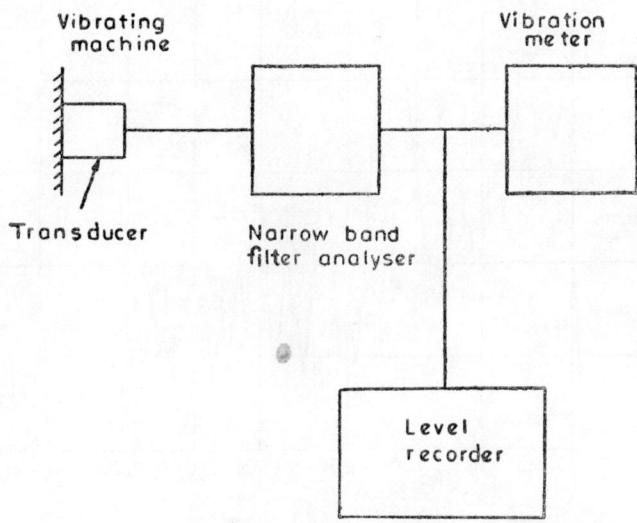

Fig. 13.7 Condition monitoring based on narrow band analysis

operating speed is 170 cps and the critical speed of the system is 217 cps. The major components present are harmonics of 73.3 cps and sum and difference frequencies of the rotational frequency 166 cps and 73.3 cps. 73.3 cps corresponds to $0.44 \times$ rev of the rotor, thus showing half frequency whirl predominant in the system. The spectrum is shown in Fig. 13.10.

13.3 Noise Spectrum

A machine part vibrating in one of its normal modes produces pure tone sound at the frequency of vibration. The energy of vibration is emitted in the form of sound, and the sound pressure produced by a vibrating surface is proportional to the velocity of vibration of the surface. Machinery noise in general is very complex as there are many moving parts, different structural elements vibrating at different frequencies, aerodynamic noise and noise of neighbouring machines and structures etc. Noise is sound, random in nature, the spectrum of which does not exhibit distinct frequency components. In some specific cases analysis of noise signature may be helpful and easier than measurement and analysis of several vibration signatures obtained at different locations on the machine.

288 Rotor Dynamics

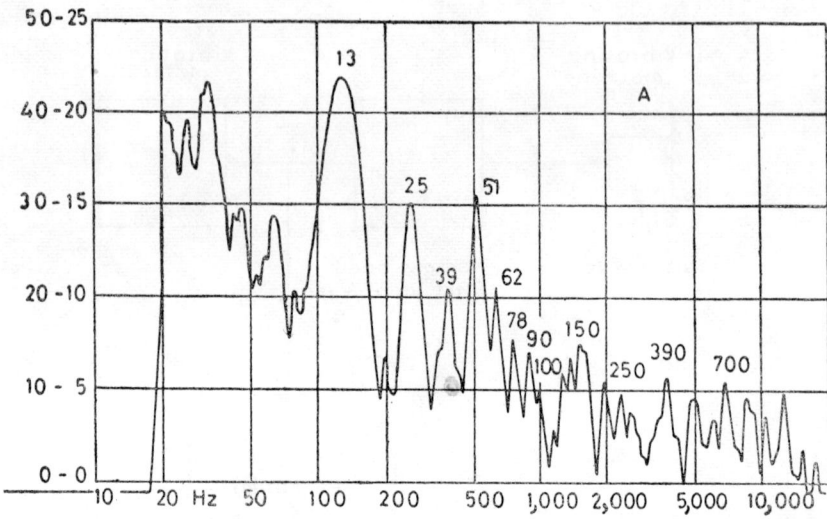

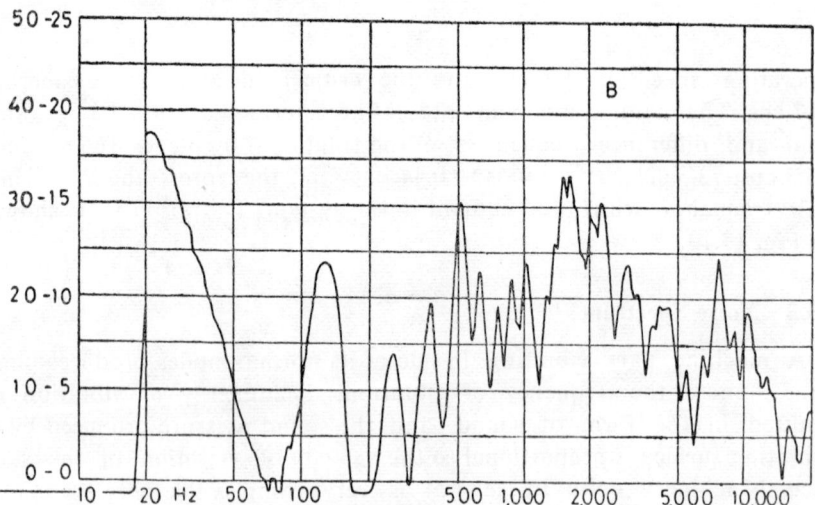

Fig. 13.8 Frequency spectrum of vibration signature of a reciprocating compressor: (A) Displacement, (B) Acceleration

An example is a gear box, where there is a continuous engagement and disengagement of gear teeth on different shafts. The noise signature will

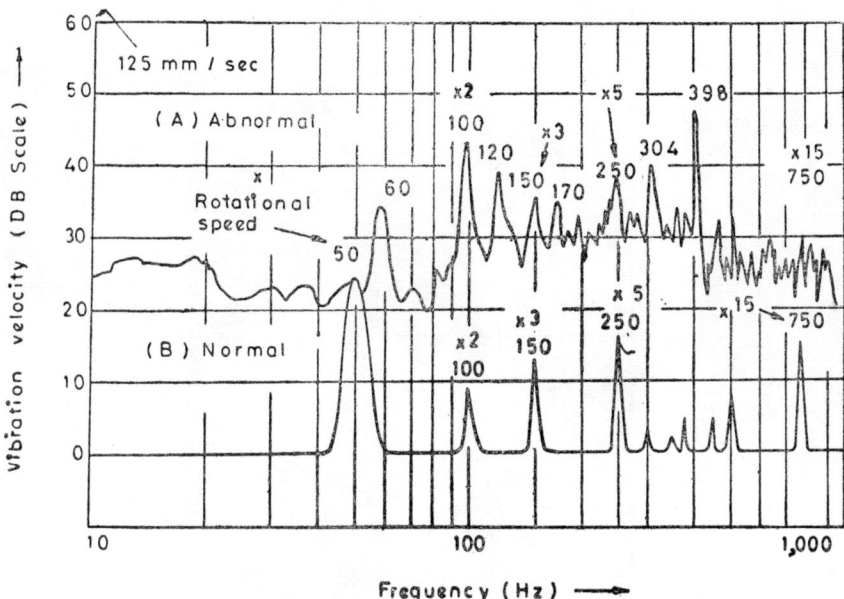

Fig. 13.9 Frequency analysis of vibration of a BFW pump during (A) abnormal (B) normal running

contain the input shaft frequency, output shaft frequency, tooth meshing frequency and their harmonics. If there is a misalignment, bending of one shaft, improper teeth etc., these frequency components will be high and a diagnosis can be easily made to rectify the improper condition. The frequency spectrum [11] of a gear box with good and bad alignment is given in Fig. 13.11. This demonstrates that noise signature can be used to detect malfunction in some machinery and put to use as a valuable tool with a little flair and discretion.

13.4 Real Time Analysis

For dynamic systems under transient conditions, real time analysis is necessary to monitor the condition and make any diagnostic analysis. A typical case study is discussed below [12].

This is concerned with a new 3.5 MW Diesel Generator set. The engine rated at 800 rpm drives a six pole generator at 1000 rpm. The generator bearings recorded unusually excessive vibration levels and attempts of rigid rotor balancing of generator rotor at 500 rpm did not improve the condition of operation at 1000 rpm. When the rotor speed was raised

290 *Rotor Dynamics*

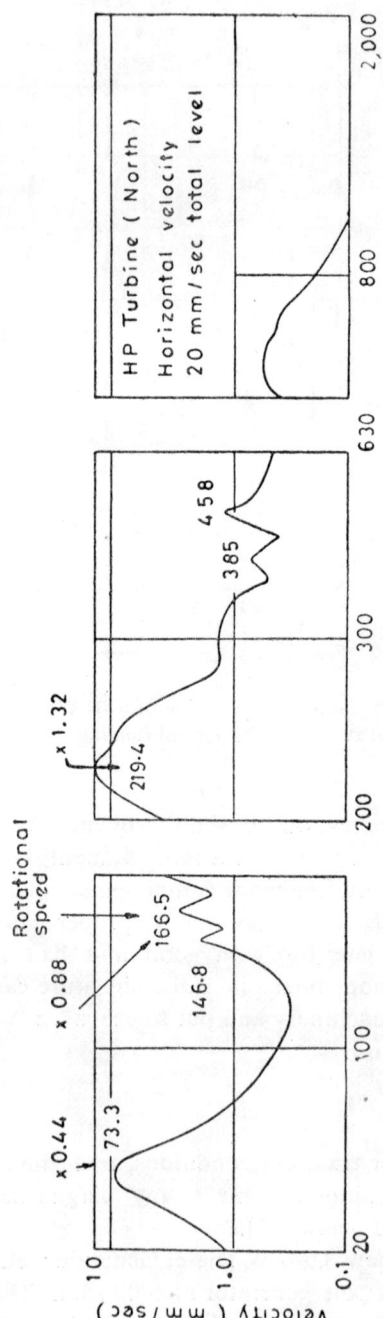

Fig. 13.10 Typical frequency analysis of vibration on turbine during a period of high vibration

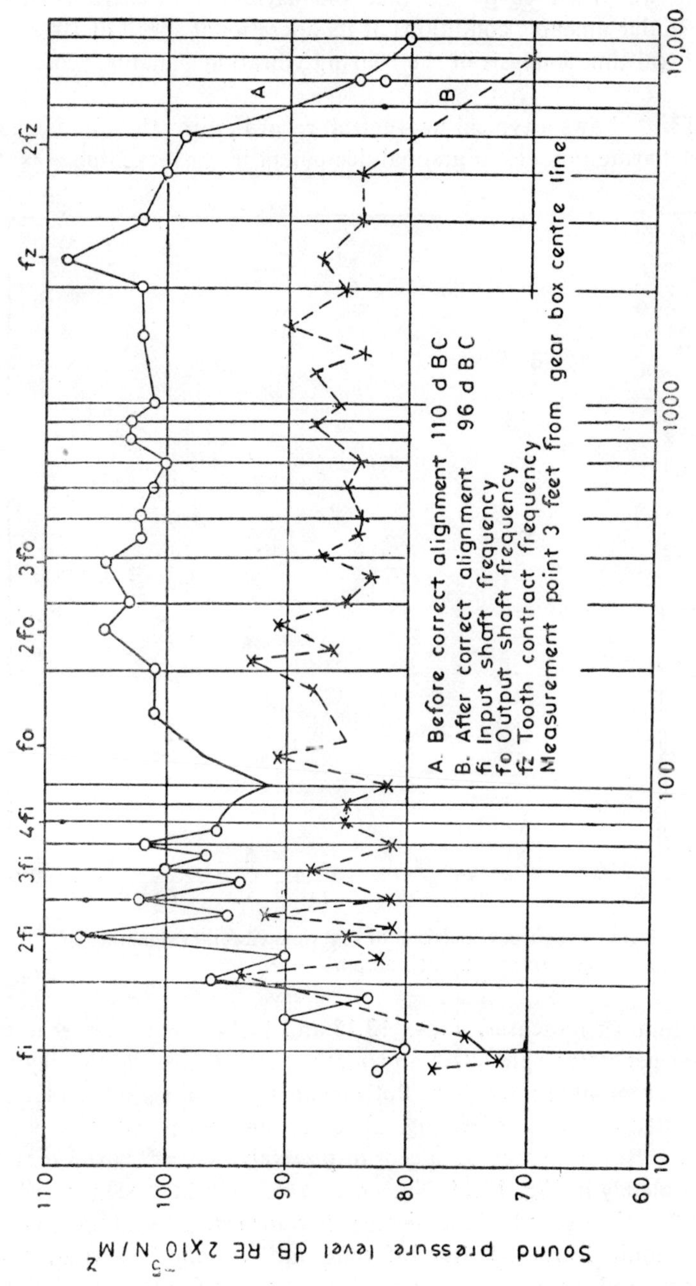

Fig. 13.11 Frequency spectrum of gearbox noise with good and bad alignment

from 500 to 1000 rpm, the vibration levels increased eightfold rather than four-fold. The critical speed was found to be 1955 rpm by Rayleigh's method. It was then suspected that the asymmetric generator shaft is operating under unstable conditions at its operational speed of 1000 rpm. A detailed real time analysis of the bearing vibration signatures confirmed this.

Figure 13.12 shows a typical uneventful record with the fundamental and higher harmonics with a gradual decrement in the amplitude as the

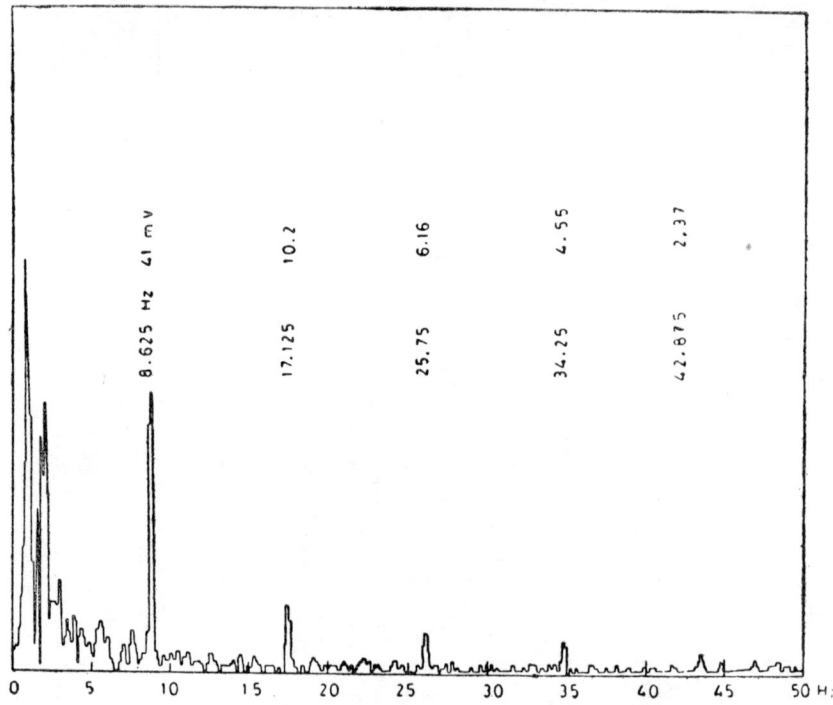

Fig. 13.12 Frequency spectrum of real time vibration signature of generator rotor at 520 rpm

harmonic number increases. Figs. 13.13 and 13.14 show the real time spectrum at generator rpm 607.5 and 735 respectively, while the engine is coasting up from 400 to 800 rpm. Both records show a signal at 1455 rpm (24 Hz) excited by third order (Fig. 13.13) of engine speed and second order (Fig. 13.14) of generator speed respectively. Other harmonics are absent completely in Fig. 13.14. The speed 1455 rpm has been attributed to torsional critical speed of the diesel generator set. The amplitude of the second harmonic decreased after 735 rpm, until around 930 rpm, which rapidly increased to almost the same level as the fundamental at about 997.5 rpm (Fig. 13.15). This shows that the critical speed of the generator rotor is 1995 rpm. Since 1995 rpm is the critical speed which is excited by the second harmonic of the generator rpm, large amplitude vibration

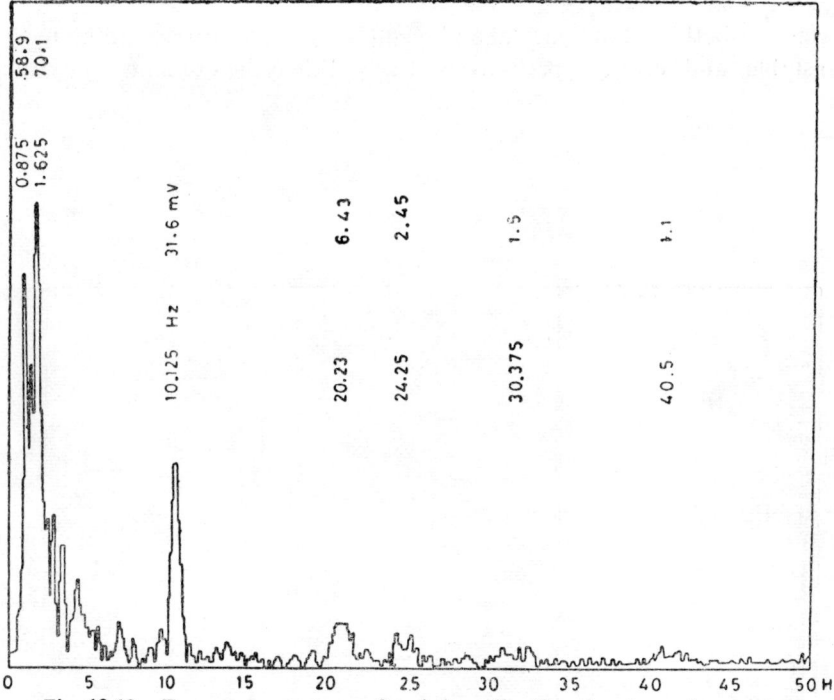

Fig. 13.13 Frequency spectrum of real time vibration signature of generator rotor at 607.5 rpm

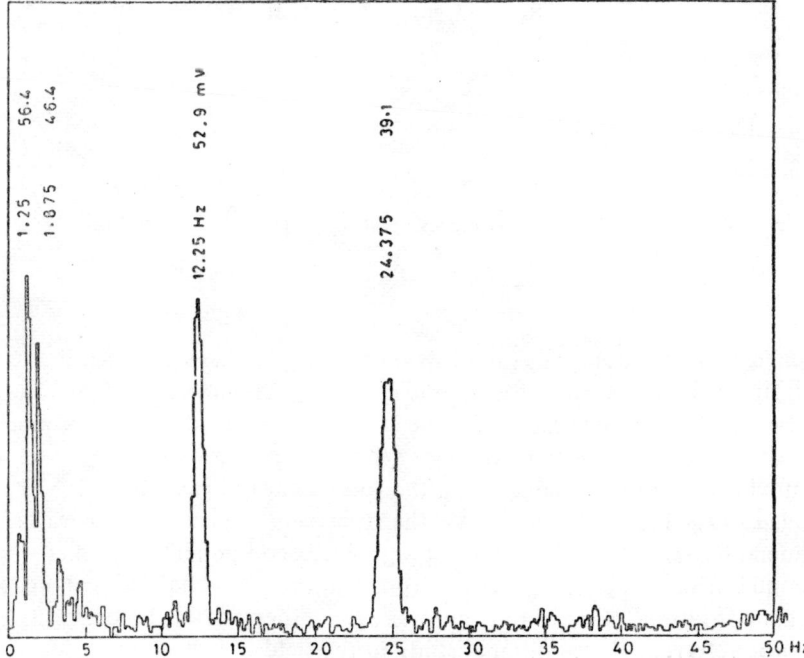

Fig. 13.14 Frequency spectrum of real time vibration signature of generator rotor at 735 rpm

occurs at both the fundamental and second harmonic corresponding to the unstable and critical speeds respectively. When the excitation is put on,

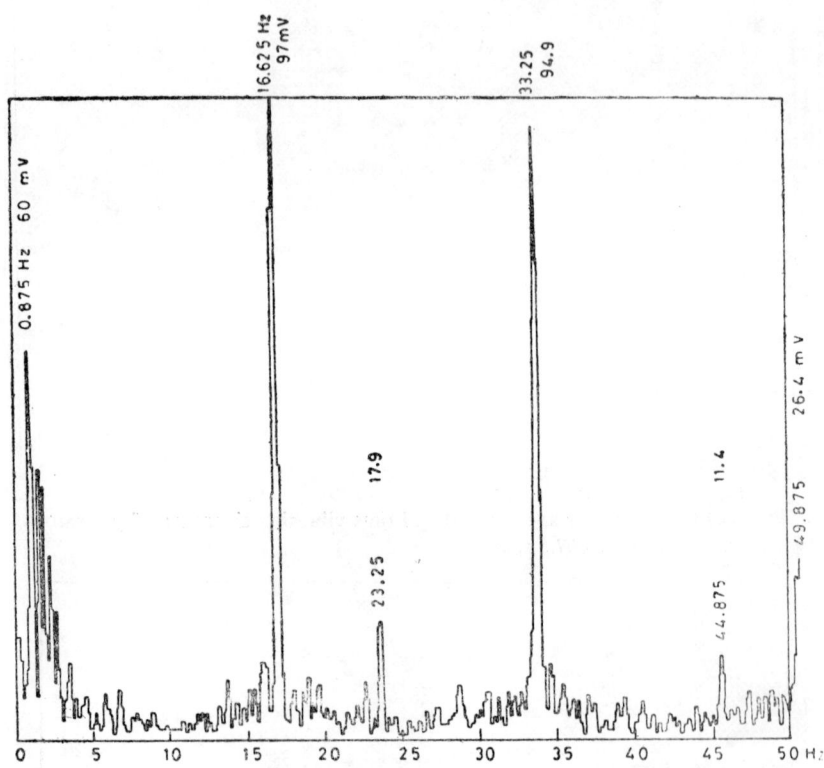

Fig. 13.15   Real time frequency spectrum of generator at 997.5 rpm

the frequency content was found to shift rapidly in a narrow band particularly between 990 and 1100 rpm (Fig. 13.16). The application of voltage causes transient conditions in the rotor which in turn cause frequency entrainment in the unstable regime. When the excitation is complete, thus removing the transient conditions, the rotor behaved steadily at 1087 rpm (Fig. 13.17). Figure 13.18 shows the spectrum when the generator is loaded at 1027 rpm with only three significant components, the first and second harmonics of the generator rpm and the torsional critical speed. For application of real time analysis, two other project reports [13, 14] may be referred. For additional study in the field of condition monitoring reference may be made to Dawson [15] and the application notes of Bruel and Kjaer [16].

Condition Monitoring Using Vibration Measurements 295

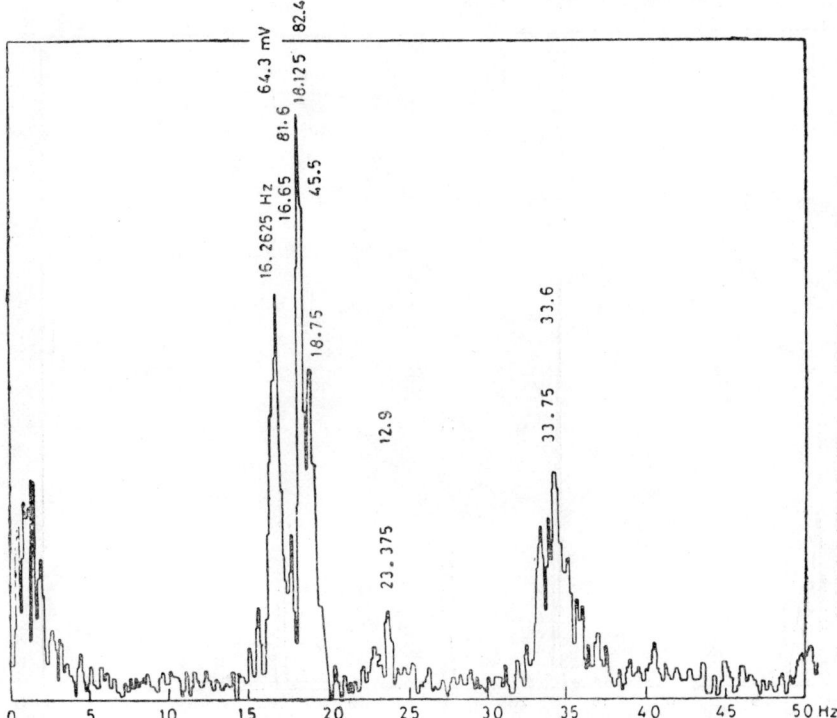

Fig. 13.16 Real time generator spectrum record during excitation

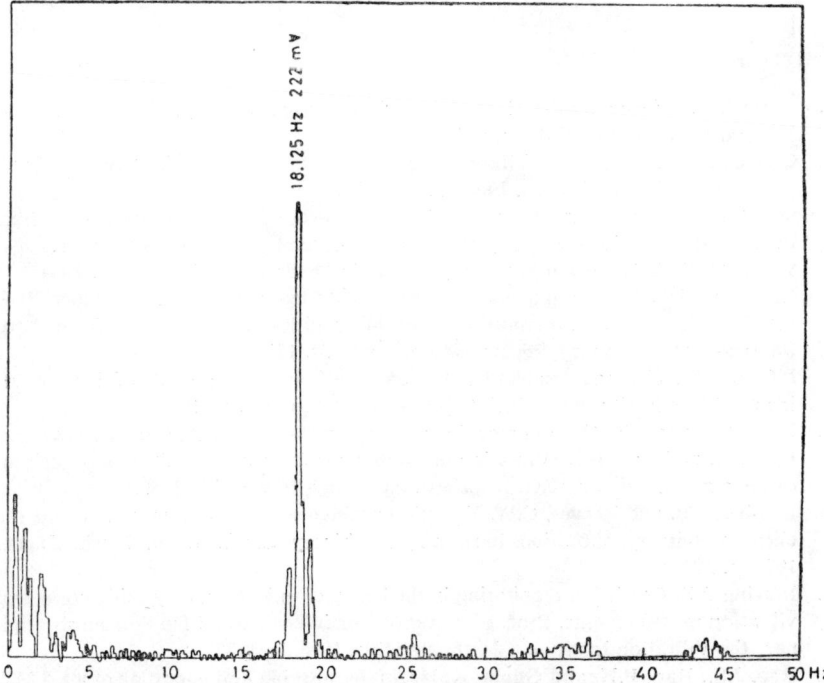

Fig. 13.17 Frequency spectrum of generator rotor at 1087 rpm

296 *Rotor Dynamics*

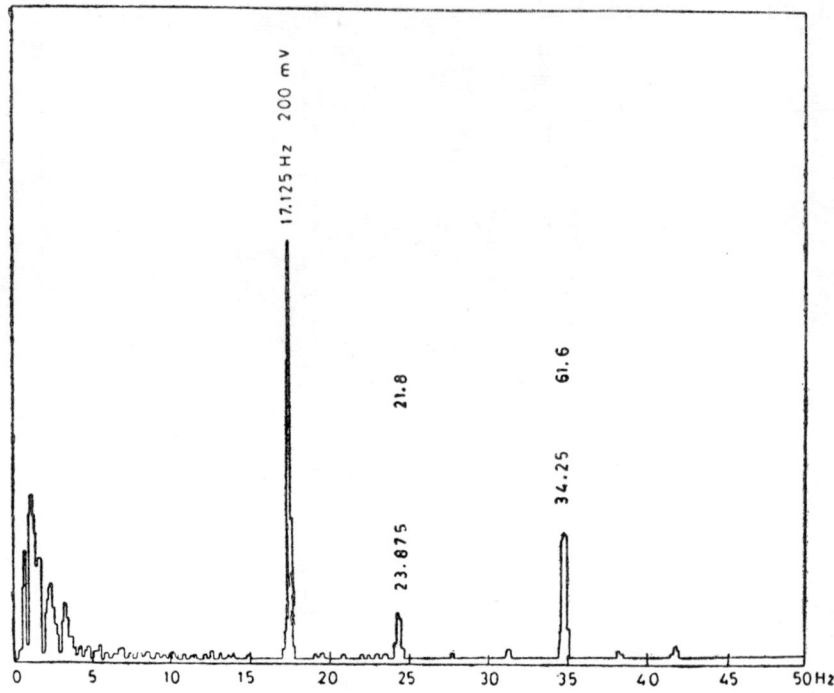

Fig. 13.18 Frequency spectrum of generator rotor at 1027 rpm

References

1. Hubbell, J. Here comes the United States' Super-speed Navy, Reader's Digest, February 1978, Indian Edition, p. 70.
2. Collacot, R.A. Mechanical failure—Diagnosis and monitoring, Chartered Mechanical Engineer, Instn. of Mech. Engrs., July 1976.
3. Sohre, J.S. Operating problems with high speed turbomachinery, Causes and correction, ASME Petroleum Mech. Engng. Conf., September 1968, Dallas, Texas.
4. VDI 2056/1964, Criteria for assessing mechanical vibrations of machines, 1964.
5. Rathbone, T.C. Vibration tolerance, Power plant Engng., Chicago, November, 1939.
6. ISO/IS 2732. Mechanical vibration of machines with operating speeds from 10 to 200 rps—Basis for specifying evaluation standards, 1972.
7. ISO/IS 3945. The measurement and evaluation of vibration severity of large rotating machines in situ, operating at speeds from 10 to 200 rps, 1975.
8. IRD Mechanalysis. General machinery vibration severity chart, Chester, UK.
9. Rao, J.S. and Rao, P.N. Vibration measurements for health monitoring of reciprocating air compressors, K.G. Khosla Compressors, New Delhi, 1980.
10. Erskine, J.B. and Reeves, C.W. Vibration problems on rotating machinery in the chemical industry, Vibrations in rotating machinery Conf., Instn. Mech. Engrs., 1976, p. 209.
11. Erskine, J.B. Condition monitoring in the heavy chemical industry using noise and vibration measurements, Proc. of preventive maintenance and fault diagnosis seminar, the Vibration Institute, Clarendom Hills, Illinois, 1977.
12. Rao, J.S., Rao, P.N. and Gupta, K. Health monitoring and condition of a 3.5 MW DG set Generator rotor, Hindustan Copper Ltd., Khetri, 1982.

13. Rao, J.S., Brake assembly problem study, STI Project 81 PN 008, WMATA, Washington, 1981.
14. Rao, J.S., Rao, P.N. and Suryanarayana, S. Diagnosis of vibration and noise in detergent, alkylate plant at IPCL, Baroda, Engineers India Ltd., New Delhi, 1979.
15. Dawson, B. Vibration condition monitoring techniques for rotating machinery. The Shock and Vibration Digest, Vol. 8, 1976, p. 3.
16. Bruel and Kjaer. Application notes on the use of vibration measurements for machinery condition monitoring.

Index

Archer dynamic magnifier 38
Asymmetric shaft 163
Asymmetric shaft at critical speed 167
Average acceleration method 181

Backward nonsynchronous whirl 85
Backward whirl 69
Backward whirl of rotors 123
Backward whirl response—Effect of hydrodynamic bearings 111
Balance quality 252
Balance quality flexible rotors 272
Balancing criteria flexible rotors 269
Balancing criteria—Rigid rotors 248
Balancing of flexible rotors 260, 263
Balancing of rotors—Static 247
Bearing damping 96
Bearing matrix 127
Bearing matrix—Dual rotor 145
Bearing stiffness 96
Beating period—Dual rotor whirl 150
Bending critical speeds 47
Blade ring 135
Branched system torsional vibration 20
Breathing action of a transverse crack 191

Causes of rotor unbalance 277
Classification of rotors 248
Coefficient of fluctuation of speed 41
Condition monitoring—Narrow band analysis 285
Condition monitoring—Octave band analysis 284
Condition monitoring—Simple measurements 281
Conical whirl 50
Conical whirl instability 204
Conical whirl of flexible rotors 206
Continuous element—Bending vibration 131
Continuous element—Torsional vibration 11
Coupled shafts 86
Crack model—Breathing action 191
Cracked shaft deflection under gravity 189
Cracked shaft response 192
Cradle balancing 254

Crank equivalent length 28
Crankshaft torsional stiffness 28
Critical speed condition 114
Critical speeds—Effect of hydrodynamic bearings 110
Critical speeds—Gyroscopic effects 84
Critical speeds—Reciprocating machines 37
Cyclic irregularity 40

Disk element 134
Dissimilar moments of area 163
Dual rotor—Junction conditions 147
Dual rotor on hydrodynamic bearings 153
Dual rotor system 142
Dunkerley's method 51
Dynamic balancing 251
Dynamic magnifier 38

Eccentricity ratio 92
Effect of shaft damping—Rotors on hydrodynamic bearings 121
Effective damping of rotors on hydrodynamic bearings 118
Effective stiffness of rotors on hydrodynamic bearings 118
Eigen value problem discrete torsional system 29
Engine excitation torque 35
Equilibrium amplitude 38
Equivalent discrete system 16
Equivalent length of crank 28
Euler's method 241
Excitation torque due to short circuits 17

Field matrix 53
Field matrix—Dual rotor 144
Field matrix modified 66
Firing angle diagram 36
Flexible rotor 48
Flexible rotor balancing—Influence coefficient method 263
Flexible rotor balancing—Modal method 260
Flexible rotor balancing criteria 269

Flexible rotor instability due to oil film forces 205
Flexible rotors—Classification 248
Flexible rotor balance quality 272
Flexible rotor conical whirl instability 206
Flexible rotor translatory whirl instability 206
Floquet's theory 163
Fluid film bearings 91
Forced vibration—Reciprocating machines 37
Forward nonsynchronous whirl 85
Forward whirl 69
Forward whirl response—Effect of hydrodynamic bearings 111
Frequency equation—Overhung shafts 59
Frequency equation—Transfer matrix 56

Geared torsional system 22
Generalized coordinate 38
Gravity orbital solution 183
Gravity solution of dissimilar shaft 175
Gravity unbalance response of cracked shaft 194
Grooved bearing coefficients 101
Gyroscopic effect coupled rotors 86
Gyroscopic effects 79

Harmonic components due to inertia 34
Hollow shaft 136
Horizontal shaft at half critical speed 167
Horizontal shaft with rectangular section 167
Hydrodynamic bearing steady performance 92
Hydrodynamic bearing transfer matrix 127
Hydrodynamic bearings 91
Hysteresis 210

Influence coefficient balancing—Flexible rotors 263
Influence coefficient balancing—Rigid rotors 257
Instability—Negative cross-coupled stiffness 214
Instability—Torsional vibrations 236
Instability due to oil film 199
Instability due to transverse crack 187
Instability of flexible rotors due to oil film 205
Instability threshold—Transfer matrix method 209
Instability zones of cracked shaft 193
Internal hysteresis 210

Jeffcott rotor 47
Jeffcott rotor in fluid film bearings 107
Jeffcott rotor with asymmetric shaft 178
Junction conditions—Dual rotor 147

Locomotive diesel engine 31
Loops in dual rotor whirl 150

Major natural frequency 217
MAN dynamic magnifier 38
Mass moment of inertia equivalent value for piston 27
Mathieu equation—Variable torsional stiffness 238
Mathieu equation 163
Matrix eigen value problem torsional vibration 29
Mechanical looseness 279
Minimum unbalance response 154
Minor natural frequency 217
Misalignment 279
Modal balancing 260
Modal expansion method 37
Modelling of reciprocating machines 25
Modelling of rotors—Bending vibration 133

Narrow band analysis 285
Newmark time marching method 181
Noise spectrum 287
Nonsynchronous whirl gyroscopic effects 85

Octave band analysis 284
Oil whip 201
Oil whirl 200
Optimum bearing design 154
Optimum values of bearing parameters 156
Orbital analysis—Transfer matrix method 219
Orbital periodicity—Dual rotor 150
Orbital response due to gravity 183
Orbital response of asymmetric shaft 183
Out of balance response 62
Out of balance response—Rigid end supports 68
Overall transfer matrix 55
Overhung rotor—Unbalance response 71
Overhung rotor gyroscopics 82
Overhung shaft 56

Plain cylindrical bearing coefficients 97
Point matrix 55
Point matrix—Dual rotor 144
Point matrix modified 66

Index

Precession of a spinning disc 80

Rayleigh's method 52
Real time analysis 289
Reciprocating unbalance 277
Rectangular ripple variation—Stiffness 168
Rectangular shaft 164
Residual unbalance—Quality 254
Resonant stress—Reciprocating machines 39
Reynolds equation 91
Rigid rotor balancing 251
Rigid rotor instability due to oil whirl 202
Rigid rotors—Classification 248
Root searching technique 13
Rotor classification 248
Rotor on fluid film bearings 107
Rotor on hydrodynamic bearings with no critical speeds 119
Rotors on hydrodynamic bearings—Effect of damping 116

Shaft with dissimilar stiffness 163
Short circuit subtransient 17
Short circuit transient 17
Side rod locomotive 237
Sommerfeld number 92
Spatial mode—Whirling 140
Stability—Variable inertia system 244
Stability index 216
Stability of asymmetric shaft 163
Stability of gravity solution 176
Stability of rigid rotor in oil bearings 204
Stability of vertical shaft—Tondl's analysis 173
State vector cosine and sine components 65
Static balancing 247
Steel mill stand 18
Step element 133
Stiffness and damping coefficients 96
Stress modal matrix 30
Strutt diagram—Rectangular shaft 165
Strutt diagram—Variable torsional stiffness 238
Subharmonic whirls of cracked shaft 193
Synchronous whirl 51
Synchronous whirl of spinning disc 81

Tangential effort 34
Tangential effort due to gas pressure 33
Tangential effort due to inertia 34
Tapered shaft 135
Threshold instability due to oil whirl 201
Time dependent transfer matrix method 220

Time marching technique 177
Tondl's analysis 171
Torsional stiffness of crankshafts 28
Torsional system—Variable inertia 239
Torsional system—Variable stiffness 237
Torsional vibration model—Diesel generator set 26
Transfer matrix—Torsional vibration 12
Transfer matrix analysis—Continuous elements 131
Transfer matrix analysis—Distributed elements 131
Transfer matrix analysis—Simple shafts 52
Transfer matrix analysis—Rotors on hydrodynamic bearings 125
Transfer matrix method—Torsional vibration 10
Transfer matrix method branched systems 21
Transient orbital response due to gravity 183
Transient orbital response due to unbalance 186
Transient response—Cantilever rotor 226
Transient response—Nonlinear bearings 232
Transient response—Rotor on flexible bearings 230
Transient response—Simply supported rotor 228
Transient response—Torsional vibration 18
Transient response—Transfer matrix method 220
Transient response by time marching solution 177
Transient response of shaft on hydrodynamic bearings 187
Translatory whirl 50
Translatory whirl instability 204
Translatory whirl of flexible rotors 206
Transverse crack 189
Turbine rotor model—Bending vibration 136
Turbo-alternator set 14
Twist angle—Basis rotor 39
Twist modal matrix 18, 30
Two plane balancing 251
Two spool rotor analysis 142

Unbalance 48, 64
Unbalance mass—Dual rotor 145
Unbalance response—Dissimilar shaft 174
Unbalance response—Effect of hydrodynamic bearings 109
Unbalance response—Effect of shaft stiffness 124

Unbalance response—Jeffcott rotors with damping 117
Unbalance response of cracked shaft 194
Unbalance response parametric studies 155
Unstable response of asymmetric shaft on bearings 188
Unstable solution—Rectangular shaft 169

Variable elasticity shaft—Stability diagram 170
Variable inertia system 239
Variable stiffness—Torsional system 237
Vertical shaft with rectangular section 166
Vibration criterion chart 282
Vibration generating mechanisms—Asymmetric shaft 280
Vibration generating mechanisms—Bending criticals 280
Vibration generating mechanisms—Mechanical looseness 279
Vibration generating mechanisms—Misalignment 279
Vibration generating mechanisms—Oil film whirl 280
Vibration generating mechanisms—Torsional criticals 280
Vibration generating mechanisms—Unbalance 276
Vibration generating mechanisms—Reciprocating unbalance 277
Vibration severity—Quality judgement 284
Vibration severity chart 283
Viscosity 91

Whirl magnification factor 50
Whirl orbit relation 69
Whirl orbits—Effect of hydrodynamic bearings 113
Whirling of a rotor with transverse crack 190
Whirling of dissimilar shaft 171
Whirling of rotors 47
Wilson dynamic magnifier 38